The Ecology of Mangrove and Related Ecosystems

Developments in Hydrobiology 80

Series editor
H. J. Dumont

The Ecology of Mangrove and Related Ecosystems

Proceedings of the International Symposium held at Mombasa, Kenya, 24–30 September 1990

Edited by
Victor Jaccarini & Els Martens

Reprinted from Hydrobiologia, vol. 247 (1992)

Kluwer Academic Publishers
Dordrecht / Boston / London

Library of Congress Cataloging-in-Publication Data

The Ecology of mangrove and related ecosystems : proceedings of the international symposium held at Mombasa, Kenya, September 24-30, 1990 / edited by Victor Jaccarini & Els Martens.
p. cm. -- (Developments in hydrobiology ; 80)
"Reprinted from Hydrobiologia vol. 247 (1992)."
Includes indexes.
ISBN 0-7923-2049-2 (hb : acid free paper)
1. Mangrove swamp ecology--Congresses. 2. Mangrove fauna--Congresses. 3. Mangrove swamps--Congresses. I. Jaccarini, Victor. II. Martens, Els. III. Series.
QH541.5.M27E25 1992
574.5'26325--dc20 92-38659

ISBN 0-7923-2049-2

Published by Kluwer Academic Publishers,
P.O. Box 17, 3300 AA Dordrecht, The Netherlands.

Kluwer Academic Publishers incorporates
the publishing programmes of
D. Reidel, Martinus Nijhoff, Dr W. Junk and MTP Press.

Sold and distributed in the U.S.A. and Canada
by Kluwer Academic Publishers,
101 Philip Drive, Norwell, MA 02061, U.S.A.

In all other countries, sold and distributed
by Kluwer Academic Publishers Group,
P.O. Box 322, 3300 AH Dordrecht, The Netherlands.

The publication of this volume was funded by the Commission of the European Communities, and sponsored by the Hydrobiological Society of Eastern Africa (HYSEA).

Printed on acid-free paper

Printed in Belgium

Contents

Hydrobiologia **247**: vii–viii, 1992.
V. Jaccarini & E. Martens (eds), The Ecology of Mangrove and Related Ecosystems.

Preface

The international symposium on *The Ecology of Mangrove and Related Ecosystems*, Mombasa, Kenya, 24-30 September, 1990, was organized by the Department of Zoology, University of Nairobi, in collaboration with the Kenya Marine and Fisheries Research Institute, and the Regional Dispatch Centre of the Regional Cooperation in Scientific Information Exchange in the West Indian Ocean, Mombasa, Kenya. It was officially opened by Professor F. N. Onyango, Principal, College of Biological and Physical Sciences, University of Nairobi. The closing address was given by Professor M. Hyder, retired Principal and Professor of Zoology, CBPS, University of Nairobi.

Scientific investigations of the mangrove and coral reef ecosystems in Kenya have been going on at least since the eartly 70's and have been gathering momentum with the involvement of the University of Nairobi, KMFRI, and British, Belgian, Dutch and Italian universities and scientific institutions. So the time was ripe to take stock of the situation and to try to bring together workers in the same field from as many tropical regions as possible. A cursory glance at the list of participants is sufficient proof of the widespread interest in the theme of the symposium. Contributions came from all the main tropical marine areas of the globe, representing twenty-three countries, in West and Eastern Africa, South and Southeast Asia, Central and South America, and Australia. The thirty papers published in this volume were selected from seventy presented at the symposium.

It is recognized that the mangroves have intimate and complex linkages with other coastal systems, notably seagrass beds and coral reefs. So, starting from the mangrove forest and its users at the interface between the terrestrial, freshwater and marine environments, the papers presented proceed to the fully submerged seagrasses and further out to the corals. The nature of the coupling between the different systems was touched on at the symposium, but stands out as a serious gap in our knowledge. Next, chemical and physical processes which constitute the mechanisms of linkage received attention. Human uses and abuses of the ecosystems formed the last major topic of the symposium.

Workshop discussions were held during the symposium in preparation for the adoption in plenary session of two declarations: *A Statement of Priorities for Scientific Research* and *A Declaration on Management Strategies*.

The following priorities for basic research were agreed: the interlinkages between the mangrove, coral reef and seagrass ecosystems; the need for more information on nearshore hydrological processes operating at the small and meso-scale and elaboration of predictive models; more work on nutrient cycling and energy flow including soil processes; biological processes and use of bioindicators.

Priorities were also set for applied research. This, it was agreed, should be directed towards management of the forest or aquatic product on a sustainable basis, taking into account the integrated nature of the ecosystems. Emphasis should be placed on minimizing perturbations to the systems and the maintenance of ecological integrity. Broad areas needing investigation were identified: methods for efficient and ecologically sound land use in the watershed adjacent to the mangal, seagrass beds and coral reefs; effect of management of the mangroves on off-shore fisheries; impact of industralization on the mangal and related ecosystems; effect of domestic and industrial effluents and management of their impact; comparative studies on destructive and non-destructive aquacultural techniques with a view to minimizing negative effects; the effect of human activities on physical, chemical and biological soil processes; research on management practices appropriate to different mangrove forest types and promotion of a biotechnological aproach in mangrove silviculture; socio-economic studies of human coastal settlements; natural products of the mangals; the impact of global climate change; and active rehabilitation of seagrass beds and coral reefs.

The *Declaration on Management Strategies* was positive, concluding that much can be done to stop the degradation of damaged mangals, seagrass beds and coral reefs and to rehabilitate them, but that new techniques must be worked out to allow use on a sustainable basis for long term economic return and for the wellbeing of coastal human settlements and a healthy environment.

To achieve the above goals the Symposium recommended the following strategies.

The establishment of **National Coastal Management Plans** including the following components: survey of both natural and human resources and their present uses; identification of the problems affecting coastal ecosystems; adjustment of legal frameworks with regard to the uses of these ecosystems and enforcement of existing management regulations; provision of technical assistance and training for the introduction of economically *and* ecologically viable activities in these ecosystems; rehabilitation of damaged areas; creation of natural reserves both for tourism and for the recuperation of natural plant and animal populations; education and creation of public awareness on the value and problems of these ecosystems.

In formulating and implementing each national Plan *all* concerned parties should be involved, i.e., the inhabitants and the outside users of these ecosystems, the policy makers, the scientists and technical experts, and also the national and international funding agencies. Use should be made of existing institutions and networks. The scientists gathered in the symposium urged the establishment or strengthening of regional and international associations of scientists to provide the scientific basis to national and regional management plans.

It is our pleasant duty as Chairman and Secretary of the Symposium Organizing Committee to thank the many sponsors who generously provided financial support for the Symposium: the African Biosciences Network for providing the seed funds; the US National Research Council; the Commission of the European Communities; the Australian International Development Assistance Bureau; the Netherlands Ministry of Agriculture, Nature Management and Fisheries; United Nations Environment Programme, Forestry Department; UNESCO – ROSTA; and the Third World Academy of Sciences. Special thanks are due to the Commission of the European Communities for providing funds covering the publication costs of this volume. A debt of gratitude is owed to *Hydrobiologia* (Kluwer Academic Publishers) and the Chief Editor, Professor H. Dumont of University of Ghent (Belgium), for agreeing to publish the Proceedings as a special volume in their series Developments in Hydrobiology. We would also like to thank the members of the Executive Committee for their help in the organization of the Symposium. Finally, we wish to thank the University of Nairobi in the persons of its previous Vice-chancellor, Professor P. M. Mbithi, Professor F. N. Onyango, Principal, College of Biological and Physical Sciences, and Professor R. W. Mwangi, Dean, Faculty of Science, for their support. Thanks should also go to Professor G. M. O. Maloiy who negotiated the initial funding with the African Biosciences Network.

V. Jaccarini & E. Martens

1. D. Onyango
2. J. Kasyi
3. C. Amadou
4. J. Francis
5. J. Kathupotha
6. G. C. Trono
7. V. Jaccarini
8. C. A. Muhando
9. A. Thorhaug
10. A.-F. Woitchik
11. J. L. Agatsiva
12. L. Solorzano
13. J. Azariah
14. E. Slim
15. N. O. Isirimah
16. M. Hemminga
17. P. O. Gang
18. M. A. Amakiri
19. M. D. Amarasinghe
20. R. W. Mwangi
21. M. Vannini
22. S. Gunasekaran
23. J. P. Shunula
24. M. Vuai
25. K. Deges
26. J. D. Woodley
27. B. Musyoki
28. N. Visser
29. A. M. Ellison
30. A. Robertson
31. A. H. Moallim
32. F. van Avermaat
33. V. Selvam
34. A. G. Tejan
35. A. Sasekumar
36. J. K. Muraya
37, U. Helbing
38. K. Naskar
39. L. Vergeer
40. D. Prasad
41. J. Rasowo
42. J. Mutere
43. W. M. Kudoja
44. J. G. Mutwiri
45. A. Yanez-Arancibia
46. M. D. Fortes
47. L. Kannan
48. E. Coppejans
49. P. Akiwumi
50. S. Abdulrahman Issa
51. G. van der Velde
52. G. Wamukoya
53. E. Wolanski
54. J. Hatton
55. F. W. Fong
56. F. M. Chale
57. M. A. Ngoile
58. M. Mastaller
59. N. Revisa
60. E. Okemwa
61. E. Martens
62. K. Delbeke
63. R. Mohanraju
64. M. Vincx
65. T. Done
66. D. van Gansbeke
67. J. Schrijvers
68. M. J. Ntiba

Hydrobiologia **247**: x–xii, 1992.
V. Jaccarini & E. Martens (eds), The Ecology of Mangrove and Related Ecosystems.

List of participants

ABDULRAHMAN ISSA, S., Commission for Lands & Environment, P.O. Box 811, Zanzibar, Tanzania

AGATSIVA, J. L., Remote Sensing, Department of Resource Surveys and Remote Sensing, P.O. Box 47146, Nairobi, Kenya

AIKEN, K., Department of Biological Sciences, Florida International University, Tamiami Campus, Miami, FL 33199, USA

AKIWUMI, P. K., Oceans & Coastal Areas Programme, Activity Centre (OCA/PAC) United Nations Environment Programme, P.O. Box 30552, Nairobi, Kenya

ALLENG, G., Department of Zoology University of the West Indies, Mona, Kingston 7, Jamaica

AMADOU, C., UNDP/FAO Fuelwood Project c/o UNDP, P.O. Box 1011, Freetown, Sierra Leone

AMAKIRI, M. A., Rivers State University Science & Techn. Faculty of Agriculture, P.M.B. 5080, Port Harcourt, Nigeria

AMARASINGHE, M. D., National Aquatic Resources Agency, Regional Research Centre, Kalpitiya, Sri Lanka

ANDERSON, M., Department of Biological Sciences, Florida International University, Mamiami Campus, Miami, FL 33199, USA

ANYIENI, Z. M., Permanent Presidential Commission on Soil Conservation & Afforestation, P.O. Box 46191, Nairobi, Kenya

AZARIAH, J., Department of Zoology University of Madras, Guindy Campus, Madras – 600 025, India

AZARIAH, H., CAS in Botany, University of Madras, Guindy Campus, Madras – 600 025, India

BACON, P. R., Department of Zoology, University of the West Indies, Mona, Kingstone 7, Jamaica

BALASUBRAMANIAM, S., Department of Botany, University of Peradeniya, Peradeniya, Sri Lanka

BORGER, R., Ecology & Systematics Laboratory, Free University of Brussels, Pleinlaan 2, B-1050 Brussels, Belgium

CARBY, B., Department of Biological Sciences, Florida International University, Tamiami Campus, Miami, FL 33199, USA

CHALE, F. M., Dept. of Zoology & Marine Biology, University of Dar Es Salaam, P.O. Box 35064, Dar Es Salaam, Tanzania

CHONG, V. C., Department of Zoology, University of Malaya, 59100 Kuala Lumpur, Malaysia

COPPEJANS, E., State University of Ghent, Laboratory of Morphology, Systematics & Ecology of the Plants, K. L. Ledeganckstraat 35, B-9000 Ghent, Belgium

COUTO, M., Faculty of Biology, University Eduardo Mondiane, C.P. 257, Maputo, Mocambique

D'CRUZ, R., Department of Zoology, University of Malaya, 59100 Kuala Lumpur, Malaysia

DARO, M. H., Ecology & Systematics Laboratory, Free University of Brussels, Pleinlaan 251, B-1050 Brussels, Belgium

DEGEORGES, A., Regional Environment Advisor, U.S. – AID, P.O. Box 30261, Nairobi, Kenya

DEGES, K., State University of Ghent, Institute of Zoology – Marine Biology, K. L. Ledeganckstraat 35, B-9000 Ghent, Belgium

DELBEKE, K., Kenya-Belgium Project Marine Science Kenya Marine & Fisheries Research Inst., P.O. Box 81651, Mombasa, Kenya

DEN HARTOG, C., Laboratory of Aquatic Ecology, Catholic University of Nijmegen, Toernooiveld, 6525 ED Nijmegen, The Netherlands

DONE, T., Australian Institute of Marine Science AIMS, PMB 3 Townsville MC, Queensland 4810, Australia

ELLISON, A. M., Department of Biological Sciences, Mount Holyoke College Clapp Laboratory, South Hadley, MA 01075-1484, USA

Farnsworth, E. J., Department of Botany University of Vermont, 120 Marsh Life Sciences Building, Burlington, VT 05405, USA

FONG, F. W., Department of Zoology, University of Malaya, Field Studies Centre, 59100 Kuala Lumpur, Malaysia

FORTES, M. D., Marine Science Institute, University of the Philippines, Diliman, Quezon City 1101, Philippines

FRANCIS, J., Institute of Marine Sciences, University of Dar Es Salaam, P.O. Box 668, Zanzibar, Tanzania

GAMBREL, R. P., Department of Soli Science, Rivers State University of Science, P.M.B. 5080, Port Harcourt, Nigeria

GANG, P. O., Technical Officer, Department of Resource Surveys and Remote Sensing, P.O. Box 47145, Nairobi, Kenya

GEETS, G., Dept. Zool., Kenyatta Univ., P.O. Box 43844, Nairobi, Kenya

GHERARDI, F., Department Animal Biology & Genetics, University of Florence, Via Romana 17, Firenze 50125, Italy

GICHIRU, M., Department of Soil Service, Rivers State University of Science, P.M.B. 5080, Port Harcourt, Nigeria

GIESEN, W., Laboratory of Aquatic Ecology, University of Nijmegen, Toernooiveld, 6525 ED Nijmegen, The Netherlands

GORDON, V., Department of Biological Sciences, Florida International University, Tamiami Campus, Miami, FL 33199, USA

GORISSEN, M. W., Laboratory of Aquatic Ecology, Catholic University of Nijmegen, Toernooiveld, 6525 ED Nijmegen, The Netherlands
GUNASEKARAN, S., Department of Zoology, University of Madras, Guindly Campus, Madras – 600 025, India
HALL, K. C., UNDP/FAO Fuelwood Project c/o UNDP, P.O. Box 1011, Freetown, Sierra Leone
HARRY, F. O. C., Department of Soli Science, Rivers State University of Science, P. M. B. 5080, Port Harcourt, Nigeria
HATTON, J. C., Faculty of Biology, University Eduardo Mondiane, C. P. 257, Maputo, Mozambique
HEIMSATH, A., German Agency for Technical Cooperation, Kilifi Water and Sanitation Project, P.O. Box 666, Kilifi, Kenya
HELBING, U., IBAMA, Av. Rio Branco 186, 490000 Aracaju-SE, Brazil
HEMMINGA, M. A., Delta Institute for Hydrobiological Research, Vierstraat 28, 4401 EA Yerseke, The Netherlands
HYDER, M., Department of Zoology, University of Nairobi, P.O. Box 30197, Nairobi, Kenya
ISIRIMAH, N. O., Department of Soli Science, Rivers State University Science & Techn., P.M.B. 5080, Port Harcourt, Nigeria
JACCARINI, V., Department of Zoology, University of Nairobi, P.O. Box 30197, Nairobi, Kenya
KANNAN, L., Annamalai University, Centre of Adv. Studies Marine Biology, Porto Novo – 508 502, Tamil Nadu, Indiua
KASYI, J. N., Department of Zoology, University of Nairobi, P. O. Box 30197, Nairobi. Kenya
KATUPOTHA, J., Department of Geography, University of Sri Jayewardenepura, Gangodawilla, Nugegoda, Sri Lanka
KUDOJA, W. M., Department of Zoology, University of Nairobi, P.O. Box 30197, Nairobi, Kenya
LARA-DOMINGUES, A. L., EPOMEX Programme, Autonomous University of Campeche, Ap. Postal 520, Clmpeche 24030, Mexico
LAREW, H., USAID – Office of the Science Advisor Agency for International Development, Room 320, SA-18, Washington DC 20523, USA
LEH, M. U., Department of Zoology, University of Malaya, 59100 Kuala Lumpur, Malaysia
LEVY, I., Department of Botany, University of West Indies, Kingstone 7, Jamaica
LITTAU, J., Department of Botany, University of West Indies, Mona, Kingstone 7, Jamaica
LLUISMA, A. O., Marine Science Institute, University of the Philippines, Diliman, Quezon City 1101, Philippines
MARTENS, E. E., Department of Zoology, University of Nairobi, P.O. Box 30197, Nairobi, Kenya
MASTALLER, M., Tournelystr. 16, D-8901 Stadtbergen, Germany
MEIJER, G. J., Stinapa, Bonaire, Netherlands Antilles
MENKIR, V., National Academy of Sciences BOSTID, HA 476, 2101 Constitution Ave., N.W., Washington, DC 20418, USA
MICHELI, F., Department Animal Biology & Genetics, University of Florence, Via Romana 17, Firenze 50125, Italy
MILLER, B., Department of Biological Sciences, Florida International University, Tamiami Campus, Miami, FL 33199, USA
MOALLIM, A. H., Data Collection & Storing Centre, Ministry of Fisheries and Marines Resources, P.O. Box 2589, Mogadishu, Somalia
MOHANRAJU, R., Centre Advanced Study in Marine Biology, Annamalai University, Porto Novo – 608 502, Tamil Nadu, India
MONDEGO, Faculty of Biology, University Eduardo Mondlane, C.P. 257, Maputo, Mozambique
MUHANDO, C. A., Institute of Marine Sciences, University of Dar Es Salaam, P.O. Box 668, Zanzibar, Tanzania
MUNGA, D., Kenya Marine & Fisheries Research Institute (KMFRI), P.O. Box 81651, Mombasa, Kenya
MURAYA, J. K., Department of Botany, University of Nairobi, P.O. Box 30197, Nairobi, Kenya
MUSYOKI, B. M., Kenya Wildlife Service Provincial Wildlife Office, P.O. Box 82144, Mombasa, Kenya
MUTERE, J., Kenya Marine & Fisheries Research Institute (KMFRI), P.O. Box 81651, Mombasa, Kenya
MUTWIRI, J. G., Kenya Marine & Fisheries Research Institute (KMFRI), P.O. Box 81651, Mombasa, Kenya
MWANGI, R. W., Department of Zoology, University of Nairobi, P.O. Box 30197, Nairobi, Kenya
NASKAR, K., Central Inland Capture Fisheries Research Institute, Barrackpore – 743 101, West Bengal, India
NATARAJAN, R., Ocean Data Centre for Water Resources, Annamalai University, Madras 600 025, India
NGOILE, M. A. K., Institute of Marine Sciences, University of Dar Es Salaam, P.O. Box 668, Zanzibar, Tanzania
NGULI, M. M., Kenya Marine & Fisheries Research Institute (KMFRI), P.O. Box 81651, Mombasa, Kenya
NTIBA, M. J., Department of Zoology, University of Nairobi, P.O. Box 30197, Nairobi, Kenya
OJANUGA, A. G., Department of Soil Science, Rivers State University of Science, P.M.B. 5080, Port Harcourt, Nigeria
OKEMWA, E., Kenya Marine & Fisheries Institute (KMFRI), P.O. Box 81651, Mombasa, Kenya
OLUOCH, A., Department of Zoology, University of Nairobi, P.O. Box 30197, Nairobi, Kenya
ONYANGO, F., College Biological & Physical Sciences, University of Nairobi, P. O. Box 30197, Nairobi, Kenya
ONYANGO, D., Nairobi University, Department of Zoology, P.O. Box 30197, Nairobi, Kenya
OSORE, M. K. W., Kenya Marine and Fisheries Research Institute, P.O. Box 81651, Mombasa, Kenya
OYIEKE, H., Kenya Marine & Fisheries Research Institute (KMFRI), P.O. Box 81651, Mombasa, Kenya

PISSIERSSENS, P., RECOSCIX-WIO Project Kenya Marine & Fisheries Research Inst., P.O. Box 95832, Mombasa, Kenya
POLK, Ph., Kenya-Belgium Project Marine Sc, Ecology & Systematics Laboratory, Free University of Brussels, Pleinlaan 2, B-1050 Brussels, Belgium
PRASAD, D., Department of Botany, University of the West Indies, Mona, Kingston 7, Jamaica
RASOLOFO, V. M., Centre National de Recherches Oceanographiques (CNRO), B.P. 68, Nosy Be 207, Madagascar
RASOWO, J., Department of Zoology, Moi University, P.O. Box 3900, Eldoret, Kenya
REESE, R., Department of Biological Sciences, Florida International University, Tamiami Campus, Miami, FL 33199, USA
REVIS, N., Ecology & Systematics Laboratory, Free University of Brussels, Pleinlaan 2, B-1050 Brussels, Belgium
ROBERTSON, S. A., Coast Forest Survey, National Museums of Kenya, .P.O. Box 162, Malindi, Kenya
RODRIGUEZ, M., Department of Biological Sciences, Florida International University, Tamiami Campus, Miami, FL 33199, USA
ROHRSTED, T., Delegation of the Commission of the European Communities, P.O. Box 45119, Nairobi, Kenya
RUWA, R. K., Kenya Marine and Fisheries Institute (KMFRI), P.O. Box 81651, Mombasa, Kenya
SASEKUMAR, A., Department of Zoology, University of Malaya, Kuala Lumpur, Malaysia
SCHRIJVERS, J., State Universtiy of Ghent, Institute of Zoology – Marine Biology, Ledeganckstraat 35, B-9000 Ghent, Belgium
SELVAM, V., Department of Zoology, University of Madras, Guindy Campus, Madras 600 025, India
SEMESI, A. K., Department of Botany, University of Dar Es Salaam, P.O. Box 35060, Dar Es Salaam, Tanzania
SHAY, E. G., National Academy of Sciences BOSTID, HA 476, 2101 Constitution Ave., N.W., Washington, DC 20418, USA
SHETE, M. H. A., East African Wildlife Society, P.O. Box 20110, Nairobi, Kenya
SHUNULA, J. P., Institute of Marine Sciences, University of Dar Es Salaam, P.O. Box 668, Zanzibar, Tanzania
SIDRAK, G., Department of Biological Sciences, Florida International University, Tamiami Campus, Miami, FL 33199, USA
SLIM, E., EEC – Mangrove Project, Delta Institute for Hydrobiological Research, Vierstraat 28, 4401 EA Yerseke, The Netherlands
SOLORZANO, L., Instituto Nacional de Pesca Letamendi 102 y La Rita, P.O. Box 5918, Guayaquil, Ecuador
TACK, J. F., Ecology & Systematics Laboratory, Free University of Brussels, Pleinlaan 2, B-1050 Brussels, Belgium
TEAS, H., Department of Biological Sciences, Florida International University, Tamiami Campus, Miami, FL 33199, USA
TEJAN, A. G.,UNDP/FAO Fuelwood Project c/o UNDP, P.O. Box 1011, Freetown, Sierra Leone
THORHAUG, A., Department of Biological Sciences, Florida International University, 555 NE, 15th St. PHH, Miami, FL 33132, USA
TRONO, G. C., Jr, Marine Sciences Institute University of the Philippines, Diliman, Quezon City 1101, Philippines
VAN AVERMAET, F., State University of Ghent, Institute of Zoology–Marine Biology, K. L. Leganckstraat 35, B-9000 Ghent, Belgium
VAN DE KERKHOF, K., Laboratory of Aquatic Ecology, University of Nijmegen, Toernooiveld, 6525 ED Nijmegen, The Netherlands
VAN DER VELDE, G., Laboratory of Aquatic Ecology, Catholic University of Nijmegen, Toernooiveld, 6525 ED Nijmegen, The Netherlands
VAN GANSBEKE, D., State University of Ghent, Institute of Zoology – Marine Biology, K. L. Ledeganckstraat 35, B-9000 Ghent, Belgium
VAN HOVE, S., State University of Ghent, Institute of Zoology – Marine Biology, K. L. Ledeganckstraat 35, B-9000 Ghent, Belgium
VAN SPEYBROECK, D., Department of Botany, University of Nairobi, P.O. Box 30197, Nairobi, Kenya
VAN'T HOFF, T., Stinapa, Bonaire, Netherlands Antilles
VANDEN BERGHE, E., Department of Zoology, University of Nairobi, P.O. Box 30197, Nairobi, Kenya
VANNINI, M., University of Florence, Dept. Animal Biology & Genetics, Via Romana 17, 50125 Firenze, Italy
VASANTHA, K., Centre Adv. Studies Marine Biology, Annamalai University, Parangipettai – 608 502, Tamil Nadu, India
VERGEER, L. H. T., Laboratory of Aquatic Ecology, University of Nijmegen, Toernooiveld, 6525 ED Nijmegen, The Netherlands
VINCX, M., State University of Ghent, Institute of Zoology, Marine Biology Section, K. L. Ledeganckstraat 35, B-9000 Ghent, Belgium
VISSER, N., Agricultural Attache, Royal Nertherlands Embassy, P.O. Box 41537, Nairobi, Kenya
VUAI, M., Forestry Department, Zanzibar, Tanzania
WAMUKOYA, G. M., KEN-U-TAN EXPORTS, P.O. Box 84384, Mombasa, Kenya
WIJNANT, J., Kenya-Belgium Project Marine Sciences Kenya Marine & Fisheries Research Inst., P.O. Box 81651, Mombasa, Kenya
WOITCHIK, A. F., EEC Project Mangrove Ecology, Free University of Brussels ANCH, Pleinlaan 2, B-1050 Brussels, Belgium
WOLANSKI, E., Australian Institute of Marine Science (AIMS), PMB 3, Townsville MC, Queensland 4810, Australia
WOODLEY, J. D., Discovery Bay Marine Laboratory, P.O. Box 35, Discovery Bay, Jamaica
YANEZ-ARANCIBIA, A., EPOMEX, Autonomous University of Campeche, Av. Agustin Melgar, Ap., Postal 520, Campeche 24030, Mexico

Hydrobiologia **247**: 1–10, 1992.
V. Jaccarini & E. Martens (eds), The Ecology of Mangrove and Related Ecosystems.

Developing management plans for the mangrove forest reserves of mainland Tanzania

Adelaida K. Semesi
Botany Department, University of Dar es Salaam, P.O. Box 35060, Dar es Salaam, Tanzania

Key words: mangroves, vegetation, distribution, management, Tanzania

Abstract

The mangroves of Tanzania are currently not being managed although they are legally gazetted as forest reserves. The management of this overexploited and under estimated natural resource rests with the Forest and Beekeeping Division in the Ministry of Natural Resources and Tourism. It is the Division's intention to conserve the mangroves of Tanzania, therefore steps are being taken to reach that goal. So far the Division, through its established mangrove project, has done an inventory of all the mangroves of mainland Tanzania with the assistance of NORAD. Aerial photography and ground checks were used to assess the state of all the mangrove reserves in the country. Literature search and a socio-economic study was also undertaken.

The inventory revealed that the area covered by mangrove vegetation is 115 901 ha while the previous records indicate about 80 000 ha. If the bare saline areas, water bodies, and salt pans which are part of the mangrove reserves are included, the area amounts to 172 879 ha for the Tanzania mainland. The mangrove reserves have already been mapped in 30 sheets. Some sheets are mapped using a scale of 1:25 000 while others are mapped in 1:50 000 scale. The maps show the vegetation types, the area of each vegetation type, and the stand density and height of each compartment.

The paper therefore presents the current status of the mangroves of Tanzania in respect to area, distribution and vegetation types. Included are the strategies being used to develop the mangrove management plans for Tanzania.

Introduction

Most mangrove areas in the world have received very little attention in respect to management. Therefore large areas of many mangroves are being depleted and degraded at an alarming rate (Linden & Jernelvöv, 1980; Mercer & Hamilton, 1984). An inadequate knowledge of the mangrove ecosystem, together with the intense pressure on the resource is the cause of the high rate of depletion of the mangroves (Saenger *et al.*, 1983). Large areas have been converted to agricultural land, prawn and fish ponds (FAO, 1982; Mackinnon *et al.*, 1986; Chua Thia Eng & Paw, 1987), salt pans (Semesi, 1988) or to other uses.

The mangrove forests of Tanzania are currently neither rationally exploited nor effectively protected. The Forestry Division has identified the formulation of a management strategy for the sustainable use and conservation of this areas as a prime task.

Description of the area

Location

Tanzania lies just below the equator, between 1° S–11° 45′ S and from 29° 21′ E–40° 25′ E. It covers an area of 938 000 km^2 (size of France, Germany and Belgium combined or six times the size of England). Tanzania embraces the Islands of Pemba, Zanzibar and Mafia and has a coastline of about 800 km.

Climate

The coastal areas are hot and humid with an average temperature ranging from 22 °C to 30 °C. Humidity is high throughout the year, up to 90% during the rainy season. The climate from June–September is pleasant, with little rain, lower humidity, and is the coolest time. The main rainy season is from March–May and in November to December the area gets intermittent short rains. Lindi and Mtwara have the lowest rainfall, less than 500 mm per annum while Mafia has the highest. Mafia, Rufiji, Kisarawe, Zanzibar, Pemba, Muheza and Tanga have more than 95% chance of 750 mm per annum. Kilwa, Bagamoyo and Dar es Salaam have 85–95% chance of 750 mm rainfall.

The country is influenced by Monsoon winds. The NE Monsoon winds (Kaskazi) blow from October–March and the SE Monsoons (Kusi) blow in April to October. The months of March/April and October/November are calmest because they are the time for Monsoon reversals. The average wind speed during NE Monsoon is 5 ms^{-1} and during SE Monsoon average at 8 ms^{-1}.

Tides

The coast of Tanzania experiences mixed semidiurnal tides. The average tidal range at spring tides is about 3.2 m and at neap tides it is about 1.1 m. The tide is of a standing wave type. The tides and the winds bring about locally generated waves which are variable in strength depending on the position whether in protected areas or in bays or exposed sites. Beach erosion is prominent in exposed sites.

Ocean currents

Tanzania is washed by a northward flowing current known as the East African coastal current. The surface waters are warm, 25–27 °C with only small annual temperature variation. The water bathing the continental shelves of Tanzania and Kenya is poor in nutrients and has a salinity of 34.5 ppt.

Continental shelves

The continental shelf of Tanzania has a total area of 18 000 km^2 with an average width of 15–25 km. The shelf is narrow from Pangani to the Kenya border but south of Pangani the shelf widens and encircles the island of Zanzibar. South of Zanzibar the edge of the shelf again draws near to the shore, and is only a few kilometres distant outside Dar es Salaam. South of Kisiju, the shelf expands to include Mafia Island and the Rufiji Delta. Coral grows extensively in shallow sections where there are also numerous outcrops. The shelf's sediments are generally sandy but become muddy near river mouths.

Drainage/hydrology

The coastal plains of Tanzania are traversed by permanent rivers, seasonal rivers and by numerous creeks. The permanent rivers flowing to the Indian Ocean are: Pangani, Wami, Ruvu, Rufiji, Matandu, Mbemkuru, Lukuledi, and Ruvuma. Rufiji is the largest river and drains 20% of the country.

Soils

The soils in the coastal area and island's are predominantly sandy and coralline with poor moisture holding capacity, extreme alkalinity and hard subsoil which results in poor drainage. Along river basins alluvium deposits are found.

Population

The total population of Tanzania is 23 174 336 (1988 census). Tanzania mainland has 22 533 758 inhabitants and the islands of Zanzibar and Pemba have 640 578 people. The growth rate for 1978–1988 was 2.8% and the mean population density is 25.3 km^2 for Tanzania mainland and 260 km^2 for Zanzibar.

Materials and methods

An inventory of the mangroves of Tanzania was carried out using aerial photographs which were taken in 1988/89. They were supplemented by ground reconnaissance and topographic maps to assess the magnitude of the forests and to classify them into different vegetation types. 30 maps were printed depicting 10 administrative blocks. A socio-economic study on coastal communities and how they use their mangrove environment was carried out (von Mitzlaff, 1990). Finally a workshop on mangrove management brought together resource persons and decision-makers in order to arrive at a consensus on a future policy concerning the management of this resource. These activities were carried out with the support of the Forestry Division and NORAD.

Results

Mangrove forest distribution

Mangroves occur along the entire coastline of Tanzania, from the Kenyan border in the North to the border of Mozambique in the South, and around the islands off the coast (Fig. 1). The mangroves of Mainland Tanzania, including those of the island of Mafia, were found to cover a total area of about 115475 ha. The largest area of mangroves was found in the Rufiji Delta. Fairly large areas were also found at Tanga, Kilwa, at the mouths of the Ruvu, Wami, Pangani and Ruvuma rivers.

The mangroves are distributed in the different administrative regions/Districts as shown in Table 1.

Mangrove vegetation

There are 8 species of mangroves trees that occur in mainland Tanzania belonging to six families: Rhizophoraceae, Sonneratiaceae, Avicenniaceae, Combretaceae, Sterculiaceae and Meliaceae as has been previously reported by Semesi (1986). Most common are *Rhizophora mucronata* and *Avicennia marina* (Table 2). Most important in terms of pole production is *Rhizophora mucronata*.

The outer zone is more often covered by *Avicennia marina* and sometimes by *Ceriops tagal* or *Lumnitzera racemosa*. The seaward zone is occupied by *Sonneratia alba* or *Avicennia marina*. Moving landwards the *Sonneratia alba* zone gives way to the *Rhizophora mucronata* zone which is almost impenetrable because of the arched aerial roots.

Due to the limitation of fresh water coming into the upper intertidal sites some locations like parts of the coast of Tanga, Bagamoyo, Kilwa, Lindi, and Mtwara, the concentration of salts in the soil may exceed the physiological tolerance of all the plant species and thus a vegetation-free sandy area is found before other non mangrove halophytes like *Sesuvium portulacastrum* and *Sueda monoica* are observed. This area is normally flooded by seawater twice monthly during the highest spring tides.

Mangrove fauna

The crabs are the most conspicuous invertebrates of the mangrove. The upper zone is inhabited by marsh crabs (*Sesarma*). In the *Ceriops* and *Bruguiera* zone, *Uca* spp (the fiddler crabs) are the most dominant. The creeks are inhabited by *Uca*

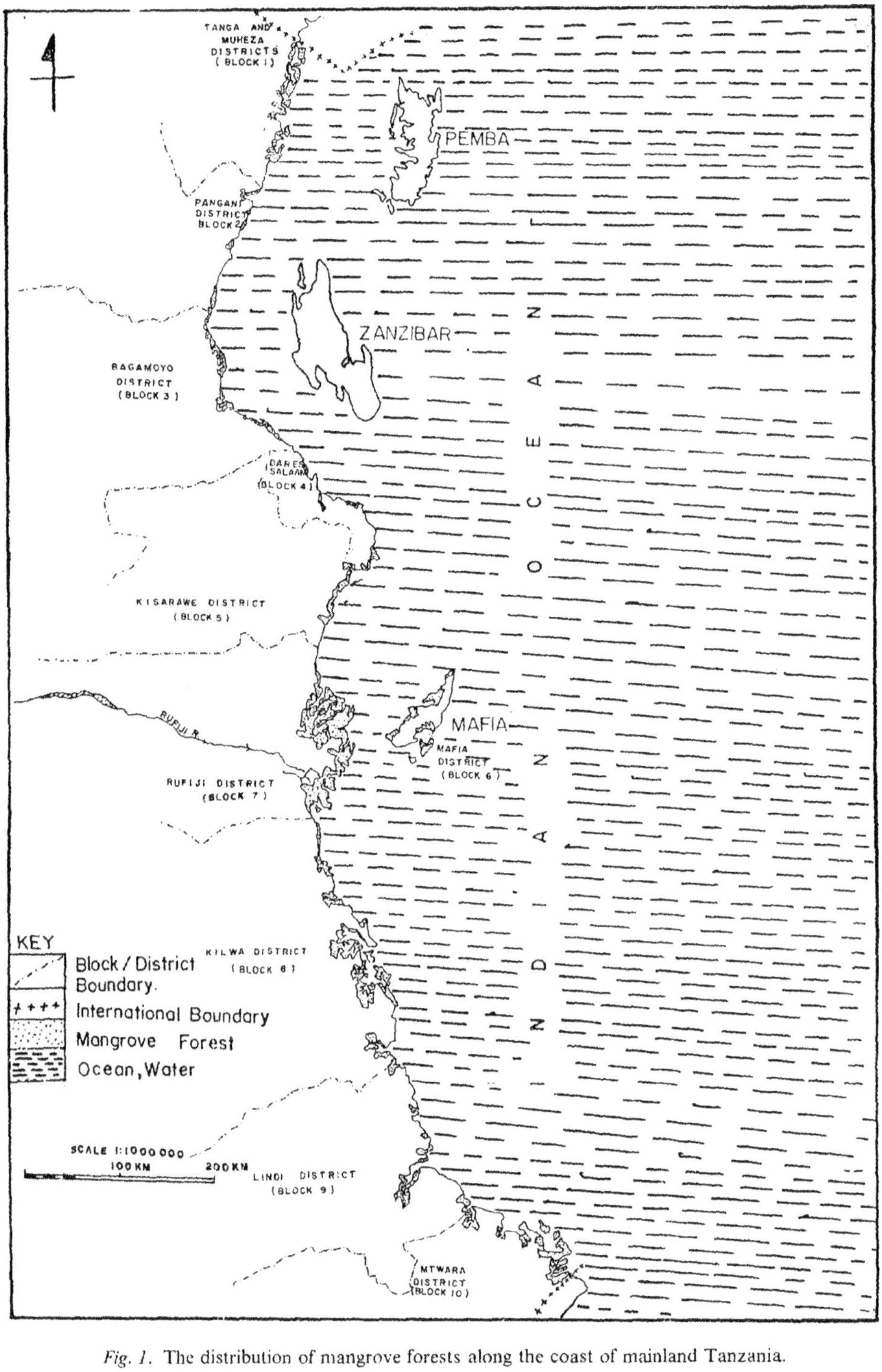

Fig. 1. The distribution of mangrove forests along the coast of mainland Tanzania.

Table 1. Mangrove forest area in the different districts in Tanzania.

District/region forest records	Area quoted in inventory (ha)	Area found in 1989 (ha)
1. Muheza and Tanga	9,996.1	9,403.3 (12,931.3)*
2. Pangani	600.0	1,755.6 (3,035.4)*
3. Bagamoyo	1,499.0	5,635.8 (9,183.0)*
4. Dar es Salaam	–	2,168.2 (3,213.3)*
5. Kisarawe	3,498.0	3,858.3 (6,050.3)*
6. Rufiji	40,470.0	53,254.8 (67,612.2)*
7. Mafia	404.7	3,472.9 (4,365.0)*
8. Kilwa	13,994.5	22,438.7 (36,747.5)*
9. Lindi	1,416.0	4,546.5 (7,300.8)*
10. Mtwara	7,996.9	8,941.5 (13,249.1)*
Total	79,875.2	115,475.6 (163,688.9)*

* The figure in the brackets include also the area of water in the creeks, saline bare areas, and salt pans.

and *Scylla serrata*. On sandy portions soldier crabs (*Dotilla* spp) are seen. *Dotilla fenestrata* is restricted to the East African coast.

The molluscs *Potamides decollatus* and *Littorina scabra* are the most common species found in the mangrove formation north of Tanga. The mangrove snails like *Terebralia*, *Cerithidea* and *Nerita* are important. Barnacles and the spiny oyster (*Crassostrea cucullata*) are very common on stems of mangroves especially on *Sonneratia* trees. On the leaves of mangroves especially those of *Bruguiera*, *Sonneratia* and *Ceriops* are seen species of *Littorina*.

Bees are visitors to the flowers of the mangroves, the honeydew of which attracts the weaver or tailor ants, *Oecophyllia* which weave the leaves together to form a nest and are most common in the *Ceriops* zone in Muheza.

Several demersal fish species inhabit the creeks and the mud skipper lives in the mangrove mud.

Mangroves are also the feeding and hunting places for a number of birds: herons, egrets, ibis, sea eagles, and several species of waders and Kingfishers. The insect and crab eating kingfisher, *Halcyon senegaloiides* is restricted to the mangroves while the other birds are also found in other habitats. The area between the Rufiji delta and Mafia island contains large numbers of waders and terns during the winter in northern Europe (Danish-Tanzania ICBP expedition to Tanzania from January–March 1988). The most

Table 2. The relative composition of Mangrove vegetation types found along the coast of Tanzania.

Classification	Area ha	%
Rhizophora dominant with occurrences of *Avicennia, Ceriops, Sonneratia, Bruguiera, Heritiera* and/or *Xylocarpus*	55549.9	48
Sonneratia almost pure stands	1223.3	1
Sonneratia dominant with *Avicennia, Bruguiera* and/or *Rhizophora*	6123.2	5
Heritiera – almost pure stands	91.2	0
Heritiera dominant with *Avicennia, Bruguiera* and/or *Rhizophora*	8188.4	7
Avicennia dominant with *Rhizophora, Bruguiera, Heritiera, Ceriops* and/or *Xylocarpus*	17141.6	15
Avicennia – almost pure stands	1687.4	1
Mixture of *Avicennia* and *Cerips*	17432.7	15
Ceriops dominant with occurrences of *Rhizophora, Avicennia* and/or *Bruguiera*	8037.9	7
Total Mangrove	115475.6	100
Water in creeks	24076.0	
Clear cut areas	4435.0	
Saline, bare areas	20740.0	
Salt pans	3093.0	
Non-mangrove forest	5069.3	
Total reserve area	172888.9	

numerous species counted were Curlew sandpiper 15000; Little Stint 8000; Crab plover 5400; and Roseate Tern 5500. Lesser Black-backed Gull, Herring Gull and Caspian Tern also spend the winter in the Rufiji delta.

Hippopotamus, monkeys and wild pigs are common mammals in the mangroves. Crocodiles are found in the major rivers. However very little information is available on the fauna found in the mangroves of Tanzania and as such we cannot identify their critical habitats within the reserves.

Uses of mangroves

Mangroves are an item of trade and they are a source of employment and income for the coastal communities: the wood is used as fuel for commercial production of salt, lime and for the processing of fish. Mangroves are cut for poles for export and as local building material. Mangrove areas are used for the construction of solar evaporation pans for the production of salt and in the Rufiji Delta they are turned into rice fields. Fisheries, and the prawn catches in particular, depend on an intact mangrove ecosystem. Mangroves are the source of domestic fuelwood and of raw material for artisans and boat-builders; fishermen make traps and fish floats from the trees, some species are used for animal fodder, vegetables and others for local medicine. Rarely bees are kept in these forests.

Potentials of the mangrove resource

Presently most of the pole cutting is done illegally and the royalty fees collected in 1988, amounting to about 10 Million TShs, were only a fraction of the actual amount of revenue due to the government. 15 Million TShs were collected as revenue from the export of prawns in 1989. It has been estimated that the present prawn fishing does not net even a quarter of the possible catch in Tanzania (Siegel, 1986). In terms of revenue, but also in terms of income and employment, the marine resource has not been fully exploited.

Salt production by solar evaporation is taking place in the mangrove areas but is not run effectively and could be much higher. Similarly the potential of this resource for tourism, for commercial honey and wax production, and for any kind of aquaculture has not yet been realized at all in Tanzania.

Problems and threats to the mangrove ecosystems

Ecological issues

Though the main threat to the mangroves comes from man, there are also natural causes for their death.

– River floods may cause considerable areas of mangroves to die as a result of alteration of water level, bank erosion and diversion of water courses. Mangroves cannot tolerate long inundation and will die. This phenomenum can clearly be observed in some parts of the Rufiji Delta.

– Sand deposition from sea and land can cut off portions of mangroves from salt water causing the mangroves to die. This problem is pronounced in most parts of Bagamoyo District.

– Sea level rise (because of the global warming) would flood the present mangrove areas and kill all mangroves (Hekstra, 1989). However they may soon colonize new land.

Social and economic issues

It is mainly commercial activities which are over-exploiting the mangrove resource whereas the coastal village communities use their mangrove environment mostly in a sustainable way.

The destruction of large areas of mangrove forests is caused by business people who usually come from inland regions. They either cut the mangroves for boiling brine for the production of salt (Tanga), for lime burning (Bagamoyo, Lindi, Mtwara) or for fish processing (Pangani). They use the resource without caring about its survival or reproduction, the communities are not enumerated for the loss of their resource and almost none of the profit gained from these enterprises remains in the villages.

This is similarly true for the construction of salt

pans for solar evaporation in Tanga, Bagamoyo and Mtwara Districts. They are often placed within mangrove areas which entails the total clearing of the forest, whilst topography and soil are equally suitable for their location behind mangroves or onbare patches.

Pole cutting, if done in an unplanned way and on a large scale, can reduce the genetic pool and endangers reproduction.

The demand for domestic fuelwood endangers the mangroves only in densely populated areas around towns. If the inland resource is already exhausted, people turn to mangroves as a second choice for fuelwood.

Agricultural activities interfering with the mangrove environment take place in the Rufiji Delta only and this quite recent development has a particular historical background. The practice of shifting cultivation for rice farming is becoming a serious threat to the mangroves.

The mangroves are also endangered by oil pollution and industrial pollution, by excessive siltation because of man's degradation of catchment areas, by herbicides coming from inland via the rivers, and by the construction of dams or major irrigation schemes.

The awareness of the overall importance of this resource is low amongst the people in general. They value it rather for its direct monetary return than its many indirect benefits.

Institutional issues

Human activities threatening the mangrove environment should be controlled by the respective government bodies. There are several institutional problems which hamper this control:

- No overall authority exists which effectively coordinates conflicting issues like salt licenses and land titles concerning the Division of Forestry, the Division of Lands and the Ministry of Water, Energy and Minerals simultaneously.
- Lack of adequately trained personnel, very low numbers of field personnel, lack of transport and infrastructure renders actual control through patrolling impossible.
- The knowledge on the mangrove ecosystem and of activities or technologies related to it is inadequate and dispersed to different institutions (e.g. Forestry, Fisheries). No institutionalized channels of communication have been established to overcome this problem.

Policy issues

The management policy which consist of control and prohibition only, without having the means and capability to carry it out (lack of funds, personnel, transport and infrastructure), has affected the management of the mangroves in Tanzania. On top of this it has not considered traditional interests of people in their resource or their involvement in its management. The National ban on cutting mangroves spelled out in 1987 has poisoned the relationship between people and government bodies even further.

Legal issues

Despite the ban, licenses for cutting poles and for the construction of salt pans are issued. Consequently the users of the mangrove resource, be it villagers or business people, are confused about the actual legal situation and their rights. Also they lack any power to lobby for their interests.

The traditional rights of the coastal village communities to use their resource for their small local needs have been disregarded although they have never been a threat to the mangroves.

Limitations and constraints encountered during management planning

The following limitations and constraints have to be recognized:

- No national strategy for conservation exists which could provide a framework for the formulation of the mangrove management plan.
- Limited knowledge on the resource concerning ecological, botanical and zoological matters is a major obstacle for planning.
- Low salaries, lack of incentives, transport and equipment are omnipresent and inextricable problems which hamper motivation and contribute to low efficiency.
- Laws do not necessarily guarantee protection

and lack of forest patrols and law enforcement is the main hindrance in resource protection.

Identified management objectives and strategies

The purpose of the management plan is to identify a programme and its objectives and strategies in order to assist the government in its endeavour to rationally use and conserve the mangrove resource of Tanzania. It will be taken as a basis for a plan of operations. The overall development goal is to enhance the contribution of mangrove ecosystems to the economy of the country and the objective of the management programme is the rational utilization of mangrove ecosystems on a sustainable basis.

To achieve this objective the following short term objectives have to be realized:

- to conserve mangrove areas with essential protective functions
- to maintain habitat for fauna and flora of this unique ecosystem
- to optimize the combination of direct and indirect uses of mangrove areas
- to meet needs of communities living in and depending on mangrove environment
- to conserve representative areas for biodiversity, tourism, research and education
- to maintain mangrove gene pools
- to improve the institutional set up
- to support research activities

Management strategies

The strategies to achieve the above mentioned objectives are geared towards the institutional as well as the operational level. First and foremost the capabilities of the Forest Division will be strengthened and enlarged. A new section within the Forest Division will have the overall authority for the management of the mangrove resource.

The number of field staff will be drastically increased, (mostly to be recruited from coastal villages) and will be provided with appropriate training in all matters concerning mangroves (management, extension work, harvesting etc.). Their equipment, transport and other infrastructure will be improved. Training for high level personnel will also be provided.

Lack of coordination between the many different government institutions directly or indirectly involved in issues concerning mangroves has been identified as a major hindrance to successful management.

The governmental institutions directly or indirectly involved in issues concerning mangroves (Division of Lands, Fisheries Division, Forest Division, Ministry of Water, Energy and Minerals, Ministry of Local Government) and potential resource persons will be represented in a steering committee which is directly linked to the Director of Forestry and the coordinator of a National Mangrove Management Project. This steering committee, as the linking body to the ministries, will have to deal with the conflicting interests and responsibilities of their institutions and agree on a hierarchy of decision-making and communication. Problems arising from unclarified legal issues have to be solved before any new policy can be formulated.

In order to find a consensus on such disputed legal issues the respective government bodies concerned will be brought together by the Director of Forestry to discuss and decide on the necessary steps to be taken. This decision will be widely publicized in order to ensure the different government institutions as well as the users of the mangrove resource know about their rights and responsibilities.

The participation of the coastal people is of paramount importance for the success of the management of the mangrove resource. Only if it is an asset to them they will defend it against illegal exploiters. Different approaches will be used to seek their participation:

- local people will be involved actively in the identification of zones for different uses near their villages. Not only has this merit as a kind of awareness campaign, but it will also give the villagers a chance to voice their interests.
- certain areas will be allocated to villages for their local use (firewood, raw material).

- other areas will be allocated to individual villagers or groups for commercial use, i.e. they will have preferential access to licenses for the exploitation of the resource (pole cutting, salt production).
- training on the management by rationally exploiting the resource will be available for those who are engaged in it. This extension service has to be provided by the Mangrove Forest Officers.
- villagers will be given support in developing alternative economic opportunities. The Beekeeping Division will encourage and assist village women and men in the establishment of a honey and wax industry, in its processing and marketing. The Fisheries Division will assist in improving the situation of the artisanal fishermen through the supply of gear or through loans, through training and through improving technologies of processing fish (the latter concerning women mainly).
- complementary measures, like establishing fuelwood plots for villages or individuals, will receive every possible support.
- infrastructure improvements will benefit the villages as well as the mangrove management.
- villagers will be employed as Mangrove Forest Guards
- contracts will be made between the villagers or groups of villagers involved and the Forest Division with regard to their rights and obligations of exploiting and protecting the resource. Ultimate authority lies with the Forest Division who can withdraw these rights if the obligations are not met.
- through the establishment of Users' Advisory Boards villagers and other commercial users will be given an opportunity to voice their interests.

Zoning as a management strategy

The mangrove forests are of very different qualities and occur under varying conditions. In order to optimize their direct and indirect uses they have been categorized in four zones.

Zone I are all forests which will receive *prime protection*.

Zone II are all ecologically stable areas with sufficient regeneration potential. They are considered productive forests suitable for controlled harvesting.

Zone III are degraded areas which will be closed for some time to allow *recovery and rehabilitation*.

Zone IV are areas marked for '*development*' of different kinds.

Conclusion

Ecological changes, uncontrolled exploitation by man, uncoordinated and inadequate institutional measures, deficiencies in policies and an unclear overall policy have all contributed to the present situation, where the survival of the mangrove forests of Tanzania is in doubt. There is urgent need for planned, coordinated action, there is need for a new policy and there is need also for an organizational set-up which can meet these tasks.

Acknowledgements

I would like to thank the Norwegian Agency for International Development (NORAD) for financial support and the University of Dar es Salaam for granting me permission to work on the mangrove project. Many thanks should also go to messrs A. Mttoi, S. Kipangula, A. Mtweta for the airphoto interpretation, to Mr. Håkan Sjöholm, Coordinator of Catchment Forest Project for his valuable discussions. Dr. Von Mitzalaff, Dr. K. Howell, Dr. F. Urasa, Dr. A. Kivaisi and Mangrove Workshop Participants must be thanked for their critical comments.

References

Chua Thia. Eng & J. N. Paw, 1987. Aquaculture Development and Coastal Zone Management in S.E. Asia: Conflicts and Complementary. Proc. of the Fifth Symp. on Coastal and Ocean Management. 2: 2007–2019. Elsevier Scientific Publishing Company, Amsterdam.

FAO, 1982. Management and Utilization of Mangroves in Asia and the pacific. FAO Environment paper No. 3. Rome: FAO.

Hekstra, G. P., 1989. Global warming and Rising Sea levels: The Policy Implications. The ecologist, 19 (1): 4–15.

Linden, O. & A. Jernelov, 1980. The Mangrove Swamp – An ecosystem in Danger. Ambio, 9: 81–88.

Mackinnon, J., K. Mackinnon, G. Child & J. Thorsell, 1986. Managing Protected Areas in the Tropics. IUCN Switzerland.

Mercer, E. & L. S. Hamilton, 1984. Mangrove ecosystems: Some economic and natural benefits. Mar. Resour. UNESCO, 20: 14–19.

Von Mitzalaff, U., 1990. Coastal communities in Tanzania and their mangrove environment. A socio-economic study prepared for NORAD-Dar es Salaam, 60 pp.

Saenger, P., E. J. Hegerl & J. D. S. Davis (eds), 1983. Global status of mangrove ecosystems. Commission on Ecology Papers Number 3. Gland Switzerland: IUCN. The Environmentalist 3. Supplement No. 3.

Semesi, A. K., 1986. Zonation and vegetation structure of mangrove communities in Tanzania. In: Mainoya J. R. & Siegel P. R. (eds), Proc. of a Workshop on Save the Mangrove Ecosystems in Tanzania. 21–22nd February, University of Dar es Salaam. 174–181.

Semesi, A. K., 1988. Status and utilization of mangroves along Tanga, Tanzania. In: Mainoya J. R. (ed.), Proc. of a Workshop on Ecology and Bioproductivity of the Marine Coastal Waters of Eastern Africa. 18–20th January, University of Dar es Salaam. 174–181.

Hydrobiologia **247**: 11–16, 1992.
V. Jaccarini & E. Martens (eds), The Ecology of Mangrove and Related Ecosystems.

Structure and species distribution in Coringa mangrove forest, Godavari Delta, Andhra Pradesh, India

Jayapaul Azariah, Hilda Azariah,* S. Gunasekaran & V. Selvam
*Department of Zoology, University of Madras, Guindy Campus, Madras – 600 025, India; *CAS In Botany, University of Madras, Guindy Campus, Madras 600 025, India*

Key words: Coringa mangrove, structure, distribution, atmospheric pollutants, industrial impact

Abstract

Coringa mangrove forest is located in the Godavari delta, Andhra Pradesh, India. The mangrove community consisted of more than 13 species of mangrove and other plants in the present study area. The following three dominant mangrove plants, *Avicennia marina*, *Excoecaria agallocha* and *Sonneratia apetala* were found to be present on the banks of a major channel of the Godavari river running through the forest. The structure and species distribution of mangrove, in the Channel Nagathana Kalava has been described. The area behind the belt consisting of *Acanthus ilicifolius* and *Myriostachya wightiana* is generally colonized by *E. agallocha* and *A. marina*. The zone has been called the *Avicennia* and *Excoecaria* zone. Adjacent to this zone species like *Aegiceras corniculatum* and *A. officinalis* were the common species. In the flat clayey soil, *Suaeda maritima* was found to grow. In areas of high elevation, devoid of inundation of tidal seawater during the high tidal period, species such as *M. wightiana* and *Acanthus* were found to colonize both the banks of the channels.

An analysis of species diversity, indicated a definite trend in the distribution of mangrove from the mouth of the estuarine region to the inland waters.

The levels of atmospheric pollutants such as sulphur dioxide (SO_2), oxides of nitrogen (NO_x), ammonia (NH_3) and suspended particle matter (SPM) were within the legal limits.

Introduction

A survey of the global status of mangroves may show that vast areas are being destroyed either intentionally or as a secondary result of other activities (Saenger *et al.*, 1983). The mangrove of India also is no exception to such an uncontrolled exploitation. The mangrove vegetation of India have an area of approximately 356 500 ha of which Krishna-Godavari mangrove complex comprises an area of approximately 12 800 ha (Sidhu, 1963). The Coringa mangrove forest, which is a part of the Godavari mangrove forest, is located on the southern coast of the Kakinada town in Andhra Pradesh, where the present investigation was carried out. The Kakinada port is a natural harbour and an all weather port which can provide protection from severe weather conditions. Hence, it is possible to handle ships for more than 300 days per year. Currently, loading and unloading are carried out around the clock. Recently, the Government of Andhra Pradesh has developed a master plan to convert the Kakinada port into a major port so as to make it a gateway for trade and export. The proposed developmental plan involves the construction of three deep water berths, extension of break water to 3.35 km, dredging of access canal and provision of ancillary facilities such as road, approach bridges and other navigational aids. The project, costing Rupees 1 335.8 million, is expected to be completed by April 1995 (Indian Express, 1990).

A second major developmental activity is the building of the fertilizer complexes namely, Godavari Fertilizers and Chemicals Ltd (Capital cost of Rupees 1 080 million, work completed) and the Nagarjuna Fertilizer and Chemicals Ltd (capital cost of Rupees 6 700 million-work nearing completion). These two fertilizer factories are expected to meet the fertilizer demands within a 200 km range from the factory. In view of such a rapid pace of industrialization in the coastal zone of Kakinada, a detailed study on the environmental impact assessment was made on a multidisciplinary basis. In an earlier paper the results of the analysis of the mangrove vegetation within a 15 km radius of the fertilizer complex have been given (Azariah *et al.*, 1990). The present paper examines the structure of the mangrove community and its distribution in the Coringa forest, which is located on one of the deltaic branches of the northern zone of the Godavari delta, so as to provide quantitative baseline information on the current status of the ecosystem together with a survey of the current levels of atmospheric pollutants such as sulphur dioxide, oxides of nitrogen, ammonia and suspended particulate matter.

Description of site

Mangrove forests are seen in places like Coringa, Yanam, Kandikuppa, Kakinada and Sacramento which are in the East Godavari delta. The Coringa river originates at Yanam which is about 6 km on the upper reach of Godavari river. This tributory along with Gadeur forms a part of the Godavari estuarine system. Both the tributaries open into the southern side of the Coringa bay. The bay is named after Coringa (Koriangi, a well known ancient port). A group of islets are located in the vicinity of Coringa bay. Among these, Hope Island, Bhairavapalend series and Coringa and tributaries, Gaderu and Vasavakalava are noteworthy (see Fig. 1).

Materials and methods

The present study was carried out in the Nagathana Kalva (canal) which extends for a distance of 10 km from the Coringa river. The area is swampy with knee deep mud.

A series of quadrats of 5 × 5 m (Sidhu, 1963) were laid at an interval of 1 km along the 10 km long canal. The number of trees in each quadrat was counted and the data were tabulated. Using certain indices of species structure in community, data on the index of dominance, species diversity, richness of species were calculated. Besides, data on the relative density, relative frequency and relative dominance were calculated. Details of the formulae used are given in an earlier paper (Azariah *et al.*, 1986). A few micrometereological assessments were carried out to obtain baseline information on the level of sulphur dioxide, oxides of nitrogen, ammonia and suspended particulate matter. The details of the analysis are given in an earlier paper (Azariah *et al.*, 1990).

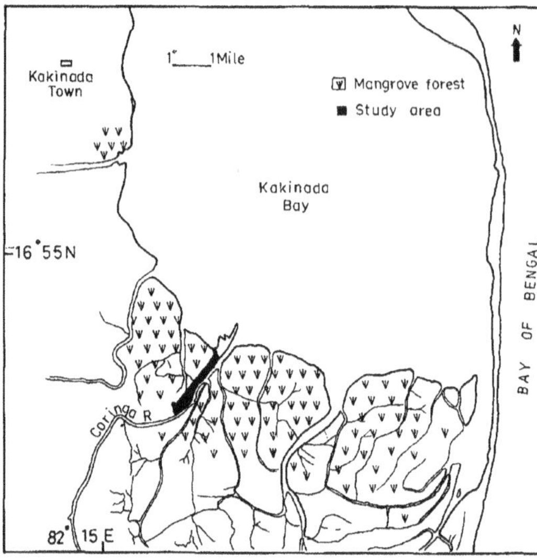

Fig. 1. Map showing the location of the Coringa mangrove forest and the study site.

Results

Concentration of atmospheric pollutants

A biweekly sampling of atmospheric pollutants were made using a high volume sampler. The studies were carried out from December 1988 – May 1989 (Table 1).

The SO_2 concentration varied from zero to 10 $\mu g\ m^{-3}$ over a period of 6 months. The values didn't exceed the maximum admissible limit of 80 $\mu g\ m^{-3}$. The NO_x values varied from zero to 98 $\mu g\ m^{-3}$. The maximum value was observed during the month of September 1988, which may indicate that there were occasional outburst of oxides of nitrogen. It may be seen that the levels of NO_x exceed the maximum admissible limit (80 $\mu g\ m^{-3}$). The concentration of ammonia varied from zero to 28 $\mu g\ m^{-3}$ against the admissible limit of 30 $\mu g\ m^{-3}$. It can be seen that the levels of ammonia were nearing the critical concentration. The total suspended particulate matter (SPM) varied between 31 and 137 $\mu g\ m^{-3}$, which is well below the admissible limit of 200 $\mu g\ m^{-3}$. The dust fall is low, the value being 9.8 metric tonnes km^{-2} per month.

A comparison of data between Coring village and Kakinada neighbourhood was made. It can be seen that the levels of the SO_2 were higher in the Coringa mangrove forest than at Kakinada neighbourhood during the summer season, i.e., March to May. It can be noted that the level was higher at Kakinada during the post monsoon season (December to February). A similar pattern may be seen with respect NO_x. The concentration of ammonia and SPM was higher during both the summer and post monsoon months. A similar pattern was observed with respect to SPM (See Fig. 2).

Nature of environmental pollutants in the vicinity of Kakinada (Table 1)

Level of sulphur dioxide

In an earlier study, it has been found that the SO_2 concentration did not exceed the limit throughout the year (Azariah *et al.*, 1990). Notable variations in its concentration were found between the summer and North East monsoon (June to August) seasons at Kakinada township with the values of 1.9 and 4.4 $\mu g\ m^{-3}$ respectively. During the summer the concentrations of was minimal in the entire vicinity of Kakinada, the value being 1.9 $\mu g\ m^{-3}$ (Table 1).

NO_x

The concentration of NO_x exceeded the admissible limit in the Kakinada town but it remained within the admissible limit in the coastal zone during the summer (March to May) season. A relationship between NO_x and human activity

Table 1. Concentration of atmospheric pollutants ($\mu g\ m^{-3}$) in Coringa forest and Kakinada neighbourhood

Period	SO_2	NO_x	NH_3	SPM
Coringa:				
Dec 88–Feb 89	1.3 ± 2.4	10.6 ± 5.7	14.4 ± 8.1	64.9 ± 23.8
	(0–3.0)	(2–20.2)	(0–28)	(31–97)
Mar 89–May 89	3.6 ± 2.9	36.0 ± 27.0	13.0 ± 4.7	80.5 ± 31.6
	(0–10)	(0–98)	(0–19)	(38–137)
Kakinada neighbourhood:				
Dec 88–Feb 89	2.76 ± 2.83	46.9 ± 30.5	31.0 ± 17.3	104.03 ± 52.8
	(0–20)	(0–330)	(0–180)	(24–294)
Mar 89–May 89	1.9 ± 2.26	45.1 ± 36.2	9.3 ± 9.46	106.86 ± 56.8
	(0–12)	(0–276)	(0–81)	(28.–302)
Maximum admissible limit	80 $\mu m\ m^{-3}$	80 $\mu g\ m^{-3}$	30$\mu g\ m^{-3}$	200 $\mu g\ m^{-3}$

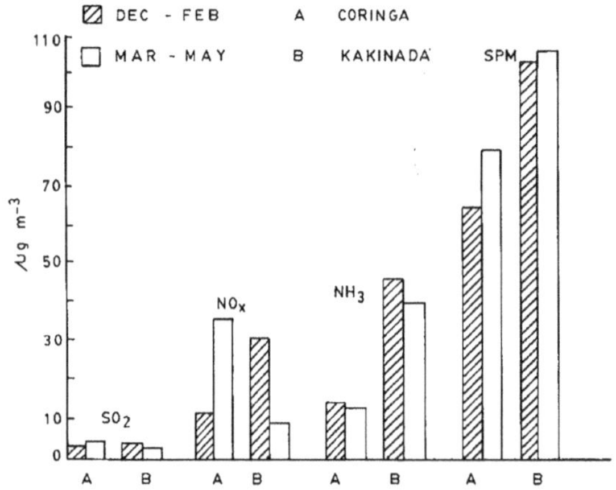

Fig. 2. Concentration of atmospheric pollutants in the Kakinada town and Coringa mangrove forest.

was found. The concentration was higher in places where there were more human activities in terms of handling nitrogen based fertilizers (loading and unloading). During the summer months, the coastal zone registered a value ranging between zero and 40 $\mu g\ m^{-3}$ and it varied from 2 to 330 $\mu g\ m^{-3}$ during the post monsoon season i.e. December to February.

NH_3

The ammonia concentration exceeded the admissible limit, at all times, except in the coastal zone during the summer season. The maximum concentration, 180 $\mu g\ m^{-3}$, was found during the post monsoon season in the Kakinada and lower levels (0 to 8.2 $\mu g\ m^{-3}$) were recorded during the summer season in the coastal zone.

SPM

The concentration of suspended particulate matter exhibited a direct relationship with human activity. It's level was high both during the post monsoon and summer seasons, the values being 24 to 294 $\mu g\ m^{-3}$ and 28 to 302 $\mu g\ m^{-3}$ respectively.

Analysis of mangrove ecosystem

The number of plants recorded in 10 quadrats along the distance of 10 km are given in Table 2. The vegetation in the mangrove swamp consisted

Table 2. Structure of Coringa Mangrove Community, East Godavari

Sl. no.	Species	Quadrat number										Total number of individuals	Frequency (%)	Relative density (%)	Relative frequency (%)
		1	2	3	4	5	6	7	8	9	10				
		Number of individuals													
1.	*Excoecaria aquallocha*	25	9	38	28	8	17	22	24	8	9	188	100	30.62	17.54
2.	*Avicennia marina*	6	27	7	14	4	11	4	12	11	6	102	100	16.61	17.54
3.	*Avicennia officinalis*	–	–	–	–	2	6	7	3	9	7	34	60	4.4	8.77
4.	*Aegiceras corniculatum*	–	14	3	4	–	2	–	–	4	–	27	50	5.54	10.53
5.	*Acanthus ilicifolius*	–	–	–	–	–	–	6	17	17	23	63	40	10.26	7.02
6.	*Sonnaretia apetala*	–	–	–	–	–	–	–	4	8	9	21	30	3.42	5.26
7.	*Derris trifoliata*	–	–	–	–	–	–	4	6	2	3	15	40	2.44	7.02
8.	*Ipomea tuba*	–	–	–	–	–	–	3	4	3	6	16	40	2.61	7.02
9.	*Cyperus rotundus*	–	–	–	–	–	–	33	22	–	–	55	20	8.96	3.51
10.	*Clerodendrum inerme*	–	–	–	–	–	–	–	–	2	3	5	20	8.96	3.51
11.	*Ceasalpinia nuga*	–	–	–	–	–	–	–	–	1	3	8	20	0.81	3.51
12.	*Myriostachya wightiana*	–	–	–	–	–	–	–	–	4	2	6	20	1.30	5.26
13.	*Suaeda maritima*	62	12	–	–	–	–	–	–	–	–	74	20	12.05	3.51
	Total	93	62	48	46	14	36	79	93	71	72	614			

of 13 species of plants out of which 11 species belonged to the mangroves. Secondly, their distribution varied in a distinct pattern from the mouth of river towards the inland water.

The distribution of *Avicennia officinalis* Lam. was restricted towards inland waters whereas *Excoecaria agallocha* Lam. and *Avicennia marina* (Forsk.) Vierh. were found along the entire stretch of the river i.e. from the coast to the inland waters. Between *E. agallocha* and *A. marina* the dominance of the former was seen throughout the study area. However, *E. agallocha* was more thickly distributed towards the mouth of the river. On the other hand for the latter species, although distributed throughout the study area, the magnitude of colonization was relatively less than *E. agallocha*.

The zone of mangrove can be divided into two areas. First, the *Avicennia* area which is located near the inland water and the second region dominated by *E. agallocha*. This region was found more towards the ocean front. A feature of interest is that, when the number of *E. agallocha* increased the number of *Avicennia* decreased. It is also interesting to note that *Avicennia officinalis* was not distributed within the area of 4 km from the mouth of the river. It is likely that this species may prefer a inland habitat with lesser magniture of salinity variation. Whereas *A. marina* did not exhibit such a relationship. A non mangrove species *Ipomaea tuba* Linn. which was closely associated with mangroves showed a discontinuous distribution. Nine species (item No. 5 to 13 in Table 2) showed a salinity related distribution in that they were represented in regions the river, where there may be wide fluctuations in salinity, and were recorded only in quadrat numbers 6 to 12. The quadrats 1 to 3 were located near the mouth which was characterised by the presence of neritic waters throughout the year due to its proximity to the sea whereas quadrat numbers 7 to 12 were located in the inland area where fresh water condition existed during the monsoon season and brackish water regime found during the post monsoon and summer seasons (Sarma & Ganapti, 1971). *Myriostachya wightiana* Hook was found in habitats which were located more toward the inland waters. On the other hand, *Suaeda maritima* Dumort was recorded in the first two quadrats which were laid very close to the sea coast.

In general, the following three distinct zones of distribution can be made.

Zone I: *Excoecaria agallocha* – *Suaeda maritima* zone, which extends 1 to 2 km from the river mouth.

Zone II: A middle zone in which a mixed community of *Excoecaria*, *Avicennia* and *Aegiceras corniculatum* L. may be found. This zone extends upto about 6 km from the river mouth.

Zone III: Was represented by *Excoecaria*, *Avicennia* and *Cyperus* which were found at a distance of 7–8 km from the sea. In this zone a mixed community of *Excoecaria*, *Avicennia* and *Acanthus* may be recognized.

It is interesting to note that *Suaeda maritima* was found only near the swampy regions of the coastal zone along with *Excoecaria* and *Avicennia*. Another interesting feature is the occurrence of the *Rhizophora apiculata* Blume and *Lumnitzera racemosa* Willd. But these were not present within the 10 quadrats. *Rhizophora apiculata* was represented by a single tree, which was located in the middle stretch of the study area. Whereas *L. racemosa* was found to be sparingly distributed.

The species diversity varied from 0.92 to 3.13 with the highest level towards the inland waters. Similarly, an ascending value in species richness was recognized from the mouth of the river towards the inland water. Where *Excoecaria* and *Avicennia* dominated, as in the case of first 3 km, the richness index was poor (1.02 to 1.67) whereas it ranged between 4.85 and 5.4 in the region of freshwater regime. The species evenness index was relatively high in areas of low salinity (Table 3).

Discussion

Figure 2 compares the concentration of atmospheric pollution between Kakinada town and its neighbourhood and the Coringa forest. The variation in the atmospheric pollution is a function of

Table 3. Species diversity, species richness and species eveness values in different quadrats

Indices	Quadrat number									
	1	2 Coast line	3	4	5	6	7 Inland water	8	9	10
Species diversity	1.15	1.81	0.92	1.26	1.42	1.69	2.52	2.69	3.13	2.95
Species richness	1.02	1.67	1.19	1.20	1.75	1.93	3.16	4.05	5.40	4.85
Species eveness	0.73	0.90	0.58	0.76	0.90	0.84	0.90	0.85	0.90	0.89

the influence of monsoon accompanied by the direction of wind speed. The behaviour of SO_2 and NO_x was similar in that the concentration of these components were higher in Coringa during the summer season when the wind was blowing in the SE and SW directions. It is likely that wind borne SO_2 and NO_x may be concentrated at the Coringa reserve forest. On the other hand, the concentration NH_3 and SPM was higher at the Kakinada town which is due to wide spread human activity and the use of nitrogenous fertilizers.

The species structure in the mangroves of Coringa appears to vary from site to site. Rao *et al.* (1963) and Venkatesan (1966) reported the occurrence of 26 species in the neighbourhood of Kakinada. Of which *Suaeda maritima*, *Heliotrophium curassaricum* L., *Fimbristylis cymosa* R. Br. and *Aeluropus lagopoides* (L.) Trin. occur in considerable numbers. In the Coringa forest, Blasco (1975) recorded the occurrence of *Excoecaria agallocha*, *Avicennia officinalis*, *Phoenix paludosa*, *Derris* sp., *Dalbergio* sp. and *Ceasalpinia* sp. Later Sidhu (1963) recognized three major forest communities. They are tidal ever green, semi-evergreen and tidal decidous. He recognized pure *Avicennia* community, mixed *Avicennia* community and *Aegiceras* mixed community. He also recognized a total number of 20 species in his studies carried out the at Coringa Coup No. 2 and 4 (coup = area alloted by forest authorities for clear felling) and various region within the Coringa reserves such as Gundhuhalla, Varavacalea and Padavakpalam. An interesting and contradictory feature between the earlier and the present study is the variation and the dominance of species within the community. In the above areas of study, *Avicennia* was the dominant species, whereas in the present study, *Excoecaria* was found to be the dominant species and no pure stand of *Avicennia* was recognized at any place. Another feature of difference is the association of *Avicennia* with *Acanthus*.

References

Azariah, J., H. Azariah, P. Pandion & S. Gunasekaran, 1990. Coastal wetlands of Kakinada, Andhra Pradesh. A case study in environmental impact assessment. In Wetland ecology and management, Bharatpur (in press).

Azariah, J., P. Banth, H. Azariah & V. Selvam, 1986. Impact of urbanization on the status of mangrove swamps of Madras. In D. D. Hook *et al.*, (eds), The Ecology and Management of Wetlands, Croom Helm, London: 225–233.

Blasco, F., 1975. Mangroves of India. Inst. Fr. Pondicherry. Tran. Sect. Sci. Tech., 15, 175 pp.

Indian Express, 1990. Kakinada Port-Gateway for trade and industry. March, 2: 1–15.

Rao, T. A., K. R. Agerwal & A. K. Mukherje, 1963. An ecological account of the vegetation of Rameswaram Island. Bull. Bot. Surv. India, 5: 301–323.

Saenger, P., E. J. Hegal & J. D. S. Davie, 1983. Global status of Mangrove ecosystems. Commission on Ecology, Paper No. 3, IUCN Gland. 88 pp.

Sarma, D. V. R. & P. N. Ganapti, 1971. Hydrography of the Coringa river of the Gautami-Godavari estuarina system. J. mar. biol. Ass. India. 13: 234–248.

Sidhu, S. S., 1963. Studies on the mangrove of India. I. East Godavari region. Indian Forestar. 89: 337–351.

Venkatesan, K. R., 1966. The mangroves of Madras state. Indian Forestar, 92: 27–34.

Hydrobiologia **247**: 17–27, 1992.
V. Jaccarini & E. Martens (eds), The Ecology of Mangrove and Related Ecosystems.

Structural properties of two types of mangrove stands on the northwestern coast of Sri Lanka

M. D. Amarasinghe[1] & S. Balasubramaniam[2]
[1]*National Aquatic Resources Agency, Crow Island, Colombo 15, Sri Lanka;* [2]*Department of Botany, University of Peradeniya, Peradeniya, Sri Lanka*

Key words: mangrove ecosystems, community structure, floristic composition, complexity index, estuarine mangals, fringing mangals

Abstract

Mangrove stands in Puttalam lagoon and Dutch bay, two interconnected lagoons situated on the northwestern coast of Sri Lanka have been classified broadly into two groups, i.e., estuarine and island/mainland-fringing stands. Structural diversity of six mangrove stands, representing the two types was studied in terms of floristic composition, density, basal area, mean stand diameter, tree height, standing above-ground biomass and leaf-area index. *Rhizophora mucronata* and *Avicennia marina* were the dominant species. Higher mean stand diameters for the mangrove stands in Puttalam lagoon indicated greater maturity than the estuarine mangrove stands in Dutch bay. Nevertheless, estuarine stands in Dutch bay were structurally more complex (complexity indices 8.11–22.7) than the island/mainland-fringing mangrove stands (complexity indices 1.38–6.78). Higher number of species present in the estuarine mangrove stands is the major element that contributes to the higher values for the complexity indices for those stands. This appears to mask the contribution of stand-age to the complexity of a mangrove stand. Therefore complexity indices alone may not be used to explain adequately the structural diversity among mangrove stands.

Introduction

Mangrove ecosystems are open systems, which exchange matter and energy with the adjacent marine and terrestrial ecosystems. However, they are found to differ in their 'energy signature' or the sum of all forces which dictates the types of organisms that will survive and the speed of ecological processes (Cintron *et al.*, 1985). Cintron *et al.* (1978), Lugo *et al.* (1973) and Cintron & Schaefer-Novelli (1984) have shown that the magnitude and periodicities of forcing functions such as tides, nutrients, hydro-period and stresses such as cyclones, drought, salt accumulation and frost may largely determine the 'energy signature' of a mangrove stand and therefore the floristic and faunistic composition as well as the community structure.

In the present study, the structural characteristics of major types of mangrove stands in northwestern Sri Lanka have been quantified with a view to assessing whether they can be used for classification of mangrove communities in the Sri Lankan context.

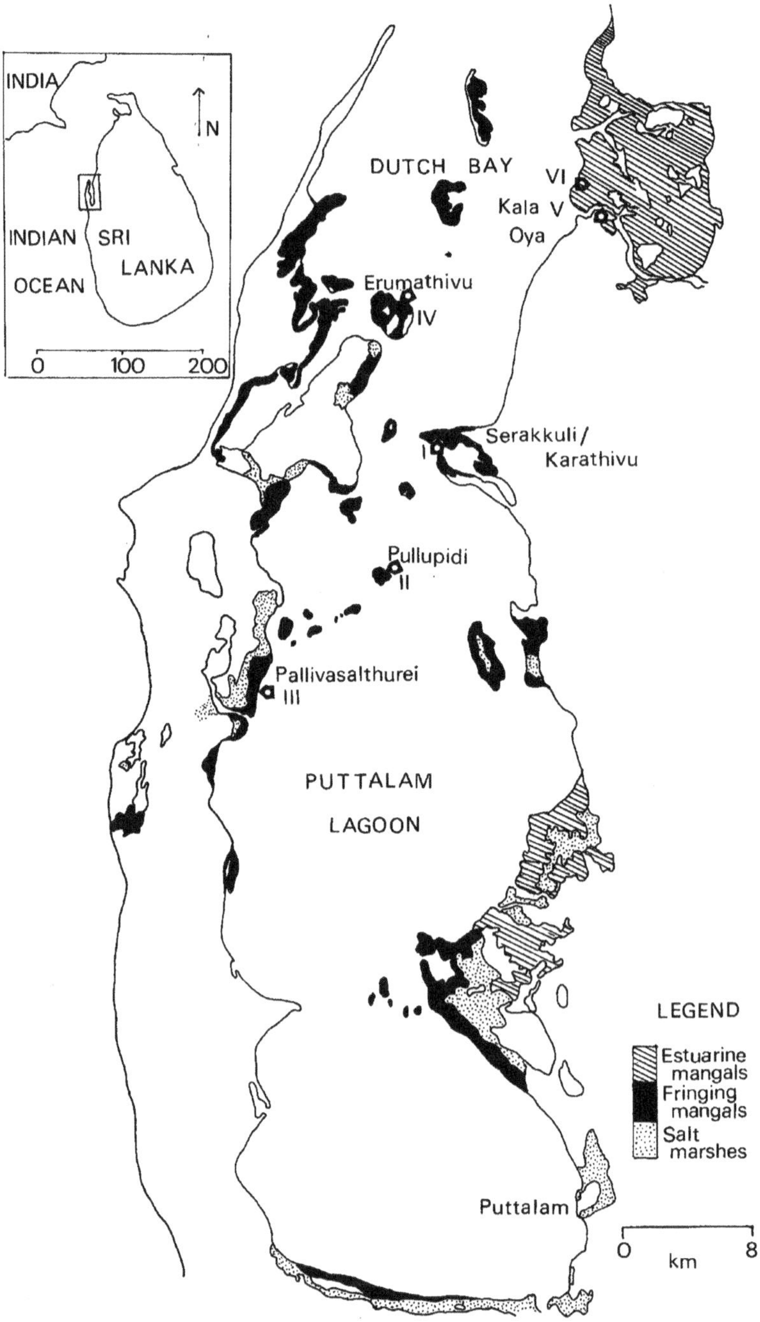

Fig. 1. Distribution of mangroves in the study area and the location of the study sites. I – Serakkuli; II – Pullupidi; III – Pallivasalthurai; IV – Erumathivu; V and VI – Kala Oya.

Study area

Puttalam lagoon and Dutch bay, together with Portugal bay (location – 8° 15′ N, 79° 50′ E; Fig. 1) form the largest brackish water body in Sri Lanka. Total water surface is approximately 46000 ha. Mangroves cover nearly 1200 ha and 2100 ha of inter-tidal land around Puttalam lagoon and Dutch bay respectively (Kanakaratne *et al.*, 1983).

Among the few river and creek systems associated with these lagoons, Kala Oya is the longest with the largest river basin. A semi-diurnal tidal pattern occurs in the lagoon and the maximum tidal range is 79 cm (Perera & Siriwardena, 1982). Surface salinities vary between 0.00 ppt (at Kala Oya estuary) and 45.00 ppt.

A dry climate prevails (Fig. 2) and the average annual rainfall (100–110 cm) is seasonal with two distinct dry seasons, i.e., January–March and July–September. Maximum rainfall is in the period from October to December by inter-monsoonal convectional rains.

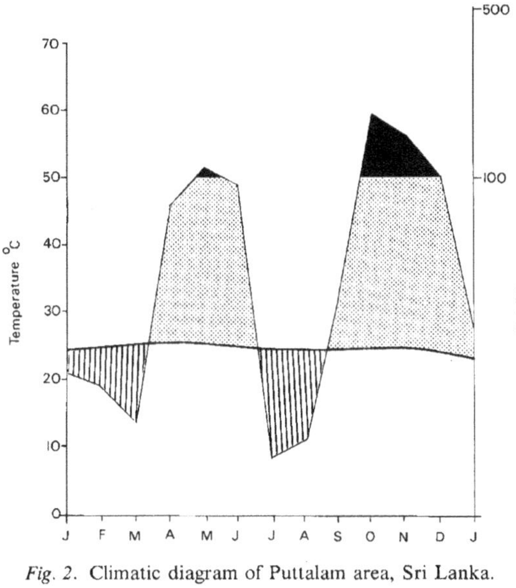

Fig. 2. Climatic diagram of Puttalam area, Sri Lanka.

Materials and methods

Using auxiliary information on physiognomy such as species present, vigor of the stand, relative location etc., mangals (the area covered with mangrove plants) of the study area have been classified according to Lugo & Snedaker (1975) and the method modified by Cintron *et al.* (1985). Representing the two types of mangals in the area i.e., estuarine and island/mainland fringing (Amarasinghe & Perera, 1984), six study sites were selected (Fig. 3).

Floristic composition

Mangrove areas on the mainland were visited over-land while the islands and Kala Oya estuary were accessed by boat. Specimens of the species encountered were identified using descriptive literature and taxonomic keys (Watson, 1928; Dassanayake & Fosberg, 1980; Blasco, 1984) as well as by comparing them with the specimens in the National Herbarium at Peradeniya, Sri Lanka. Enumerated species were categorized into true mangrove and mangrove associated species (Saenger *et al.*, 1983).

Community structure

Belt transects of 10 m in width and 50 m in length, subdivided into 10 m × 10 m plots were laid perpendicular to the shoreline to collect structural features of the mangals. Except at Kala Oya estuary and Erumathivu island (Fig. 3), only one transect in each site was employed for the purpose.

Vertical distribution of each stand was determined and profiles were prepared.

Methods described in detail by Cintron & Schaefer-Novelli (1984) for the measurement of stand density, basal area, mean stand diameter, tree height and leaf-area index were adopted. Trees which are equal to or exceeding 2.5 cm in diameter at breast height (dbh) were measured for the purpose.

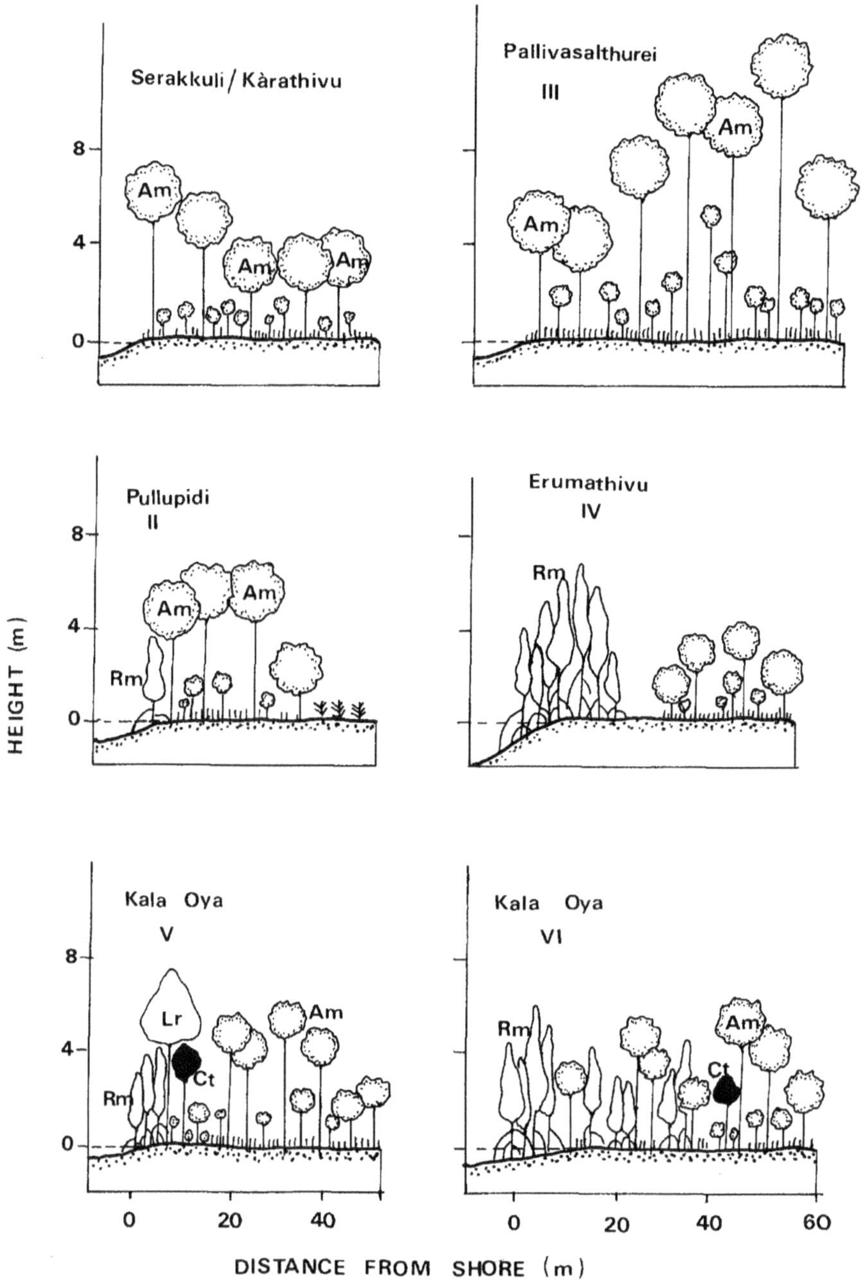

Fig. 3. Vegetation profiles of study sites. Am – *Avicennia marina*; Rm – *Rhizophora mucronata*; Lr – *Lumnitzera racemosa*; Ct *Ceriops tagal*.

Relationship between dbh and tree biomass for *Rhizophora mucronata* and *Avicennia marina*

As *R. mucronata* and *A. marina* are the dominant species in the study area these two species were chosen to determine dbh and tree biomass. Thus, the relationships were obtained using the method described by Cintron & Schaefer-Novelli (1984) and Day *et al.* (1987). Girth at breast height (gbh) of the trees in the four permanent transects (10×50 m^2) in the mangals at Erumathivu island and Kala Oya estuary (2 transects per site) were measured in order to determine the range of gbh for *R. mucronata* and *A. marina* in these two mangals. These two sites were particularly chosen for biomass studies as litterfall and gbh increment measurements (to estimate the net primary productivity, which will be discussed in a subsequent paper) were obtained only from these two mangals which were presumed to represent estuarine and fringing mangals. Ten trees of *R. mucronata* (representing the whole range of dbh) from each study site (total of 20 trees), nineteen trees of *A. marina* from Erumathivu and ten trees of the same species from Kala Oya mangals were selected for harvesting. These trees were cut as close above-ground as possible. The canopy i.e., branches, twigs and leaves (flowers and fruits were absent at the time of harvesting), was stripped and the weight of each component was obtained. Samples from each component (about 500 g were collected and the fresh weight of each sample was measured to the nearest 0.1 g in the field, using a triple beam balance. Tree trunks were cut further into convenient lengths of about 2 m, and the weight of each length was obtained with a spring-balance. A few sections from a few lengths were cut as samples and fresh weights were obtained. Fresh weight of the prop roots of harvested *R. mucronata* trees were obtained and weighed samples were brought to the laboratory to determine the dry weight to wet weight ration (conversion factor). Wet weights measured in the field were converted to dry weights using this conversion factor. These estimated dry weights for each component were summed up to obtain dry weight of trees. Slopes of the regression relationships between dbh and total biomass and the biomass of other components of the plant for each species obtained from the two sites were compared using Student's t-test and found to have no significant difference at 5% level between the two sites, thus all the data for a species in the both sites were pooled and the relationship between dbh and biomass was computed for each species.

Allometric relationships between dbh and biomass were not calculated for other species encountered in the transects, as they often occur in low densities. A conservative estimate of biomass for such species was obtained by assigning the lowest biomass to the corresponding dbh. The lowest biomass values were selected out of the biomass values calculated for *R. mucronata* and *A. marina* for the corresponding dbh.

Standing above-ground biomass

Standing above-ground biomass of the stands were calculated by determining the frequency distribution of the diameters in each stand and estimating the biomass of each diameter class separately, using the computed regression equation for each species. The average biomass of a diameter class obtained thus, was multiplied by the number of stems in that class and the results for all the diameter classes were summed to obtain the total biomass of the trees in the transect.

Table 1. List of mangrove species enumerated in Puttalam lagoon and Dutch bay mangals.

1. *Acanthus ilicifolius* L.
2. *Aegiceras corniculatum* (L.) Blanco
3. *Avicennia marina* (Forsk.) Vierh.
4. *Bruguiera cylindrica* (L.) Blume
5. *Bruguiera gymnorhiza* (L.) Lam
6. *Ceriops tagal* (Perrot.) C.B. Rob.
7. *Cynometra iripa* Kostel.
8. *Exoecaria agallocha* L.
9. *Lumnitzera racemosa* Wiild.
10. *Rhizophora apiculata* Blume.
11. *Rhizophora mucronata* Lam.
12. *Scyphiphora hydrophyllaceae* Gaertn.
13. *Sonneratia alba* J. Smith
14. *Xylocarpus granatum* Koenig.

Table 2. Relative density (Rde), dominance (Rdo), frequency (Rfr) and importance value (Imv) for the major mangrove species (Based on the trees with diameter > 2.5 cm). Roman numerals in the parentheses indicate the location in Fig. 3. A.m. – *Avicennia marina*; R.m. – *Rhizophora mucronata*; L.r. – *Lumnitzera racemosa*; C.t. – *Ceriops tagal*.

Mangal type/location		A.m.	R.m.	L.r.	C.t.	Other
Riverine mangals						
(Dutch bay)						
Kala Oya (V)	Rde	65.70	25.90	1.40	2.10	4.90
	Rdo	88.53	8.85	0.39	0.69	1.54
	Rfr	33.33	25.00	8.33	16.67	16.67
	Imv	187.56	59.75	10.12	19.46	23.11
Kala Oya (VI)	Rde	13.50	52.30	28.80	4.50	0.90
	Rdo	27.18	62.75	2.96	7.00	0.11
	Rfr	36.36	27.27	9.09	18.19	9.09
	Imv	77.04	142.32	40.85	29.69	10.10
Fringe mangals						
(Puttalam lagoon)						
Karathivu/Serakkuli (I)	Rde	100.00	–	–	–	–
	Rdo	100.00	–	–	–	–
	Rfr	100.00	–	–	–	–
	Imv	300.00	–	–	–	–
Pallivasalthurai (III)	Rde	100.00	–	–	–	–
	Rdo	100.00	–	–	–	–
	Rfr	100.00	–	–	–	–
	Imv	300.00	–	–	–	–
Pullupidi (II)	Rde	89.20	6.50	4.30	–	–
	Rdo	99.42	0.24	0.34		
	Rfr	60.00	20.00	20.00	–	–
	Imv	248.62	26.74	24.64	–	–
(Dutch bay)						
Erumathivu island (IV)	Rde	36.20	63.80	–	–	–
	Rdo	36.20	63.80	–	–	–
	Rfr	50.00	50.00	–	–	–
	Imv	122.40	177.60	–	–	–

Complexity index

Complexity index (CI, Holdridge *et al.*, 1971) was calculated as follows:

$$CI = \text{Number of species} \times \text{stand density} \times \text{stand basal area} \times \text{stand height} 10^{-5}$$

Results

Floristic composition

Out of the 14 true mangrove species encountered in the study area, *R. mucronata* and *A. marina* were the dominant. True mangroves, mangrove

Table 3. Regression coefficients used to estimate the above-ground standing biomass. The equations are in the form, $\log_e Y = A + B \log_e X$ with Y = dry weight (g), X = dbh (cm) and r^2 = coefficient of determination.

Species	Plant component	A	B	r^2
R. mucronata	Leaves	5.408	1.241	0.597
	Branches	4.262	2.103	0.781
	Trunk	4.339	2.520	0.956
	Prop roots	5.359	2.024	0.863
	Total biomass	6.247	2.064	0.945
A. marina	Leaves	3.984	1.825	0.816
	Branches	4.074	2.299	0.807
	Trunk	5.148	2.044	0.906
	Total biomass	5.551	2.153	0.922

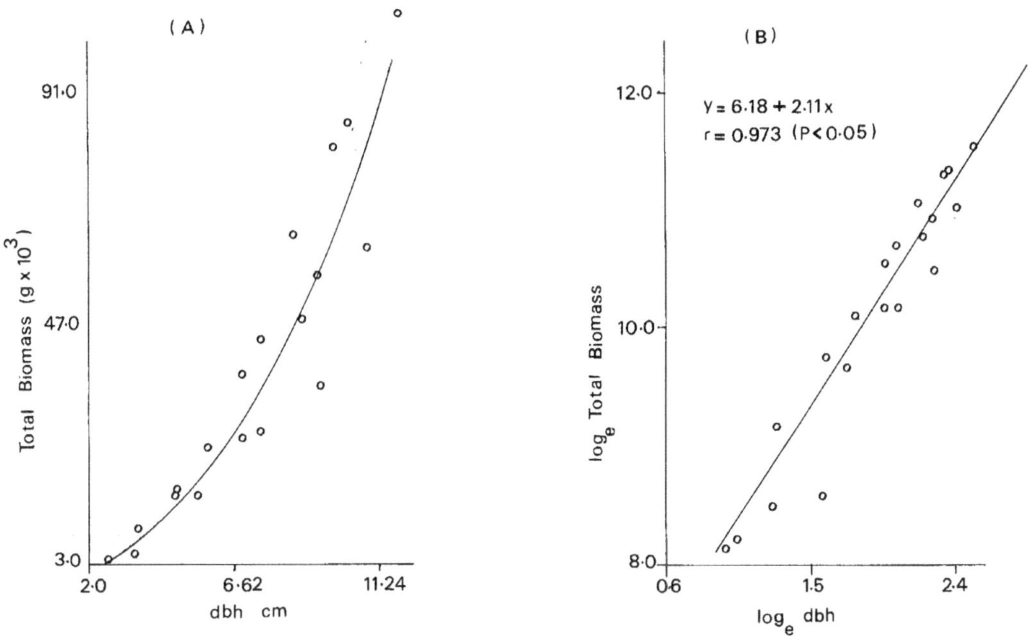

Fig. 4. (A) Curvilinear and (B) log-log relationships between total biomass and dbh for *R. mucronata* in Dutch bay.

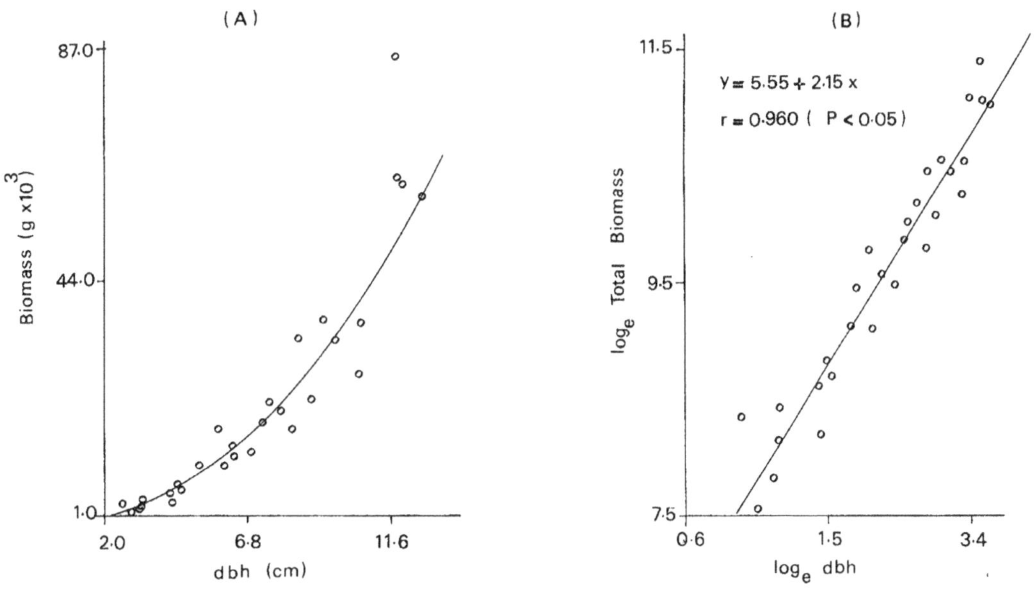

Fig. 5. (A) Curvilinear and (B) log-log relationships between total biomass and dbh for *A. marina* in Dutch bay.

Table 4. Structural characteristics of riverine and fringing mangals of Puttalam lagoon and Dutch bay (units on a per ha. basis. Trees measured at dbh). Roman numerals in parentheses indicate the location in Fig. 1.

Type & location of the mangal	No. of species	No. of stems/trees of the stand				Stand density (No. ha^{-1})	Basal area		Stand basal area (m^2)	Mean stand diameter (cm)	Stand height (m)	Complexity index	Above-ground biomass (t ha^{-1})
		> 2.5 cm		> 10 cm			> 2.5 cm	> 10 cm					
		Stems	Trees	Stems	Trees								
Riverine mangals													
Dutch bay													
1. Kala Oya (V)	7	4000	3075	650	525	3600	10.07	9.92	19.99	8.41	4.4	22.16	85
2. Kala Oya (VI)	5	4960	2550	230	200	2750	10.37	2.74	13.11	7.79	4.5	8.11	57
Fringe mangals													
Puttalam lagoon													
(Mainland habitats)													
1. Karathivu/Serakkuli (I)	1	833	200	667	600	800	1.31	42.46	43.77	26.40	7.2	2.87	240
2. Pallivasalthurei (III)	1	438	125	300	325	450	1.06	28.66	29.72	29.00	10.3	1.38	193
3. Pullupidi Island (II)	3	2167	767	867	767	1534	5.92	28.36	34.28	16.87	4.3	6.78	172
Dutch bay													
(Island habitats)													
4. Erumathivu Island (IV)	2	4420	1980	230	220	2200	8.15	3.35	11.4	8.12	3.9	1.95	71

associated and salt marsh species enumerated in Puttalam lagoon and Dutch bay are listed in Table 1. Vegetation profiles of the study sites are depicted in Fig. 3. In Puttalam lagoon mangals, *A. marina* had a higher importance value (Cintron & Schaefer-Novelli, 1984) than in the Dutch bay mangals and the reverse was true for *R. mucronata* (Table 2).

Community structure

Structural features are presented in Table 4. Stand densities of the mangals in Dutch bay are considerably higher than those in Puttalam lagoon and a reverse trend occurs with regard to stand basal area and standing above-ground biomass. Complexity indices for estuarine mangals are greater than those for the fringing mangals.

Relationship between dbh and biomass for *R. mucronata* and *A. marina* are graphically presented in Figs 4 & 5. Regression coefficients used to estimate above-ground biomass are presented in Table 3.

Discussion

Floristic composition and zonation

As in many other mangals in the dry coastal regions in Sri Lanka, *R. mucronata* and *A. marina* are the major constituent species of the mangals of Puttalam lagoon and Dutch bay. *A. marina* is the most dominant species in the mangals in Puttalam lagoon with relative dominance values ranging from 60 to 100 and importance values ranging from 248.62 to 300. Ability of *A. marina* to tolerate a wide range of environmental conditions (Wells, 1982) such as salinity and air temperature, may favour its distribution, not only in frequently inundated areas but also in infrequently flooded upper inter-tidal areas. *R. mucronata* is more dominant than *A. marina* in the estuarine habitats of Kala Oya.

Most of the fringing habitats, especially those in Puttalam lagoon are occupied by a single species, frequently by *A. marina*. According to Cintron *et al.* (1985), the most constant conditions of salinity and nutrients as well as the exposure to expectedly higher kinetic energy regimes which are relatively unfavourable to support a large number of species may restrict the number of species occurring in fringing habitats. In the estuarine mangals, nutrients and salinity may oscillate widely and the kinetic energy of water flow may anticipated to be lower than that of fringing mangals. Moreover, during periods of flood, considerable tidal and water flow energies may be dissipated in the riverine (estuarine) sites (Cintron *et al.*, 1985). Although such periods are relatively short, they may favour mangrove establishment more than the constant tidal motion in the fringing mangrove habitats, especially in a low-tidal environment. Thus as a result of the oscillating conditions and more favourable nutrient balances, estuarine mangals may be able to support more species than fringing mangals.

General zonation of species within a mangal reflects the gradient in flooding and soil characteristics of the mangrove species. Kenneally (1982) listed frequency of tidal flushing, soil type, soil salinity, drainage plant and animal interactions as the major factors that appear to determine the survival of individual species and thus zonation within a mangal. In the study area where a dry climate and a small tidal amplitude prevail, the frequency of tidal flushing in the upslope may be significantly low and as the terrain is flat, soils are poorly drained thus creating a soil salinity gradient inland. More salinity tolerant mangrove species such as *A. marina* and *Excoecaria agallocha* as well as the salt marsh species occupy such landward portions while known less tolerant *R. mucronata* occupies the water-front zones of the inter-tidal land. Seasonal precipitation and high evapo-transpiration rates may have largely contributed to the development of salt marshes and salt flats which characteristically occur around Puttalam lagoon where freshwater influx is apparently low.

Structural parameters

Tree densities of estuarine mangals are higher than those of the fringe mangals, owing to the higher number of smaller trees (diameter less than 2.5 cm) in the estuarine habitats. Fringing mangals on the mainland (around Puttalam lagoon) are the least densed while those on the islands record intermediate values. These trends indicate that the stands in Puttalam lagoon are more matured than those in Dutch bay.

Stand basal area, mean stand diameter, stand height and standing above-ground biomass values calculated for these mangrove stands also manifest a similar trend. The mainland-fringing mangrove stands in Puttalam lagoon seem to be the most matured, followed by the island fringing mangals in the same area. The least matured stands are the estuarine mangals at Kala Oya estuary. In general, the presence of fewer number of small trees and higher number of large trees in the mangals in Puttalam lagoon imply that these mangals are more matured than those in Dutch bay. This may imply that the coastal areas in Puttalam lagoon are geomorphologically more stable than in Dutch bay.

Estuarine mangals manifest a greater structural complexity (complexity indices 8.11–22.7) than the mainland and island fringing mangals (complexity indices 1.38–6.78). This may be attributed to the prevailing favourable flushing patterns as well as the nutrient and salinity regimes which contribute to the 'energy signature' (Cintron *et al.*, 1985) in the estuary.

However, the low complexity index-values for the mangrove stands in Puttalam lagoon, which have resulted mainly due to the low number of species present, appear to mask their relative ecological importance. Therefore complexity index may not adequately indicate the structural complexity and functional capacity of a mangal comprised of a few species.

Acknowledgements

This study was supported by the National Aquatic Resources Agency and the assistance given by the staff of the Regional Research Centre at Kalpitiya is gratefully acknowledged. Our thanks are also extended to Dr. Yara Shaefer-Novelli at Universidade De Sao Paulo, Brazil for her critical comments on the manuscript.

References

Amarasinghe, M. D. & W. K. T. Perera, 1984. A preliminary survey of the peripheral vegetation communities around Puttalam lagoon and Dutch bay, Sri Lanka. Proc. 41st Ann. Sess. SLAAS (abstract only).

Blasco, F., 1984. Taxonomic considerations of the mangrove species. In S. C. Snedaker & J. C. Snedaker (eds), The mangrove ecosystems: research methods. UNESCO, Paris; France: 81–90.

Cintron, G., A. E. Lugo, D. J. Pool & G. Morris, 1978. Mangroves of arid environments in Puerto Rico and adjacent islands. Biotrop. 10: 110–121.

Cintron, G. & S. Y. Schaefer-Novelli, 1984. Methods for studying mangrove structure. In S. C. Snedaker & J. C. Snedaker (eds), The mangrove ecosystem: research methods. UNESCO, Paris; France: 91–113.

Cintron, G., A. E. Lugo & R. Martinez, 1985. Structural and functional properties of mangrove forests. In W. G. D'Arcy & A. M. D. Correa (eds), The botany and natural history of Panama: La botanica e historia natural de Panama IV series, Monographs in systematic botany, V 10, Missouri Botanical Gardens, St. Louis, Mo; USA: 53–56.

Dassanayake, M. D. & F. R. Fosberg, 1980. A revised handbook to the flora of Ceylon, Vol. I. Amarind Publishers, New Delhi; India: 508 pp.

Day, J. W. Jr., W. H. Corner, F. Ley-lou, R. H. Day & A. M. Navarro, 1987. The productivity and composition of mangrove forests, Laguna de Terminos, Mexico. Aquat. Bot. 27: 267–284.

Holdridge, L. R., W. C. Grenko, W. H. Hatheway, T. Liang & J. Tosi, 1971. Forest environment in tropical life zones; a pilot study. Peragamon Press, New York: 747 pp.

Kanakaratne, M. D., W. K. T. Perera & B. U. S. Fernando, 1983. An attempt at determining the mangrove coverage in Puttalam lagoon, Dutch bay and Portugal bay, Sri Lanka, using remote sensing techniques. Proc. 4th Asian Conf. Remote Sensing, Colombo; Sri Lanka; P-15-1-9.

Kenneally, K. F., 1982. Mangroves of Western Australia. In B. F. Clough (ed.), Mangrove ecosystems in Australia: structure, function and management, Australian Institute of Marine Sciences; Australia: 95–110.

Lugo, A. E., G. Evink, M. M. Brinson, A. Broce & S. C. Snedaker, 1973. Diurnal rates of photosynthesis, respiration and transpiration in mangrove forests in south Florida. In: The role of mangrove ecosystems in the maintenance of environmental quality and high productivity of desirable fisheries. Centre for Aquatic Sciences.

Lugo, A. E. & S. C. Snedaker, 1975. Properties of a mangrove forest in southern Florida. In G. E. Walsh, S. C. Snedaker & H. J. Teas (eds), Proc. International Symposium on Biology and Management of Mangroves, IFAS, University of Florida; USA: 170–212.
Perera, W. K. T. & P. P. G. S. N. Siriwardena, 1982. Topography of Puttalam lagoon. J. Inland Fish. (Sri Lanka) 1: 97–104.
Saenger, P., E. J. Hegerl & J. D. S. Davie, 1983. Global status of mangroves. IUCN Commission on Ecology, Pap. No. 3.
Watson, J. D., 1928. Mangrove forests of the Malay peninsula. Malaya Forest Records 6: 1–275.
Wells, A. G., 1982. Mangrove vegetation of northern Australia. In B. F. Clough (ed.), Mangrove ecosystems in Australia: Structure, Function and Management. Australian Institute of Marine Sciences: 57–78.

Hydrobiologia **247**: 29–36, 1992.
V. Jaccarini & E. Martens (eds), The Ecology of Mangrove and Related Ecosystems.

The current status of mangroves along the Kenyan coast: a case study of Mida Creek mangroves based on remote sensing

P. O. Gang & J. L. Agatsiva
Department of Resource Surveys and Remote Sensing, Ministry of Planning and National Development, P.O. Box 47146, Nairobi, Kenya

Key words: mangroves, Kenya, remote sensing, species composition and distribution

Abstract

Mangroves form a unique ecosystem of limited extent covering an area of about 53000 ha along the Kenyan coast which need protection from overexploitation. Background information on the areal extent and status of these mangroves is limited and makes their protection and management difficult.

A model study has been carried out on the Mida Creek mangroves based on a double sampling approach starting with SPOT multispectral satellite imagery followed by ground checks to provide information on species composition, density and distribution of mangroves. The utilization of mangroves and uses around and within the mangroves were determined.

During this study, seven mangrove species were identified in Mida Creek, namely: *Avicennia marina*, *Bruguiera gymnorrhiza*, *Ceriops tagal*, *Rhizophora mucronata*, *Sonneratia alba*, *Xylocarpus granatum* and *Lumnitzera racemosa*. This report briefly outlines the present utilisation of mangrove species. It is recommended that a multidisciplinary management plan be developed in order to conserve and manage the mangroves of Kenya on a sustainable yield basis.

Introduction

Mangroves are a group of highly evolved halophytes occupying the intertidal zone in estuaries, lagoons and coastal mud flats in tropical and subtropical zones. While they have proven ability to function in poorly oxygenated water-logged, saline habitats, a seasonal flushing with freshwater is necessary for their continued survival. There are some which prefer a daily tidal wash while others find their optimum conditions in areas subject to only seasonal high tides (UNDP/UNESCO, 1984). This determines the distribution of mangrove species within a given community. Mangroves have been exploited by coastal people for their numerous natural products. These include poles for building, vegetable tannins, pulp, fuelwood and medicine (UNEP, 1988).

The Kenyan coastline has a total of about 52980 ha of mangrove forest (Doute *et al.*, 1981). The bulk of these forests occur in the intertidal areas where submarine ground water discharge (SGD) or seepage occur rather than in estuaries (Ruwa & Polk, 1986).

Background and problem

The lack of appraisal of the status of Kenyan mangrove forest has been a limitation on the formulation of an adequate management and conservation policy. Quite often, the exploitation of mangroves has occurred in the past unnoticed or unquantified, as has been previously reported by Doute *et al.* (1981) who based their study on the use of Landsat Multispectral Scanner (MSS) imagery. The study is now under review (Agatsiva *et al.*, 1990) to up-date it at the same time giving more details which were never achieved in the 1981 study.

The lack of up-to-date and accurate data poses planning problems on conservation and management of such mangrove species. It has been noted that mangroves and their related ecosystems have in the past undergone drastic changes without much attention being paid to their ultimate fate. When exploiting a renewable resource, the concept of sustainable development should be borne in mind. Conflicting land-uses such as fishing, cutting of mangroves for different uses, salt mining and various marine activities are likely to exhaust the mangrove resources.

The Mida Creek mangroves (Gang, 1988) have been considered in an effort to design an appropriate methodology for having the Kenyan mangroves rapidly and cheaply assessed. The recommendations made may enable them to be properly conserved, managed and where exploited, to be appropriately harvested on a sustainable yield basis. Mida Creek was preferred because the area has a variety of mangrove types as well as different regeneration stages and some dense old stands. There are also several land use practices in and around the creek which may also influence the mangrove ecosystem along the coast. The broad recommendations made on this area are therefore likely to represent a wider spectrum of information concerning all mangroves along the Kenyan coast.

The objectives of the study are therefore

(i) To document the extent and status of the mangroves of Mida Creek.
(ii) To assess the suitability of SPOT multispectral satellite imagery in providing baseline data for the monitoring and management of mangrove areas.
(iii) To prepare a checklist of plants growing in close proximity to the mangroves both within and without the mangrove area in Mida Creek.
(iv) To give possible suggestions and recommendations that may be applied for sustainable utilisation of the mangroves.

The study area

Location of the study area

Mida Creek is located in Kilifi district on the north coast of Kenya and is approximately 25 km south of Malindi town. The area lies within approximate bounds of longitude 39° 58′ E and latitude 3° 20′ S (Fig. 1).

Climate

The rainfall pattern is bimodal. The long rains fall in April to June, with a peak in May. The short rains fall from October to December. The area receives an annual rainfall of about 900–1100 mm (District Development Plant 1989/83).

Soil types

Soils of the mangrove swamps are poorly drained, deep and excessively saline, olive to greenish grey loam to clay and often with sulfidic material (District Environmental Assessment Report, 1985).

Land use

Agriculture is one of the land use types in the area. Tree crops occupy more than one third of the area under cultivation. The most important of which are cashewnuts, mangoes and coconuts. Coconuts are very much valued due to the role they play in providing daily requirements such as cooking fat, woodfuel, building materials, local beer and hair oil. Cashewnuts are sold to the processing plant based in Kilifi. Mangoes are sold to the village, hotel industry and traders in Mombasa.

Fishing is another important activity in the study area. Most of the catch is harvested by individual fishermen using their own boats and gear following traditional methods. They are often seen in their small boats in the creek which is surrounded by mangroves.

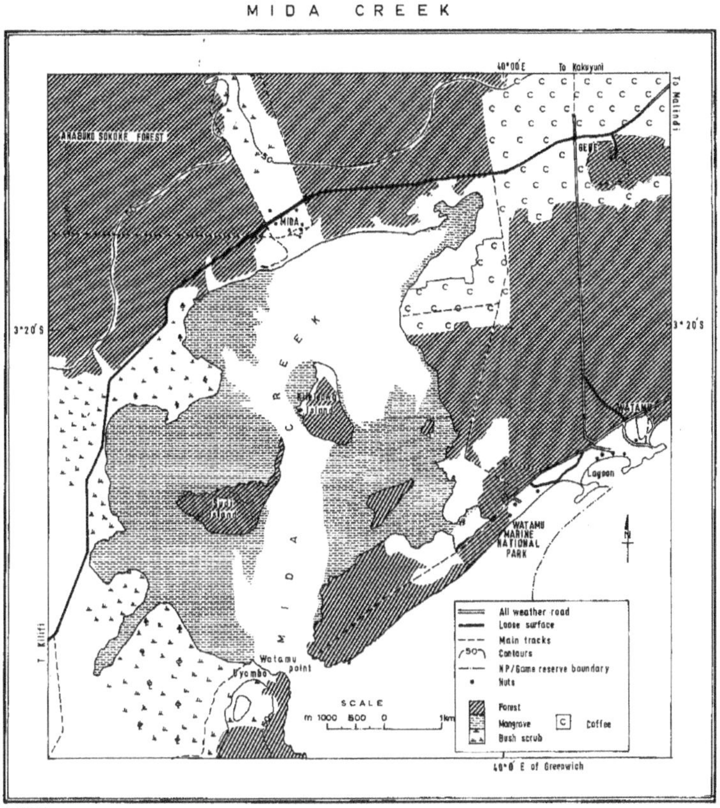

Fig. 1. Study area map.

Methodology

The methodology used was based on a double sampling approach which include the following:

(i) Use of remote sensing to provide initial stratification of the area into major cover types and ground-truthing of the satellite image.

(ii) Use of ground checks for species identification and categorisation into various cover types.

Remote sensing

Remote Sensing is broadly defined as a process of gathering information about objects from a distance based on the electromagnetic energy reflected or emitted by that object. Satellite remote sensing data can provide natural resource managers and planners with periodic information on the extent, location and status of the resources. Such information types are used in the day to day decision making on the implementation of management policies such as conservation and resource allocation.

A whole range of remote sensing techniques could have been used to map the area but weather conditions did not allow all these techniques to be used and their results compared for the area of interest. The region is under cloud cover for much of the year. However, SPOT 1 multispectral imagery was obtained for the area from Toulouse – France. SPOT multispectral imagery KJ 146 357 (scale 1:50 000) which has a spatial resolution of 20 by 20 m was the ideal imagery for mapping cover types because different mangrove species and zonation could be easily identified from the image. This was done by use of a transparent overlay put over the image and placed on a light table where all the distinguishable features were delineated. The differentiation was based on colour intensity, chroma and hue in addition to other image characteristics and interpretation elements including pattern and texture among others as described by Lillesand & Keifer (1979).

A topographic map at the scale of 1:50000 (sheets No 193/1 and 193/3 from survey of Kenya) was used to associate various manifestations in the image with contents of the map referring to land use and also for orientation and navigation during field surveys.

Ground measurements

In each stratum obtained, representative stands were sampled to obtain information regarding the mangroves on:

(i) Mangrove Species
(ii) Height class
(iii) Crown cover
(iv) Density of species
(v) Diameter at breast height (dbh).

Representative stands were selected within each stratum, using 50 m by 50 m quadrats. The total count method was used as described in Muchoki (1988). All mangrove trees 0.7 m high and above were measured and quantitative attributes such as height, dbh and crown cover were recorded. Visual observation was also done to determine the species being exploited. This included actual counting of stumps of the already cut mangroves and counting the dhows transporting mangrove poles to a collection yard at Uyombo. Finally an attempt was made to prepare a vegetation checklist of plants growing in close proximity to mangroves. The methods used were those used by Kenya Rangeland Ecological Monitoring Unit (KREMU) in its Rangeland Monitoring attempts (Kuchar, 1979) and those used by botanists elsewhere in compiling lists of plants by occurrence in different habitats.

Results and discussion

The above double sampling approach yielded good results on the mangrove species distribution and uses. Seven major mangrove species were identified (Table 1).

The SPOT image delineated had the following cover classes (Fig. 2).

Avicennia marina
This is found in the landward fringe of the Creek. The underground is mainly covered by pneumatophores. This covers 160 ha (2.9%).

Avicennia/Ceriops/Rhizophora
This is characterised by species which have discontinous canopy formations and generally in poor state of health, covering 145 ha (2.6%).

Rhizophora/Ceriops
Dominated mainly by *Rhizophora mucronata* and *Sonneratia alba* at the edge of the creek and *Ceriops tagal* with few *Bruguiera gymnorrhiza* and *Xylocarpus* spp. The undercanopy consists of

Table 1. Mangrove species found in the study area.

Mangrove species	Standard name in Swahili	Family
Avicennia marina (*Forsk.*) Vierh	Mchu or Mtu	Avicenniacea
Bruguiera gymnorrhiza (L.) Lam	Muia	Rhizophoraceae
Ceriops tagal (perr.) C. B. Robinson	Mkanda	Rhizophoraceae
Lumnitzera racemosa Wild.	Kikanda or Mkanda-Mwitu	Combretaceae
Rhizophora mucronata Lam.	Mkoko	Rhizophoraceae
Sonneratia alba Sm.	Mlilana or Mpia	Sonneratiacea
Xylocarpus granatum Koen.	Mkomafi	Meliaceae
Xylocarpus sp.		Meliaceae

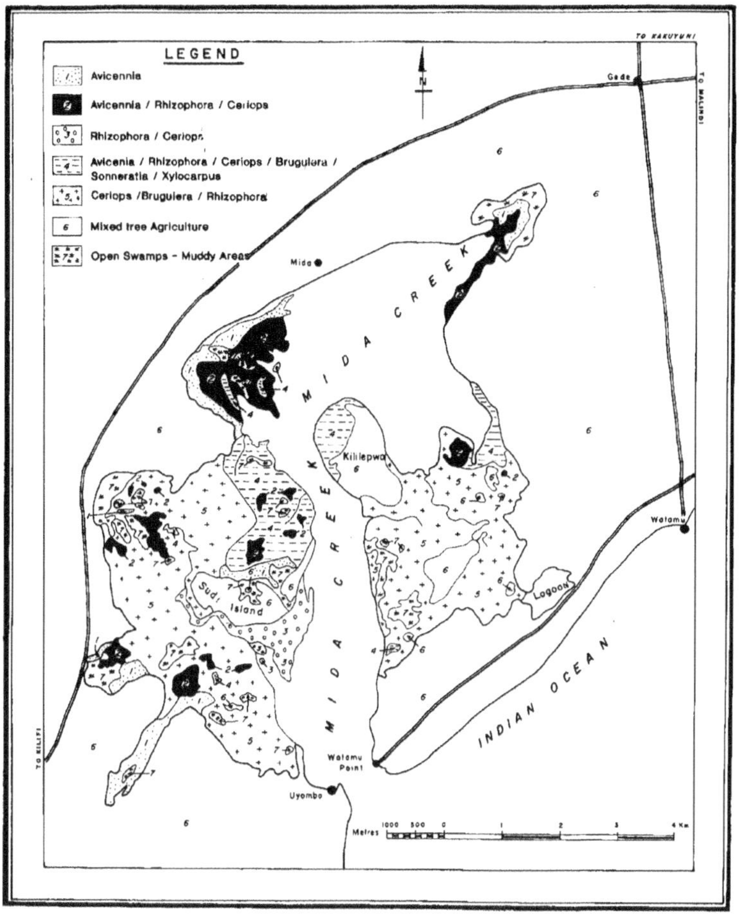

Fig. 2. Classification of Mida Creek mangroves and surrounding cover types based on SPOT imagery.

Rhizophora and *Ceriops* seedlings growing vigorously. Cutting in this zone is selective and rather intensive. This category covers 67.5 ha (1.2%).

Avicennia/Rhizophora/Ceriops/Bruguiera/Sonneratia/Xylocarpus

This category is characterised by very mature stands of majority of the species. There are very-tall *Bruguiera*. Cutting of *Ceriops* and *Xylocarpus* is most intensive probably due to their straight poles. *Avicennia* is very scarce and least utilised. This unit covers 237.5 ha (4.2%).

Ceriops/Bruguiera/Rhizophora

This is dominated by *Ceriops tagal* with very high regeneration of both *Rhizophora* and *Ceriops* on the outer fringes. From the land seawards, this zone has few *Avicennia marina* while *Rhizophora mucronata* mixed with *Sonneratia alba* dominate the edge of the creek. This covers 890 ha (15.9%).

Mixed tree agriculture

This occurs in the hinterland surrounding the creek, Sudi and Kilelepwa islands found in Mida creek. Tree crops found in this zone are mangoes, coconuts and cashewnuts. There is also natural vegetation growing together with tree crops. This includes *Thespesia danis* Olive and *Azadrachta indica* A. Juss. This class covers 3805 ha (68%).

Open swamps/muddy areas

These are swampy areas that no longer receive flooding except during high tides, immediately within the vicinity of these areas, small halophytic plants such as *Salicornia* sp. and *Sueda* sp. are dominant. The category covers 292.5 ha (5.2%).

Satellite imagery (SPOT image) proved to be very useful in identifying mangrove areas and other land use types in close proximity. Different mangrove zonations are easily identifiable using tonal differences in the image. It is not possible to identify mangroves to species level in the image but tonal differences interpreted were confirmed to represent different mangrove types during field surveys. It was not possible also to detect the cutting of the mangroves from the satellite imagery because the cutting is selectively done and it is not concentrated at one spot. Actual counting in the sampling plots indicated a cutting intensity of thirty poles per 50 × 50 m plot. Table 2 shows some of the mangroves species identified in the Mida Creek using the total count approach. The rest of the plants in the surrounding areas are indicated in Table 3. The cutting intensity decreases away from the landward fringes probably due to accessibility and convenience. The forest department controls the cutting by issuing licenses and ensures that poles are cut according to the quantity specified, and suspends further cutting in some areas until regeneration is restored. There are no afforestation activities in the creek but rather the mangroves regenerate naturally, growing into exploitable poles after 3–6 years. No quantitative measurements were carried out on regeneration but it was observed during field surveys that species like *Ceriops tagal* and *Rhizophora mucronata* have several seedlings actively growing. This may suggest that the forest department need to regulate the mode of harvesting in order to ensure sustainable exploitation of the mangroves in this creek.

Table 2. Total number of species obtained through the total count method from five sampling plots randomly chosen in Mida Creek.

Mangrove	Plots					
	1	2	3	4	5	Total
Ceriops tagal	46	107	7	47	2	209
Xylocarpus granatum	7	–	12	1	–	20
Bruguiera gymnorrhiza	2	1	–	2	1	6
Avicennia marina	1	–	–	–	82	83
Rhizophora mucronata	–	32	27	23	1	83
Sonneratia alba	–	–	–	–	5	5

Table 3. Plants growing in close Proximity to mangroves in Mida Creek.

Species	Family
Aeva lanata	Amaranthaceae
Anisotes parviflora	Acanthaceae
Aristogeitonia monophylla	Euphobiaecae
Asparagus africanus Lam.	Liliaceae
Azima tetracantha Lam.	Salvadoraceae
Cremaspora trifloa	Rubiceae
Cyperus tuberosus Rottb.	Cyperaceae
Canaralia carthartica Thouars	Papilionaceae
Dactyloctineum germinatum Hack	Gramineae
Derris spp.	Papilionaceae
Digitaria milanjiana (Randle) Stapt	Gramineae
Eleaodendron schweinfurthianum Loes	Celastraceae
Enteropogon macrostachysus (A. Rich) Benth	Gramineae
Euclea schimperi (A. D. C.) Dandy	Ebenaceae
Flacourtia indica (burn, f) Merr.	Flacourtiaceae
Fimbristylis triflora (L.) K. Schum	Cyperaceae
Fuirena ciliaris (L.) Roxb.	Cyperaceae
Grewia vaughanii exell.	Tilliaceae
Hyphaene compressa H. Wendl.	Palmae
Ipomoea pes caprae (L.) R. Br.	Convolvulaceae
Mormodica trifoliata Hook F.	Cucurbitaceae
Panicum infestum Anderss.	Gramineae
Panicum piniplium	Gramineae
Phoenix reclinata Jacq.	Palmae
Phragnites communis	Gramineae
Premna spp.	Verbanceae
Psilotrichum sericeum (Roxb.) Dal 2	Amaranthaceae
Saliconia sp.	Chenopodiaceae
Salvadora persica L	Salvadoraceae
Sesuvium portulacastrum (L.) L	Aizoaceae
Sideroxylon inerme L	Sapotaceae

The licensing policy of the forest department should be reviewed and strengthened to ensure that licenses issued to licensees are consistent in order to prevent illegal cutting of mangrove species and types which are not originally included in the license. This requirement should control the current practices in Mida creek where mangroves are cut to the edge of the creek. The cutting of *Rhizophora* sp. at the edge allows siltation since the stilt roots which hold sand leads to the death of the tree. This activity may interfere with the future of the creek since siltation may in the long run turn the creek into the beach.

Mida Creek mangroves are not much exploited for firewood. A random survey indicates that people living in the periphery of the creek prefer to collect their fuelwood from the neighbouring Arabuko Sokoke forest. Firewood is mainly obtained from the dead branches of common tree species including: *Brachystegia speciformis*, *Cynometra webberi*, *Afzelia quanzensis*, *Croton pseudopulchellus* and *maytenus undata*. Nevertheless, Mida Creek mangroves (Table 1) are heavily exploited for building and fencing poles. A visit to a collection yard at Uyombo, confirms the export of mangrove poles to the mainland. They are piled in the different classes depending on the diameter at butt end. These range from 3.8 cm to 14 cm in diameter.

Complementary studies on woodfuel/timber demand and supply may be essential to ensure sustainable utilisation of the woody biomass. These studies should be able to compute further, the attributes such as the ones shown in the Table 4 to precisely calculate the amount of biomass available to the entire creek.

Conclusion

The superior resolution of SPOT data made it possible to identify different cover types in the study area. With the complement of ground survey, species identification was possible in the area. From these results, SPOT data promise a new dimension in the use of Remote Sensing particularly in this region where rapid acquisition of data is necessary for development planning. Hopefully, with further research into the use of satellite imagery for coastal resource planning and management, it will be possible to answer some of the questions arising from such studies and refine what amounts to a very useful and time saving method of monitoring mangrove areas. Questions such as how much area of kenya is currently under mangroves, their status and exploitation patterns will be answered.

Suggestions and recommendations

Use of further satellite imagery in monitoring mangrove areas

Having proved the ability of satellite and recognizing it as a powerful aid to fast field data collection, it would be worthwhile to apply this tool in documenting the status of the mangroves in the entire Kenyan coast. A good sampling strategy incorporating aerial photography on a sample basis may rapidly enhance the results for mangrove management.

Further studies into the parameters controlling the distribution of mangroves in a given habitat

Further studies should be carried out in order to

Table 4. Summary of mangrove data in plot No. 2 near Dabaso in Mida Creek.

Name of plant	Total above ground biomass kg/ha	Utilizable woody biomass kg/ha	% of trees per species	Density	Frequency
Rhizophora mucronata	11832.0	9727.0	32.0	46.0	23
Bruguiera gymnorrhiza	1029.0	846.0	3.0	4.0	2
Ceriops tagal	24178.0	19877.0	64.0	94.0	47
Xylocarpus granatum	514.0	423.0	1.0	2.0	1

gain in-depth understanding of the parameters controlling the distribution of each mangrove species within a given habitat. Effects of such parameters as land use activities and erosion in the hinterland, salinity, temperature and depth of substrate type are important for designing advisory and protection schemes for the proper interpretation of the results from monitoring programmes.

Use of computers and digital mapping techniques
The capability to digitize SPOT data should be enhanced as well as the use of computers and digital mapping techniques in the region in order to derive data from the imagery. Certainly for any time series this would ease the problem of identifying changes to mangrove cover.

Increased awareness of mangroves values
The efforts to conserve the mangroves would benefit from a greater awareness in Governmental, Non Governmental organisations and amongst the general public of the importance of mangroves. To this end, an educational campaign should be initiated which emphasizes the values of the mangroves and their links to other elements in the ecosystem, principally seagrass beds, coral reefs and fish populations.

Silvicultural measures could potentially enhance the productivity of the mangrove forests, but at present there is very limited information available on which to base such measures. Further research should therefore be carried out to document the correct silvicultural practices that may be useful for the Kenyan coast.

A National inventory of the mangroves of Kenya should be made. The Status of the entire mangrove ecosystem should be highlighted in this inventory report. It is after this that a national mangrove management scheme can be framed and prepared by a multidisciplinary group.

Acknowledgements

Gratitude is due to Mr. D. K. Andere, Department of Resource Surveys and Remote Sensing, D.R.S.R.S., Nairobi, for his support and encouragement. I also extend my gratitude to African Wildlife foundation, regional office in Nairobi, through the fund for conservation endeavour for providing part of the funding for the survey activities in Mida Creek. I am also grateful to members of the East African Herbarium for their help with plant identification and finally to technical staff, D.R.S.R.S. for criticizing the text.

References

Agatsiva, J. L., J. P. Delsol, N. Ochanda, M. Shimba & J. M. Terresa. (unpublished results). A remote sensing approach to mapping and monitoring forest cover in Kenya.

Doute, R., N. Ochanda & H. Epp, 1981. Forest cover mapping using remote sensing techniques Kenya Rangeland Ecological Monitoring Unit, P.O. Box 47146, Nairobi, technical report No. 30, 72 pp.

Gang, P. O., 1988. Application of Remote Sensing in Landuse/Landcover assessment of Coastal Environments (Dissertation submitted to the Kenya Polytechnic, P.O. Box 52428, Nairobi, Kenya).

Gang, P. O., 1988. The distribution of mangroves forest of Kenya. A case study in Mida Creek, Sabaki and Ngomeni areas, Kenya Rangeland Ecological Monitoring Unit, Box 47146, Nairobi (unpublished results).

Kuchar, P., 1979. A methodology for sampling in order to monitor conditions, trend and plant biomass. Kenya Rangeland Ecological Monitoring Unit, Box 47146, Nairobi.

Lillesand, T. M. & R. W. Kiefer, 1979. Remote sensing and image interpretation. John Wiley and Sons Inc., New York: 528–590.

Kenya Government, Ministry of Planning and National Development 1989, Kilifi district Development plan. 1989/93, pp. 1–3.

Muchoki, C. H. K., 1988. A preliminary survey of efficiency of three methods used in studying rangeland vegetation: The quadrat method, PCQ and line transect methods, Kenya Rangeland Ecological Monitoring Unit, Box 47146, Nairobi. Technical Report No. 132. pp. 3–10.

Kenya Government, National Environmental Secretariat, 1985. District environmental assessment report of Kwale district, 6 p. (6 pp.)

Ruwa, R. K. & P. Polk, 1986. Additional information on the mangrove distribution in Kenya: some observation and remarks – Kenya J. Sci. Ser. B7: 41–45.

UNEP, 1988. Ambio, A Journal of the human environment. Royal Swedish Academy of Sciences/Pergamon Press, vol. xvii No. 3 1988: 214–216.

UNDP/UNESCO, 1984, Regional Mangrove Project RA/79/002 in Asia and the Pacific.

Hydrobiologia **247**: 37–47, 1992.
V. Jaccarini & E. Martens (eds), The Ecology of Mangrove and Related Ecosystems.

Net primary productivity of two mangrove forest stands on the northwestern coast of Sri Lanka

M. D. Amarasinghe[1] & S. Balasubramaniam[2]
[1] *National Aquatic Resources Agency, Crow Island, Colombo 15, Sri Lanka;* [2] *Department of Botany, University of Peradeniya, Peradeniya, Sri Lanka*

Key words: mangrove ecosystems, estuarine mangals, fringing mangals, litterfall, primary productivity

Abstract

Productivity studies were carried out from September, 1985 to August, 1987 in two mangrove stands, i.e. estuarine and island fringing, in Dutch bay, a lagoon situated on the northwestern coast of Sri Lanka. Net above-ground primary productivity was measured by monitoring litterfall and above-ground biomass increment. The average annual rate of litterfall in the estuarine and island-fringing mangrove stands are 588.14 g m^{-2} (approximately 6 t ha^{-1}) and 407.33 g m^{-2} (approximately 4 t ha^{-1}) respectively. The average annual rates of above ground woody growth are 614.74 g m^{-2} (approximately 6 t ha^{-1}) in the estuarine stands and 286.8 g m^{-2} (approximately 3 t ha^{-1}) in the island-fringing mangrove stands. Hence estuarine mangrove stands record a higher annual rate of above-ground net primary production (NPP; 1207.88 g m^{-2} or approximately 12 t ha^{-1}) than the fringing mangrove stands (694.22 g m^{-2}; approximately 7 t ha^{-1}). The annual rate of NPP in the water front zones of the stands (1300.47 g m^{-2} in the estuarine stands and 874.56 g m^{-2} in the fringing stands) are greater than those in the back-mangrove zones (115.28 g m^{-2} in the estuarine stands and 513.88 g m^{-2} in the island-fringing stands). These variations may be attributed to the differences in tidal flushing and influence of freshwater in the two localities.

Introduction

Tropical forest ecosystems especially mangrove forests, are among the most productive in the world in terms of gross primary productivity (GPP), (Rodin, 1975) and it has been shown that considerable variation in GPP exists among the various mangrove areas (Lugo & Snedaker, 1974; Snedaker & Brown, 1982). An important portion of this GPP is the net primary productivity (NPP) which represents the portion of the GPP that remains after carbon loss through respiration is deducted and accumulated as biomass, primarily in wood, leaves (above-ground NPP) and roots (below-ground NPP). Shedding of plant parts, i.e. leaves, floral parts, fruits, twigs, branches and bark, known as litterfall, is an important source of organic detritus, which support economically important detrital marine food webs (Odum & Heald, 1975). Indicating this ecological role of mangals, a positive relationship between the extent of mangals and marine produce (e.g. prawns) has been observed in a number of tropical areas (Martosubroto & Naamin, 1977; Turner, 1977).

As tidal flushing, salinity and nutrient regimes as well as substrate characteristics appear to influence primary productivity (Carter *et al.*, 1973), its magnitude may vary from one mangrove area to another. As such, present study was undertaken to understand the productivity of estuarine and fringing mangrove ecosystems associated with a dry coastal zone lagoon in North Western Sri Lanka where a low tidal amplitude prevails.

Study area

Dutch bay is a part of the largest brackish water body in Sri Lanka (Location – 8° 15′ N, 79° 50′ E). Description of the study area, map of Dutch bay and the location of study sites are given in Amarasinghe & Balasubramaniam (this volume).

Mangroves cover nearly 2100 ha of the intertidal areas of Dutch bay and the mangals at Kala Oya estuary are the most extensive (nearly 1800 ha; Kanakaratne *et al.*, 1983). A dry climate prevails and the average annual rainfall (1000–1100 mm) is seasonal with two distinct dry seasons i.e., January–March and July–September, coinciding with the north east and south west monsoons. Maximum rainfall is in the period from October to December by inter-monsoonal convectional rains (Amarasinghe & Balasubramaniam, this volume). Surface water salinity at the sampling sites in Dutch bay varies from 0.00‰ to 39.00‰ (Fig. 1).

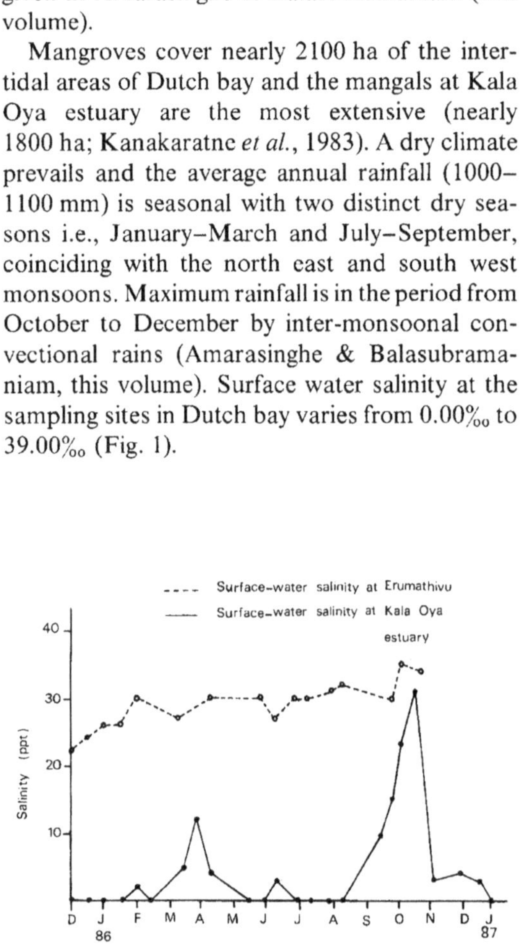

Fig. 1. Monthly variation in salinity from December 1985 – January 1987 at the study sites.

Materials and methods

Study sites

Study sites (Fig. 2) were selected so as to represent the major types of mangals i.e., estuarine mangals (Kala Oya estuary), fringing mangals (Erumathivu island). At both sites, *Avicennia marina* and *Rhizophora mucronata* were the dominant mangrove species and the latter is found occupying the water-front zone and the floodways within the mangals while the former species occupies a land-ward position (back-mangrove areas). Kala Oya mangals are floristically richer and structurally more complex than the fringing mangals at Erumathivu island (Amarasinghe & Balasubramanium, this volume). *Ceriops tagal, Lumnitzera racemosa* and *Aegiceras corniculatum* occur sporadically in the back-mangrove areas in the estuary.

Litterfall

Using auxiliary information, i.e., presence of different mangrove species and associations, it was possible to divide the mangals into sub-populations or zones (strata), each of which was assumed homogeneous internally. In both the study sites, two major zones or strata were distinguishable. Number of traps to collect litter, for each stratum or zone was determined on the approximate proportion of ground area of each zone and a total of 38 traps, 20 at Kala Oya (7 in the water-front *R. mucronata* zone and 13 in the mixed back-mangrove zone) and 18 at Erumathivu (8 in the water-front *R. mucronata* zone and 10 in the *A. marina* back-mangrove zone) were used for the purpose.

Traps of 1 m^2 in area, made out of wooden frames and nylon mesh were placed randomly in each zone, along the 10 m wide belt transects which have been laid perpendicular to the shoreline/riverbank across the major zones of the mangals upto 50 m inland, at the two study sites (2 such transects per site). Traps were installed at a level higher than the highest tide level, using stakes/poles and nylon ropes.

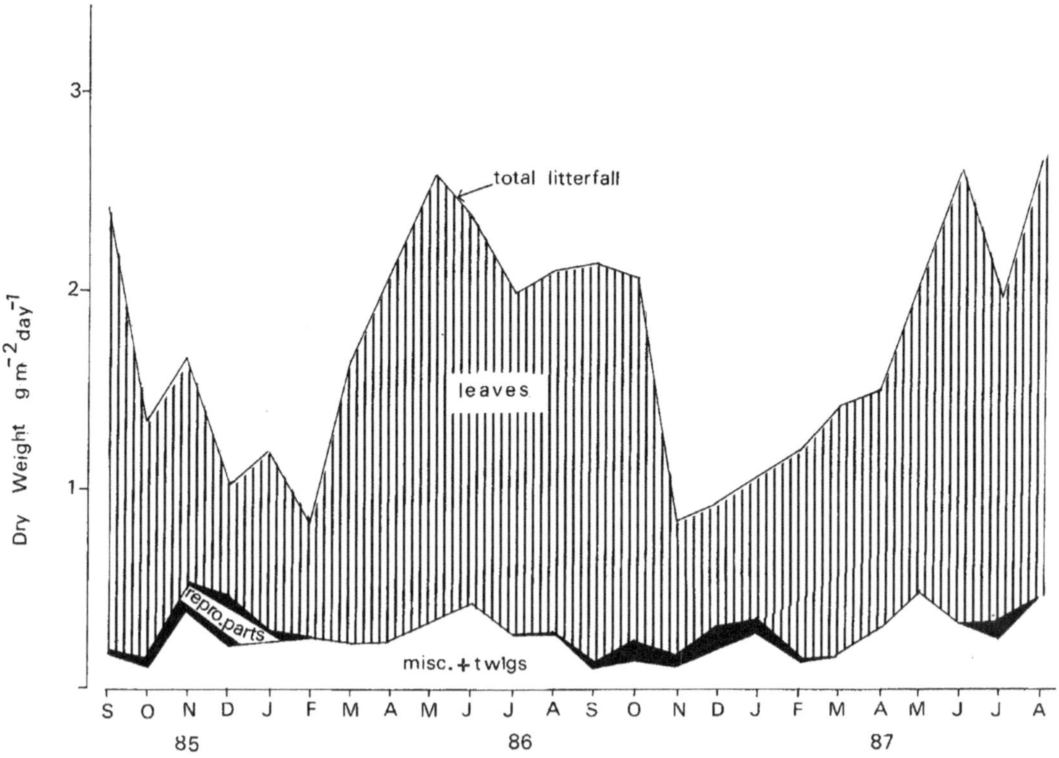

Fig. 2. Variation in total litterfall and components in the *Rhizophora mucronata* zone of the riverine/estuarine mangals at Kala Oya estuary.

Litter (trapped) was collected fortnightly, for a period of two years from September 1985 to August 1987. Collected litter was brought into the laboratory in plastic bags, sorted into species and litter of each species into plant components, i.e., leaves, stipules, floral parts, fruits/seedlings, branch wood and miscellaneous items and they were dried at 70 °C to constant weight.

Above-ground woody biomass production

Diameter at breast height (dbh) of all the trees with a dbh equal to or exceeding 2.5 cm, in the 10 m × 10 m subplots of the four permanent belt transects, were measured in August 1985 and remeasured in October 1987 at Erumathivu and March 1988 at Kala Oya. The initial and final biomass of the trees in the plots were estimated using the following regression equations (Amarasinghe & Balasubramanium, this volume).

For *R. mucronata*

$$\log_e (\text{total biomass}) = 6.247 + 2.64 \log_e (\text{dbh})$$

For *A. marina*

$$\log_e (\text{total biomass}) = 5.551 + 2.153 \log_e (\text{dbh})$$

Results

Litterfall

The average annual rates of component and total litterfall in the two mangals are presented in Ta-

Table 1. Averages litterfall rates ± standard deviation (g m^{-2} yr^{-1}) by component in the major zones of the mangals at Kala Oya.

	Leaves	Stipules	Repro. parts	Branches	Debris	Total
R. mucronata zone						
1985–1986	535.63	57.57	14.30	22.49	14.24	644.23
	± 18.24	± 2.24	± 2.11	± 2.25	± 0.85	± 18.49
1986–1987	505.81	52.11	13.25	18.42	13.75	603.34
	± 17.14	± 2.36	± 1.33	± 2.02	± 0.87	± 18.23
Whole period	19.22	54.84	13.78	20.46	14.00	623.79
	± 17.36	± 2.26	± 1.73	± 2.10	± 0.84	± 18.03
Mixed mangrove zone						
1985–1986	428.92	11.78	33.03	27.80	37.62	539.15
	± 12.32	± 0.58	± 3.29	± 3.19	± 1.73	± 11.83
1986–1987	445.38	13.37	51.76	17.28	38.01	565.80
	± 11.01	± 0.69	± 3.33	± 1.51	± 1.15	± 10.03
Whole period	437.15	12.58	42.40	22.54	37.82	552.48
	± 11.45	± 0.63	± 3.31	± 2.48	± 1.44	± 10.78

Table 2. Average litterfall rates ± standard deviation (g m^{-2} yr^{-1}) by component in the major zones of the mangals at Erumathivu.

	Leaves	Stipules	Repro. parts	Branches	Debris	Total
R. mucronata zone						
1985–1986	408.54	46.95	11.16	29.43	16.80	513.38
	± 19.06	± 2.11	± 1.68	± 4.99	± 1.26	± 22.58
1986–1987	273.68	44.61	20.59	13.30	16.35	368.53
	± 9.25	± 1.78	± 1.93	± 1.45	± 1.08	± 8.45
Whole period	341.11	45.78	16.13	21.37	16.58	440.96
	± 15.86	± 1.91	± 1.81	± 3.67	± 1.15	± 17.77
A. marina zone						
1985–1986	304.77	–	34.10	20.13	34.80	363.80
	± 16.68	–	± 4.20	± 2.38	± 1.43	± 15.87
1986–1987	300.00	–	10.17	14.48	28.92	353.57
	± 17.29	–	± 1.17	± 1.49	± 1.29	± 15.90
Whole period	302.39	–	22.14	17.31	31.86	373.69
	± 16.61	–	± 3.19	± 1.95	± 1.35	± 15.63

bles 1 and 2 and the percentage contribution of component litterfall to the total litterfall in the major zones of the two study sites are shown in Table 3. The average annual litterfall rates in the two zones at Kala Oya estuary are higher than those in the two major zones at Erumathivu mangals. In both the mangals, the average litterfall rates in the waterfront (*R. mucronata*) zones were greater than those in the back-mangrove zones.

Leaf litter comprised the largest portion (77–84%) of the total annual litterfall. Non-leaf plants (mangrove) litter accounts only for 2–10% of the total litterfall in these mangrove ecosystems.

Four-weekly totals for the litterfall indicate a marked seasonality in leaf, reproductive structures and total litterfalls in the two mangals. In all the zones, the period of high litterfall was during March–August (with minor variations), which coincides with the south west monsoon, generally

Table 3. Percentage contribution of component litterfall to the total litterfall in the major zones of the mangals at Kala Oya estuary and Erumathivu island (R.m. – *R. mucronata*; A.m. – *A. marina*);

Component	Percentage contribution			
	Kala Oya		Erumathivu	
	R.m. zone	A.m. zone	R.m. zone	A.m. zone
Leaves	83.5	79.1	77.4	80.0
Stipules	8.8	2.3	10.4	–
Reproductive parts	2.2	7.7	3.7	5.9
Branches	3.3	4.1	4.8	4.6
Debris (Miscellaneous)	2.2	6.8	3.2	8.5

the dry season for this area. The period of low litterfall coincides with the north east monsoon and the inter-monsoon months are the wet months of the year (Figs 2, 3, 4 and 5).

Variations in total and component litterfall of *R. mucronata* and *A. marina* in the two sites are presented in Figs 6 and 7. In both the sites, leaf fall follows a similar pattern with minor differences in the month of maximum leaf fall. The magnitude of litterfall of the species is higher at Kala Oya than in the mangals at Erumathivu.

Annual above-ground woody biomass production

Average basal area increment, average woody growth per stem per year and total annual growth

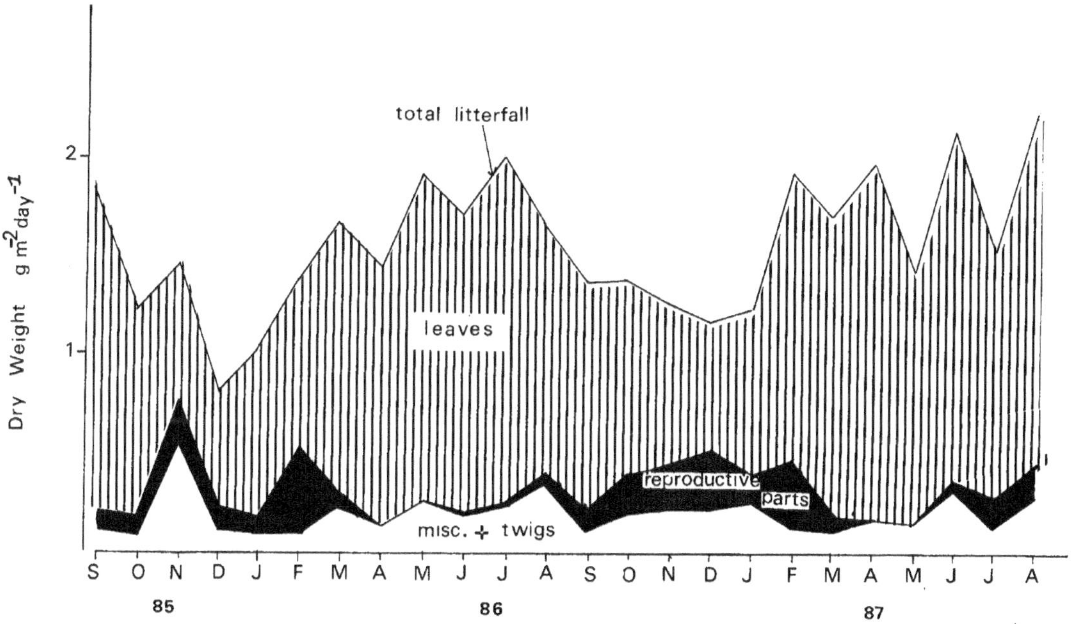

Fig. 3. Variation in total litterfall and components in the mixed mangrove (back-mangrove) zone of the riverine/estuarine mangals at Kala Oya estuary.

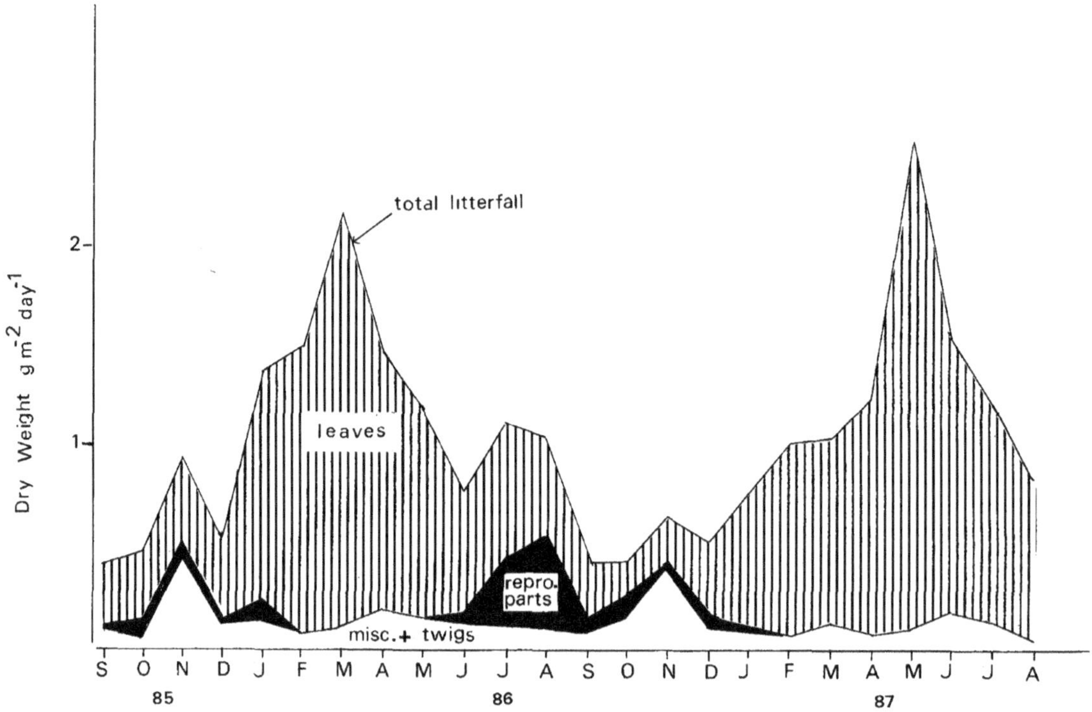

Fig. 4. Variation in total litterfall and components in the in the *Rhizophora mucronata* zone of the fringing mangals at Erumathivu island.

of mangrove trees measured in the mangals at Kala Oya estuary and Erumathivu island are given in Table 4.

The average rate of basal area increase and net woody above-ground production per stem for the two major mangrove species, i.e., *R. mucronata* and *A. marina* were greater at Kala Oya than at Erumathivu. In both sites, average woody growth of the trees in the water front zone was higher than that in the back-mangrove zone. At Kala

Table 4. Average basal area increment (A) in cm^2 per stem per year, average woody growth (B) in kg per stem per year and total annual growth (C) in $g\ m^{-2}\ y^{-1}$ of mangrove trees measured at Kala Oya estuary (I) from July 1985 to March 1988 and Erumathivu island (II) from July 1985 to October 1987.

Species	A		B		C	
	I	II	I	II	I	II
Water-front zone						
R. mucronata	1.68	0.89	1.33	0.82	676.68	433.60
Back-mangrove zone						
R. mucronata	1.66	–	1.24	–	210.35	–
A. marina	1.56	1.10	0.86	0.29	253.42	140.19
Other species	3.32	–	1.52	–	99.02	–

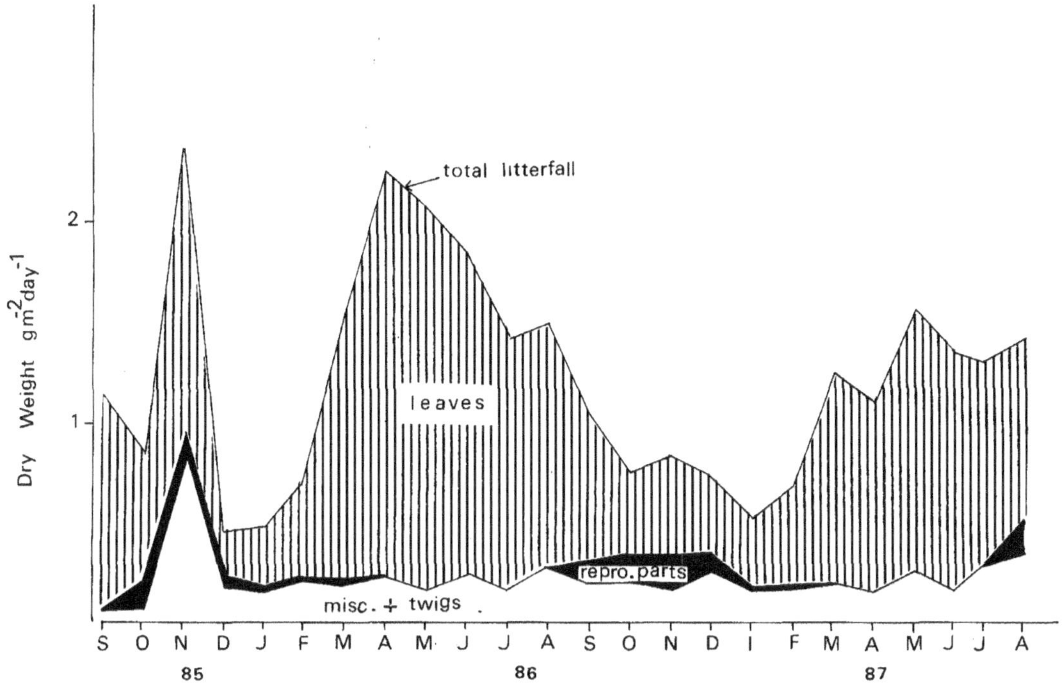

Fig. 5. Variation in total litterfall and components in *Avicennia marina* zone of the fringing mangals at Erumathivu island.

Oya, *A. marina* had a smaller basal area increase than *R. mucronata*. At Erumathivu, the trend was reversed. In both sites, increase in woody biomass of *R. mucronata* was greater than that for *A. marina* trees. Although *R. mucronata* in the mixed mangrove zone at Kala Oya had a higher basal area increase and average woody growth than *A. marina* in the same zone, its contribution to total woody growth (210.35 g yr^{-1}) was less than that by *A. marina*.

The total woody growth of the trees in the water-front zones at Kala Oya and Erumathivu were higher than those in the back-mangrove zones of the two mangals.

Total above-ground NPP

Total above-ground NPP values calculated for the major zones of the estuarine (Kala Oya) and fringing mangals (Erumathivu) are presented in Table 5. The NPP values for the two zones at Kala Oya were greater than those values calculated for the two zones at Erumathivu. At each site, the NPP for the water-front zone was higher than that for the back mangrove zone. Except in the back-mangrove zone at Erumathivu, where woody growth accounted only for about 27% of the NPP, in all other zones litterfall comprised nearly 50% of the total NPP of those mangals.

Discussion and conclusions

Litterfall

Among the different types of mangals, the highest litterfall values have been recorded from riverine/estuarine mangals (average 1100–1200 g m^{-2} yr^{-1}) and fringing mangals are reported to have a value of $906 + 89$ g m^{-2} yr^{-1}

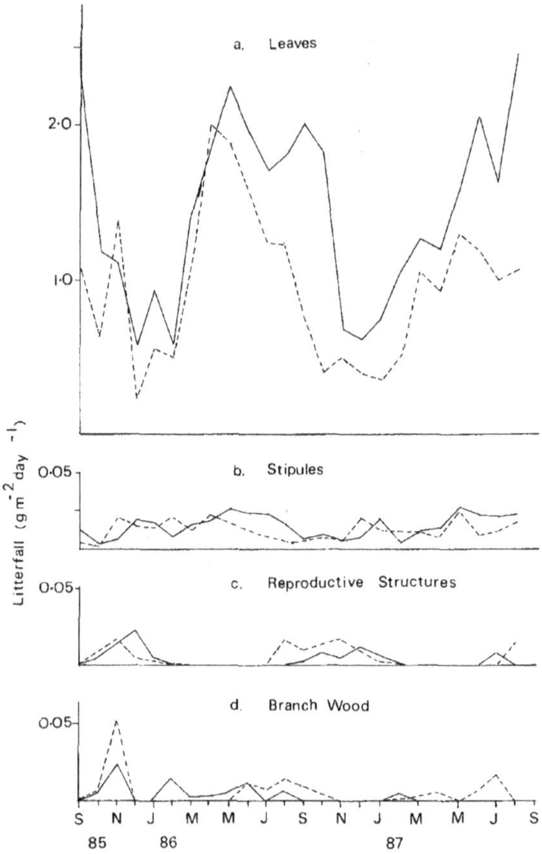

Fig. 6. Variations in litterfall by components of *Rhizophora mucronata* at Kala Oya (———) and Erumathivu (-------).

(Twilley, 1982 *cf.* Day *et al.*, 1987). The average litterfall value obtained (average for both the zones), for the estuarine mangals at Kala Oya (about 588 g m^{-2} yr^{-1}) is much lower than the average range of values reported for the estuarine mangals and this value is even lower than that reported for the fringe mangals. Litterfall rate in the fringing mangals at Erumathivu, i.e., 407 g m^{-2} yr^{-1} is much below the average range reported for fringe mangals (Cintron *et al.*, 1985). Nevertheless the total litterfall rate estimated for Kala Oya estuarine mangals (annual rainfall 100–110 cm) is comparable with that of the mangals at Siar river estuary (annual rainfall 426 cm) in Sarawak, East Malaysia (Othman, 1988). The

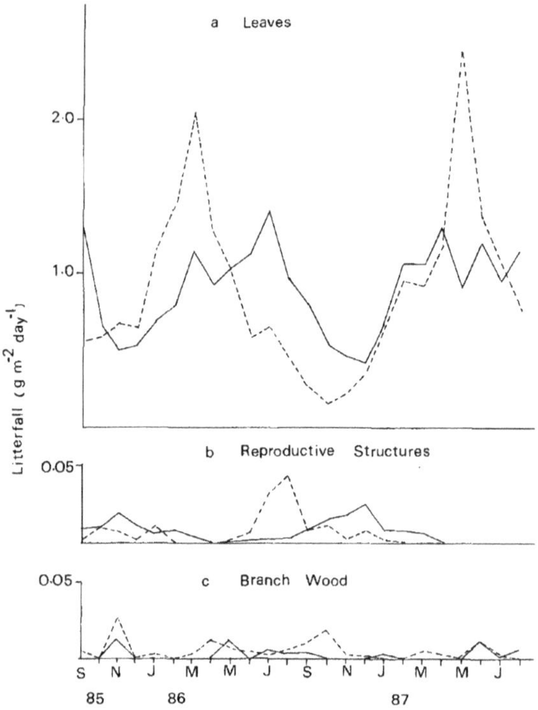

Fig. 7. Variations in litterfall by components of *Avicennia marina* at Kala Oya (———) and Erumathivu (-------).

notable similarity of these two mangrove areas is that they are dominated by *R. mucronata*. Furthermore, the annual litterfall rates for *A. marina* (at Kala Oya, about 356 g m^{-2} yr^{-1} and at Erumathivu, nearly 342 g m^{-2} yr^{-1}) are more comparable with those for the dwarf *A. marina* stands in New Zealand (Woodroffe, 1982) and *A. nitida* stands in South Florida (Snedaker & Brown, 1982), which are less than 1 m tall, in contrast to 3–4 m tall stands in this study area. This could partly be due to the differences in freshwater input and tidal amplitude. Pool *et al.* (1975) have suggested that the rate of litterfall in mangroves is related to increasing turnover of freshwater. Besides bringing nutrients required for continuous productivity of mangroves, within the mangal itself, water is important in bringing nutrients and oxygen to the root system and removing CO_2, toxic wastes and organic debris as well as in continual maintenance of the soil-salt balance

Table 5. Estimated above-ground net primary production of mangrove species in the major zones of the estuarine mangals at Kala Oya and the fringing mangals at Erumathivu island (percentage contribution of each species is given in parentheses). A – *R. mucronata* zone; B – Mixed mangrove zone; C – *A. marina* zone.

Component	Location and zone			
	Kala Oya estuary		Erumathivu island	
	A	B	A	C
Litterfall (g m^{-2} yr^{-1})				
R. mucronata	609.79 (46.9)	127.72 (11.5)	424.38 (48.5)	–
A. marina	–	355.96 (31.9)	–	341.83 (66.5)
Other species	–	30.99 (2.8)	–	–
Miscellaneous	14.00 (1.1)	37.82 (3.3)	16.58 (1.9)	31.86 (6.2)
Total	623.79 (48.0)	552.49 (49.5)	440.96 (50.4)	373.69 (72.7)
Average for mangal	588.14		407.33	
Woody growth (g m^{-2} yr^{-1})				
R. mucronata	676.68 (52.0)	210.35 (18.9)	433.60 (49.6)	–
A. marina	–	253.42 (22.7)	–	140.19 (27.3)
Other species	–	99.02 (8.9)	–	–
Total	676.68 (52.0)	562.79 (50.5)	433.60 (49.6)	140.19 (27.3)
Average for mangal	619.74		286.89	
Annual above-ground net primary production (NPP) (g m^{-2})				
R. mucronata	1286.47 (98.9)	388.07 (30.3)	857.98 (98.1)	–
A. marina	–	609.38 (54.6)	–	482.02 (93.8)
Other species	–	130.01 (11.7)	–	–
Total	1300.47	1115.28	874.56	513.88
Average for mangal	1207.88		694.22	
NPP : Litterfall	2.08	2.02	1.98	1.34

(Snedaker, 1978). Kala Oya river system is extensively utilized in the upstream areas for irrigational purposes. Eight reservoirs have been built in the upstream by damming the river and the water stored in Kala Wewa, the largest reservoir in the system, is diverted to a number of smaller tanks in the North Central Province to irrigate paddy land. Hence the volume of freshwater discharged by the river may inevitably be low. Although the nutrients in this have not been quantified, due to low freshwater inputs and associated poor sediment transportation into the estuary, it is plausible that the nutrient influx is also low. Fringing mangals at Erumathivu, situated farthest from the rivers may receive even lower input of freshwater and nutrients. Although the elevation is sufficiently low, low tidal amplitude ensue infrequent and insufficient tidal flushing may also contribute to the decreased productivity of these mangals. This also may explain the lower productivity in the back-mangrove zones than the water-front zones which subject to daily flushing.

Leaves form the largest portion of the total litterfall and hence constitutes a significant input of organic matter in both the mangals (Table 3). Although woodfall in these mangals appears to be unimportant, it may not be the reality as dead trunk fall has not been taken into account in the present study.

A general trend for peak litterfall to occur during the rainy season has been reported by a number of mangrove studies (Heald, 1971; Pool *et al.*, 1975; Leach & Burgin, 1985; Day *et al.*, 1987).

However a reverse trend was true for the litterfall in the mangals in Dutch bay. Lugo & Snedaker (1975) also have reported higher leaf fall rates during periods of extreme drought. This may be attributed to increased energetic cost of maintaining photosynthetic tissues in high-salinity conditions during droughts, thus causing shedding of leaves.

Duke *et al.* (1984) have shown that the abscission of a pair of stipules in association with the emergence of a pair of leaves, to be an accurate indicator of leaf emergence. Leaf production as measured by stipule fall of *R. mucronata* (at Kala Oya and Erumathivu) was found to be continuous throughout the study period, which suggests that the environmental conditions are favourable for leaf emergence all year round and the stress does not appear to limit the leaf production of this species. The period of slightly increased stipule fall (rates) during March–August at Kala Oya mangals coincides with higher leaf fall period. Therefore it could be assumed that the photosynthetic rates may remain constant as shown by Pool *et al.* (1975) and Lugo & Snedaker (1975). Thus this study suggests that the physiology of the trees may have an effect on the seasonality in litter production.

Above-ground woody growth of mangroves

Differences in above-ground woody growth of mangroves in the major zones at Kala Oya and Erumathivu may be attributed to the lower salinities (water and soil) and greater input of freshwater and hence nutrients in the estuarine mangals at Kala Oya. As Day *et al.* (1987) have shown, the lower rates of total woody growth in the back-mangrove zones than in the water-front zones, especially at Erumathivu may reflect the effect of stresses such as salinity and anaerobic conditions resulted from infrequent tidal flushing and also the effect of hydrogen sulphide on the biomass production of the trees in this zone. Thus more of the available energy from primary production may be used to deal with stress and less will be available to build the structure.

Above-ground NPP

Reported values of total above-ground NPP in mangrove areas are very few and they range from 380 g m^{-2} yr^{-1} for a scrub mangal in Florida (Teas, 1979) to 5475 g m^{-2} yr^{-1} for a mixed forest in Florida (Carter *et al.*, 1973). The NPP values computed for Kala Oya estuarine mangals (1208 g m^{-2} yr^{-1}) and Erumathivu fringing mangals (694 g m^{-2} yr^{-1}) fall within this range. Day *et al.* (1987) reported values of above-ground NPP of 2458 g m^{-2} yr^{-1} and 1606 g m^{-2} yr^{-1} for riverine (estuarine) and fringing mangals respectively in Mexico. The assumption that total above-ground NPP for mangroves is three times the total litterfall (Teas, 1979) is not supported by the results of the present study, in which NPP was approximately twice the amount of litterfall. Day *et al.* (1987) also have reported similar results for mangals in Mexico.

Acknowledgements

This study was supported by the National Aquatic Resources Agency (NARA) of Sri Lanka. Valuable assistance by Dr. U. S. Amarasinghe, Senior Research Officer, NARA in statistical analyses is gratefully acknowledged. Untiring assistance in the field by the staff at NARA Regional Research Centre at Kalpitiya, Sri Lanka is appreciated.

References

Amarasinghe, M. D. & S. Balasubramaniam, 1992. Structural properties of two types of mangrove stands on the northwestern coast of Sri Lanka. In: V. Jaccarini & E. Martens (eds), The Ecology of Mangrove and Related Ecosystems. Developments in Hydrobiology 80. Kluwer Academic Publishers, Dordrecht: 17–27. Reprinted from Hydrobiologia 247.

Carter, M. R., L. A. Burns, K. R. Carvinder, P. L. Dugger, D. B. Fore, D. B. Hicks, H. L. Revells, T. W. Schmidt & R. Farley, 1973. Ecosystem analysis of the big Cypress swamp and estuaries. US Environmental Protection Authority.

Cintron, G., A. E. Lugo & R. Martinez, 1985. Structural and functional properties of mangrove forests. In W. G. D'Arcy

& M. D. Correa (eds), The botany and natural history of Panama: La botanica e historia natural de Panama IV series, Monographs in systematic botany, V 10, Missouri Botanical Gardens, St. Louis, Mo; USA: 53–56.

Day, J. W. Jr., W. H. Conner, F. Ley-lou, R. H. Day & Navarro, 1987. The productivity and composition of mangrove forests, Laguna de Terminos, Mexico. Aquat. Bot. 27: 267–284.

Duke, N. C., J. S. Bunt & W. T. Williams, 1984. Observations on the floral and vegetative phenologies of North-Eastern Australia mangroves. Aust. J. Bot. 32: 87–99.

Heald, E. J., 1971. The production of detritus in a South Florida estuary. Uni. Miami Sea Grant Bulletin, No. 6.

Kanakaratne, M. D., W. K. T. Perera & B. U. S. Fernando, 1983. An attempt at determining mangrove coverage in Puttalam lagoon, Dutch bay and Portugal bay, Sri Lanka, through remote sensing techniques. Proc. 4th Asian Conference on Remote Sensing, Colombo: 1–9.

Leach, G. J. & S. Burgin, 1985. Litter production and seasonality of mangroves in Papua New Guinea. Aquat. Bot. 23: 215–224.

Lugo, A. E. & S. C. Snedaker, 1974. The ecology of mangroves. Annu. Rev. Ecol. Syst. 5: 39–64.

Lugo, A. E. & S. C. Snedaker, 1975. Properties of mangrove forests in Southern Florida. In G. E. Walsh, S. C. Snedaker & H. J. Teas (eds), Proc. Int. Symp. Biol. Management of Mangroves, IFAS, Univ. Florida: 170–212.

Martosubroto, P. & N. Naamin, 1977. Relationship between tidal forests (mangroves) and commercial shrimp production in Indonesia. Marine Research in Indonesia, No. 18: 81–86.

Odum, W. E. & E. J. Heald, 1975. The detritus based food webs of an estuarine mangrove community. Estuar. Res., 1: 265–286.

Othman, S., 1988. The structure of vegetation and rate of litter production in a mangrove forest at Siar Beach, Lundu, Sarawak, East Malaysia. A paper presented at the Symposium on Mangrove Management; Its Ecological and Economic Considerations, Bogor, Indonesia.

Pool, D. J., A. E. Lugo & S. C. Snedaker, 1975. Litter production in mangrove forests in Southern Florida and Puerto Rico. In G. E. Walsh, S. C. Snedaker & H. J. Teas (eds), Proc. Int. Symp. Biol. and Management of Mangroves, IFAS, Univ. Florida: 213–237.

Rodin, L., N. I. Bazilavich & N. N. Rozov, 1975. Productivity of the worlds main ecosystems, National Academy of Science, Washington D.C.

Snedaker, S. C., 1978. Mangroves: their value and perpetuation. Nat. Res. 14: 6–13.

Snedaker, S. C. & M. S. Brown, 1982. Primary productivity of mangroves. In C. C. Black & A. Mitsui (eds), CRC Handbook of Biosolar Resources, 1: 57–70.

Teas, H. J., 1979. Silviculture with saline water. In A. Hollaender, J. C. Aller, E. Esptein, A. E. Pietro & O. R. Zaborsky (eds), The Biosaline Concept. Plenum Press, New York: 117–161.

Turner, R. E., 1977. Intertidal vegetation and commercial yields of penaeid shrimp. Trans. am. Fish. Soc. 106: 411–416.

Woodroffe, C. D., 1982. Litter production and decomposition in the New Zealand mangrove *Avicennia marina* var. resinifera. New Zealand J. mar. Freshwat. Res. 16: 179–188.

Hydrobiologia **247**: 49–57, 1992.
V. Jaccarini & E. Martens (eds), The Ecology of Mangrove and Related Ecosystems.

The effect of coastline changes on mangrove community structure, Portuguese Island, Mozambique

J. C. Hatton & A. L. Couto
P.O. Box 257, Faculty of Biology, University of Eduardo Mondlane, Maputo, Mozambique

Key words: coastline configuration, mangrove, selective death, Mozambique

Abstract

Changes in coastline configuration and sand movements on Portuguese Island, a small island lying close to Inhaca Island 35 km east of Maputo, Mozambique have resulted in large areas of the Island being eroded away and consequent restricted tidal recharge of the mangrove. Measurements made during 1989 show that flooding of the mangrove occurred on only three occasions. Percentage mortality of the dominant species, *Ceriops tagal* (Perr.) C. B. Robinson., is high. However, at the time of the study, the co-dominant, *Bruguiera gymnorrhiza* (L.) Lam., showed no signs of stress. Halophytic conditions are indicated by an invasive spread of salt-tolerant herbs. All species of crustaceans and molluscs have disappeared from the mangrove.

Although mortality of *C. tagal* is high, the regeneration status of this species is also high, comparable to that in a regularly recharged mangrove on Inhaca Island. It is suggested that *C. tagal* propagules were shed under conditions when seedling establishment was favorable i.e. flooding. These seedlings may represent the last cohort which may also suffer stress and death unless tidal recharge is re-established.

Introduction

Mozambique has the third largest coastline in Africa, extending more than 2700 km from the Tanzanian border in the north (10° 27′ S) to the South African border in the south (26° 57′ S). Only Somalia and South Africa have a greater extent of coastline. Extensive mangrove communities are located at several sites along the coast in sheltered bays and estuaries. It is estimated that mangrove communities occupy some 850 km^2 in Mozambique, second in importance only to Tanzania (960 km^2) along the East African coast (Bennessalah, 1988).

Although mangrove communities are an important feature of the Mozambican coastline very few published data are available. General descriptions of flora and associated fauna have been given for selected sites along the coast (Cott, 1929; Jacques, 1975; Kalk, 1959; Sim, 1910). These works are of a descriptive nature or provide an inventory of plant and animal species. In contrast to the mangroves along most of Mozambique's coastline, the mangroves of Inhaca and Portuguese Islands are relatively well-documented. MacNae & Kalk (1962 and 1969) provide detailed descriptions of the taxonomy, distribution, zonation patterns and faunal associations of the Inhaca Island mangroves. Mogg (1963) carried out a preliminary investigation into the relationship between salinity and distribution of mangrove tree species on the Island. However, no investigations related to ecophysiology, mangrove dynamics or utilization patterns of the Mozambican mangroves have been carried out.

The present study represents one of several research projects related to mangrove ecology on Inhaca Island. A preliminary survey carried out by the authors in 1988 noted the presence of moribund and dead trees in the mangrove on nearby Portuguese Island which appeared to be related to impeded tidal recharge due to closure of the canal feeding into the mangrove. A comparison of aerial photographs (1958 to 1985) confirmed that marked changes in coastline configuration had taken place probably leading to this sequence of events. The present paper represents the results of a more detailed investigation into the changes in coastline configuration around Portuguese Island and the effects of these changes on the mangrove community.

Study area

Inhaca Island and nearby Portuguese Island are located 35 km east of the city of Maputo (see map, Fig. 1). Tides at Inhaca are semi-diurnal with an average amplitude of 3.3 m for springs and 1.5 m for neaps. Average annual temperature is 22.9 °C with maximum mean monthly temperature in February (26.2 °C) and mean monthly minimum in July (19.2 °C). Average annual rainfall is 874 mm with a distinct wet season during October to April and a dry season during May to September. Evaporation exceeds rainfall for all months except January and February.

Four broad vegetation types may be distinguished: intact forest, agricultural landscape with scattered trees, swamps and mangroves (Fig. 2). The main study was carried out in the mangrove on Portuguese Island with a comparative study ('non-stressed' mangrove) being made in the mangrove at Ponta Rasa, Inhaca Island.

The Ponta Rasa mangrove is a typical creek mangrove approximately 1 km in length running in a north-south direction. The banks of the canal leading into the mangrove are lined with a dense stand of *Rhizophora mucronata* Lam. The fiddler crab *Uca urvillei* (M. Edw.) is common along the banks of the canal. The central core of the mangrove is covered by a dense, mixed stand of *R. mucronata* and *Bruguiera gymnorrhiza* (L.) Lam. In this zone the soils have a relatively high organic matter content (5–6.5%) and sesarmid crabs are abundant (*Sesarma meinerti* de Man. and *S. guttata* M. Edw.). The landward edge of the mangrove comprises a less dense stand of smaller growing *Ceriops tagal* and *Avicennia marina* (Forsk.) Vierh. trees with occasional scattered individuals of *Lumnitzera racemosa* Willd. on the outermost fringes. The soils of this zone are drier and sandy with relatively low levels of organic matter (0.1–1%). Fiddler crabs (*Uca inversa* (Hoffman), *U. annulipes* (M. Edw.) and *U. gaimardi* (M. Edw.) are common faunal associates in this zone (MacNae & Kalk, 1962).

The mangrove on Portuguese Island was also, originally, a creek mangrove. This mangrove is much smaller than the Ponta Rasa mangrove and zonation patterns are less distinct. A dense stand of *R. mucronata* lines the original course of the canal. The outermost landward zone of the mangrove comprises a dense stand of *C. tagal* and *B. gymnorrhiza*. Only a few scattered individuals of *A. marina* are found in the latter zone whilst *L. racemosa* is absent altogether. The diverse mangrove fauna of crustaceans and molluscs found in the mangrove of Ponta Rasa are absent from the Portuguese Island mangrove although remains of these animals are present providing evidence of their recent disappearance.

Methods

Changes in coastline configuration were determined using aerial photographs (1:15000, Department of Geographic Surveys, Maputo) for 1958, 1967, 1985 and 1989. Changes in area of mangrove and associated vegetation types were calculated from enlarged (×4) images of these photographs. Extensive ground truthing to support the photo-interpretation was carried out in 1989.

The frequency and extent of the tidal recharge during 1989 for the Portuguese Island mangrove

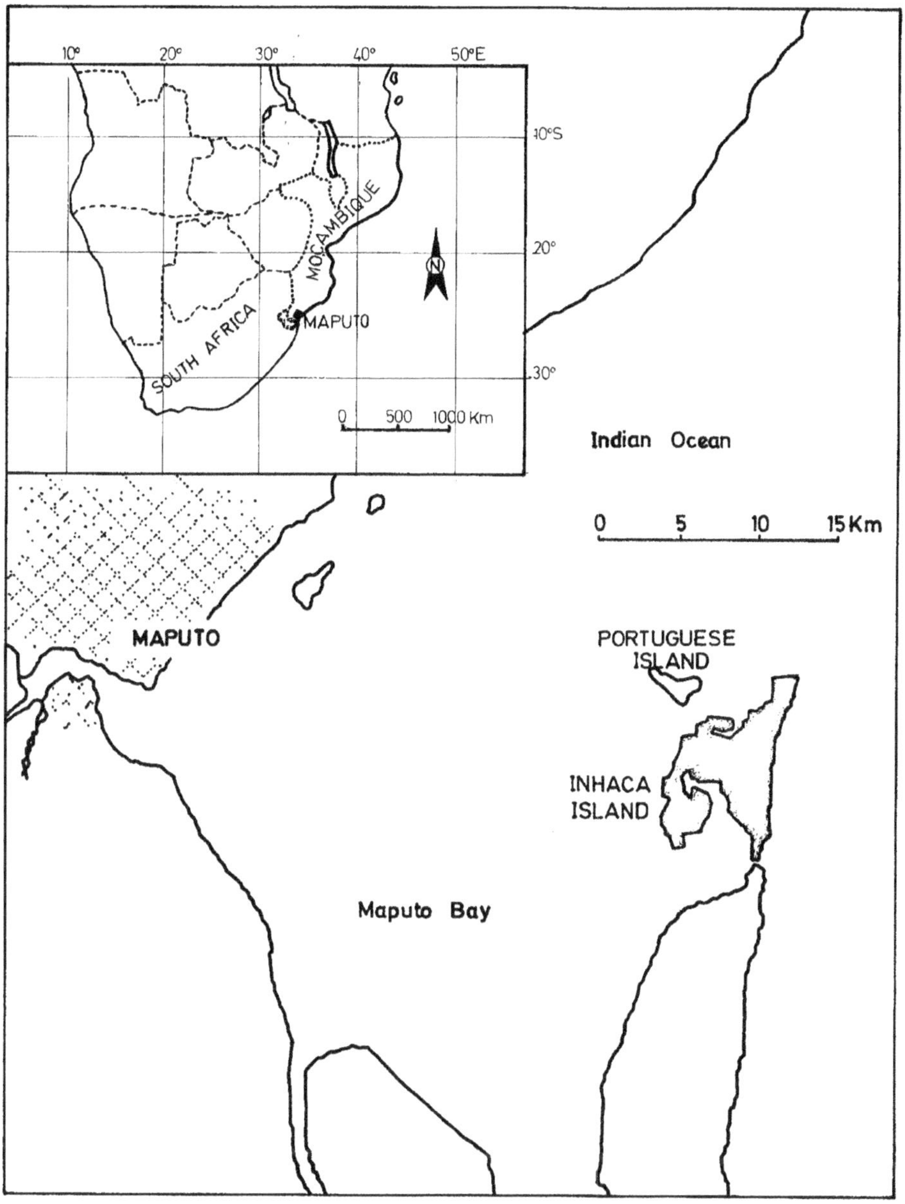

Fig. 1. The geographical setting of Inhaca Island and Portuguese Island. Inset shows the location of Maputo in southern Africa.

was determined by daily monitoring. When flooding occurred, the extent was measured by mapping on the ground with reference to permanent 'marker' trees in the mangrove.

Density (number of individuals per ha.), % frequency, % mortality and regeneration (number of seedlings per m^2) of tree species were measured in nine 5 × 5 m replicate quadrats in flooded and

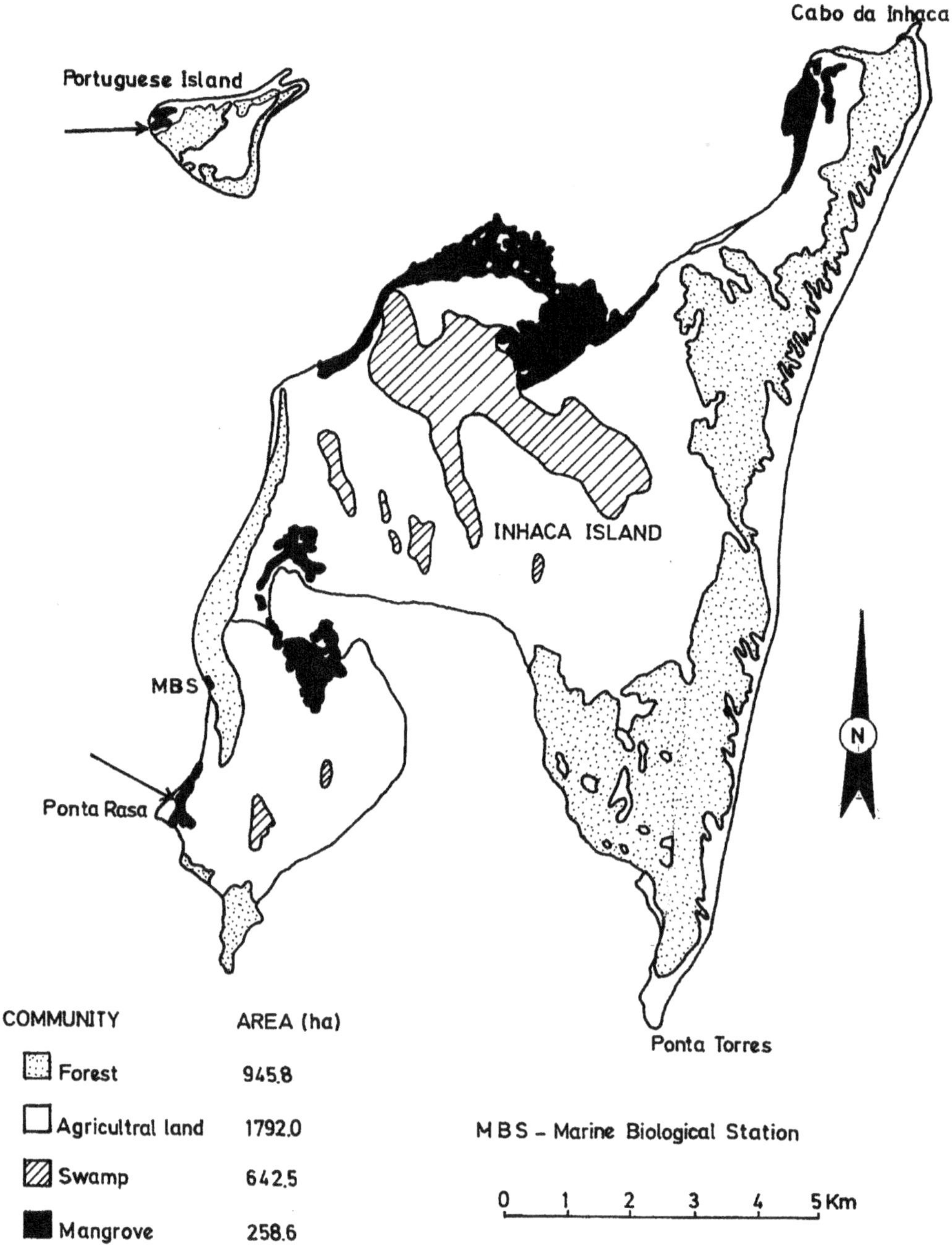

Fig. 2. Location of study sites (arrows) on Portuguese Island and Inhaca Island (Ponta Rasa). Also shown are the distribution of the four main vegetation types.

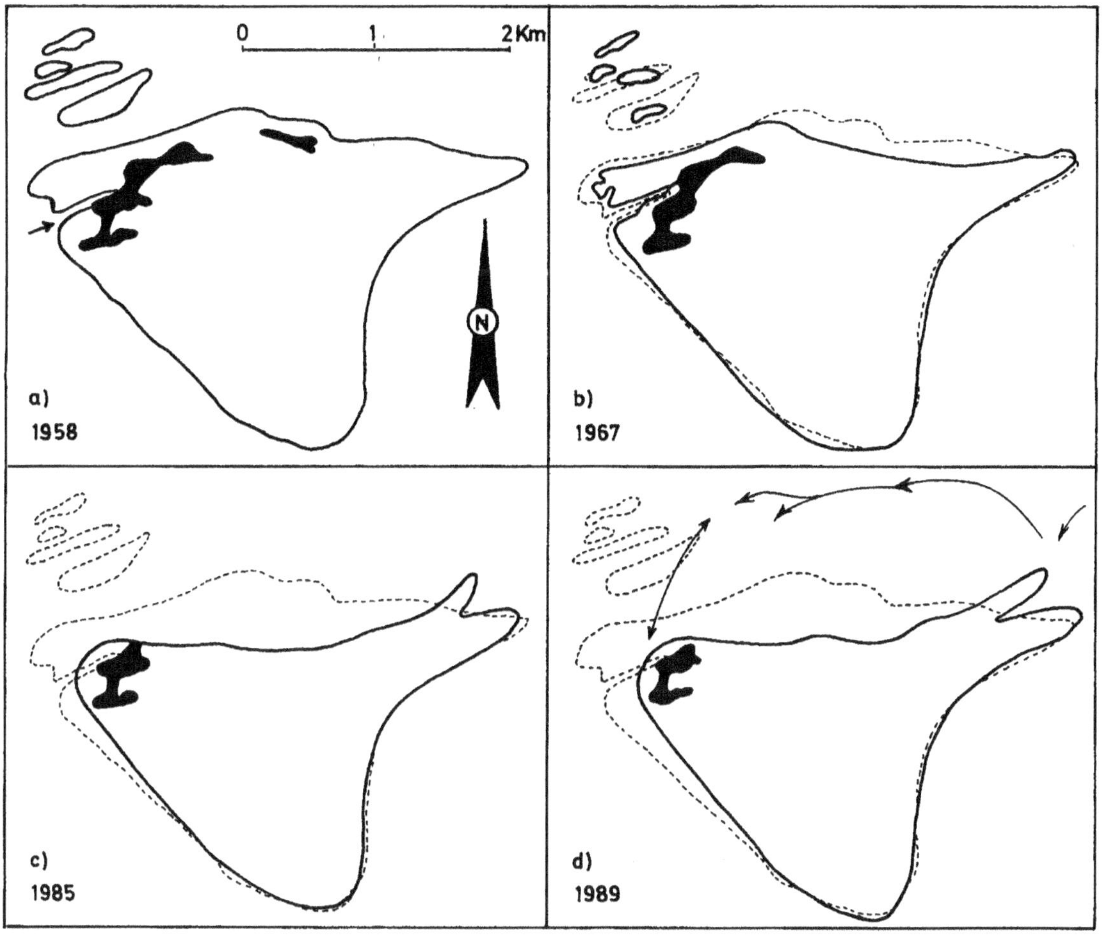

Fig. 3. Changes in coastline configuration, Portuguese Island between 1958 and 1989 (continuous line). The original 1958 contour is shown for comparison for 1967, 1985 and 1989 (dotted line). Shaded areas indicate mangrove. Arrow (1958) indicates original entrance to canal. Arrows (1989) indicate areas where new sand banks are forming.

non-flooded areas of the Portuguese Island mangrove. Herbaceous cover (%) was determined in 1 m^2 quadrats in each of the larger quadrats. Comparative studies were made in *Bruguiera* and *Ceriops* zones in the Ponta Rasa mangrove on Inhaca Island with six 5 × 5 m quadrats in each of the zones. Vegetation data were analyzed statistically by two-way t-test.

Results

Island and mangrove area changes

Aerial photographs for 1958, 1967, 1985 and 1989 show dramatic changes in the coastline configuration of Portuguese Island (Figs 3a, 3b, 3c and 3d). In 31 years much of the northern coastline has been eroded away forming a broad embayment. In the process, much of the mangrove has

Table 1. Surface area changes (m^2) of Portuguese Island and mangrove.

	Area (m^2) Portuguese Island	Area (m^2) Mangrove
1958	3027000	110150
1967	2619000	73860
1985	1519000	28430
1989	2161000	26240

also been eroded away, the canal leading into the mangrove has been closed and the remaining mangrove has become isolated from direct access to the sea by the formation of beach ridges.

In contrast to these erosion patterns, a new peninsula had formed in the north east of the Island by 1985, the process continuing until 1989. The prograding nature of this peninsula is highlighted by the formation of sand banks in an area around the northern shore, the erosion of this shore apparently having halted by 1989 (Fig. 3d).

Table 1 shows changes in area of Portuguese Island and associated mangrove.

Tidal exchange

During 1989 the Portuguese Island mangrove experienced only three periods of tidal flooding: three days in February, five days in March and four days in September. At these times (corresponding approximately to the equinoxes) the tidal amplitudes were sufficient to breach the newly-formed beach ridges on the northern shore separating the mangrove from the sea.

Although tidal recharge of the mangrove took place, the reduced strength of the tidal wash resulted in only partial immersion of the mangrove. Flooded and non-flooded areas of the mangrove corresponded to 72% and 28% respectively.

Vegetation

The density (number of individuals per ha.), frequency (%), mortality (%) and regeneration (number of seedlings per m^2) for *Ceriops tagal*, *Bruguiera gymnorrhiza*, *Avicennia marina* and *Rhizophora mucronata* in the Portuguese Island mangrove is given in Table 2. A comparison is made between flooded and non-flooded areas of the mangrove.

C. tagal and *B. gymnorrhiza* are by far the most important tree species of the mangrove. *A. marina* and *R. mucronata* occur at low densities and frequencies in both flooded and non-flooded areas. Mortality of *C. tagal* is highest with a significantly higher mortality of individuals of this species in the non-flooded areas. Mortality of the other species is low. However, regeneration of *C. tagal* is

Table 2. Density, frequency, mortality and regeneration of *Ceriops tagal*, *Brugieria gymnorrhiza*, *Avicennia marina* and *Rhizophora mucronata* in flooded and non-flooded areas in the mangrove of Portuguese Island. Standard errors about the mean are given in brackets. Differences of significance are between flooded and non-flooded areas.

	Density (no. of individuals ha^{-1})			Frequency (%)			Mortality (%)			Regeneration (seedlings m^{-2})		
	Flooded	Non-flooded	Sign	Flooded	Non-flooded	Sign	Flooded	Non-flooded	Sign	Flooded	Non-flooded	Sign
C. tagal	22444 (±3230)	19777 (±6387)	n.s.	76.2 (±5.4)	72.5 (±5.8)	n.s.	17.9 (±4.9)	56.0 (±6.2)	***	42 (±8)	28 (±6)	n.s.
B. gymnorrhiza	5933 (±1869)	4022 (±1126)	n.s.	19.8 (±5.4)	14.6 (±4.4)	n.s.	0.7 (±2.16)	0	n.s.	12 (±8)	12 (±3)	n.s.
A. marina	133 (±95)	555 (±385)	n.s.	0.1 (±0.7)	9.8 (±8.1)	n.s.	0	5.8 (±5.8)	n.s.	0	0	n.s.
R. mucronata	622 (±327)	311 (±311)	n.s.	3.04 (±1.5)	0.55 (±0.55)	n.s.	0	7.9 (±7.9)	n.s.	0	0	n.s.

n.s. = not significant; *** $p < 0.001$; $n = 9$.

Table 3. Percentage cover of herbaceous species in the Portuguese Island mangrove. Differences of significace are between floded and non-flooded areas (*** = $p < 0.001$). Standard errors about the mean are given in brackets ($n = 9$).

	Cover (%)		Sign.
	Flooded	Non-flooded	
Sesuvium portulacrastum	5.5 (± 3.6)	34.4 (± 7.6)	***
Sporobolus virginicus	5.5 (± 5.5)	33.3 (± 5.8)	***
Chenolea diffusa	0	19.4 (± 5.5)	***

highest followed by *B. gymnorrhiza*; no differences between flooded and non-flooded areas could be detected.

Three invasive herbaceous species were noted in the Portuguese Island mangrove: *Sesuvium portulacastrum* L., *Chenolea diffusa* Thunb. and

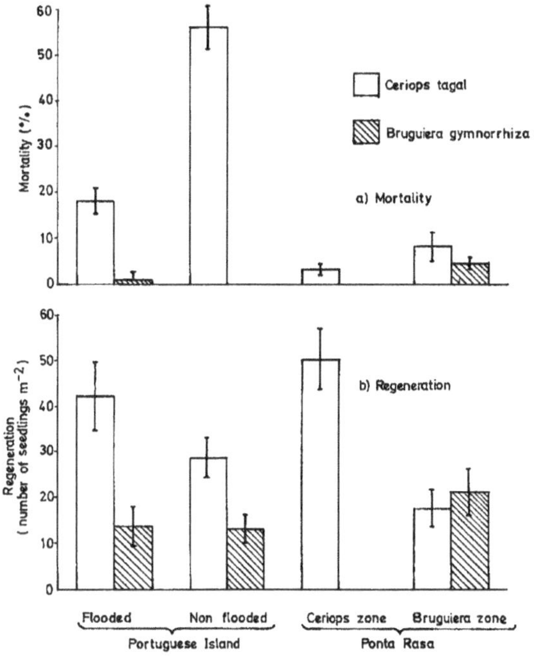

Fig. 4. Percentage mortality (Fig. 4a) and regeneration (Fig. 4b) of *C. tagal* and *B. gymnorrhiza* in flooded and non-flooded areas of the Portuguese Island mangrove and in *Ceriops* and *Bruguiera* zones of the Ponta Rasa mangrove (Inhaca Island). Standard errors about the mean are shown.

Sporobolus virginicus (L.) Kunth. Cover data (%) for these three species in flooded and non-flooded areas is given in Table 3. For all three species percent cover was significantly higher in non-flooded areas than in flooded areas.

A comparison of mortality and the regeneration status of the two dominant species, *C. tagal* and *B. gymnorrhiza*, in the 'stressed' (restricted tidal recharge) mangrove of Portuguese Island and 'non-stressed' (diurnal tidal recharge) mangrove on Inhaca Island (Ponta Rasa) is given in Figs. 4a and b. For the Ponta Rasa mangrove data is given for both a *Ceriops* zone and a *Bruguiera* zone. Although, as noted, mortality of *C. tagal* is higher in the non-flooded area as compared to the flooded areas in the mangrove of Portuguese Island (Table 2, Fig. 4a), mortality in both areas is higher than in the *Ceriops* or *Bruguiera* zones of the mangrove at Ponta Rasa. The regeneration status of *C. tagal* on Portuguese Island is similar to that found in the *Ceriops* zone at Ponta Rasa.

Discussion

It is evident that between 1958 and 1985 a strong current moving in an arc ground the north of Portuguese Island led to the erosion of much of this shoreline. In the process large areas of mangrove have been washed away with a loss of an area more than 75% of the 1958 cover (Fig. 3, Table 1). Between 1985 and 1989 new tidal forces appear to be operating resulting in deposition of sand at various locations around Portuguese Island. One of the consequences of this deposition of sand is that the Portuguese Island mangrove is isolated by beach ridges which are only rarely breached by tidal wash. Tinley (1985) has proposed that a continental shelf counter-current, moving in a direction opposite to the deep water Mozambique Current, transports sand northwards along much of the Mozambican coast. It also appears from a study of aerial photographs that the northern-most point of Inhaca Island acts as a fulcrum, sweeping sand around this point westwards towards Portuguese Island. A sand bar now extends between the two Islands and around the northern shore of Portuguese Island (Fig. 3d). These processes have probably resulted in the prograding beaches presently being formed on Portuguese Island and consequent impoundment of the mangrove.

Whatever the mechanisms leading to impoundment, by 1989 the mangrove was only occasionally flooded by tidal water and, because of reduced tidal flow, only partial immersion of the mangrove occurred. Similar impoundments and restricted tidal exchanges have been reported by Breen & Hill (1968) for the Kosi Estuary in South Africa, 120 km south of Inhaca Island, and by Gordon (1987) for the arid Dampier-Cossack region of Western Australia leading, in both cases, to mass mortality of mangrove trees.

In the case of the Portuguese Island mangrove the dominant species, *Ceriops tagal*, appears to be most affected by restricted tidal exchange. Mortality rates of *C. tagal* are much higher than individuals of the same species receiving regular exchanges of tidal wash at the Ponta Rasa mangrove on Inhaca Island (Fig. 4a). Individuals of *Bruguiera gymnorrhiza*, a codominant, do not appear to suffer the same consequences. No exact data as to the impoundment of the mangrove is available, but it probably occurred some time during 1988. *C. tagal*, therefore, appears to be most susceptible to the new adverse conditions. This is further supported by the fact that extreme and less-extreme conditions exist in the mangrove i.e. non-flooded and occasionally-flooded areas. Death of *C. tagal* is significantly higher in the non-flooded areas (Table 2). The susceptibility of *C. tagal* may be due to several factors. Along the East African coast, *C. tagal* is almost at its southern-most distributional limit (Chapman, 1976) and any environmental shift would have a proportionally larger effect on the growth of this species. Hutchings & Saenger (1987) report that the root system of *C. tagal* is particularly susceptible to sedimentation by sand, a process which may be occurring in the mangrove. The work of Breen & Hill (1968) at Kosi Bay, where *C. tagal* is rare, showed that all mangrove species, including *C. tagal* suffered death after short term restriction of tidal exchange and impoundment of fresh water. No impoundment of fresh water was observed in the Portuguese Island mangrove although it is possible that the mangrove suffered a short-term impoundment of fresh water due to rainfall which may have led to the death of *C. tagal*. Another cause of die-back could be due to increasing salinity in the mangrove. The hydrological and climatic conditions on Portuguese Island (little fresh water runoff, low rainfall and high evaporation) would probably promote salinization of the soils. An increase in salinity is further indicated by the invasive spread of halophytic herbs in areas of the mangrove not flooded by the occasional tidal recharge. Gordon (1988) reports that salinity levels of 90‰ were associated with death of *C. tagal* and other species in Western Australia. Smith (1988), in laboratory growth experiments, demonstrated that salinity levels of 60‰ resulted in death of *C. tagal*. The salinity and ground water conditions in the Portuguese Island mangrove are presently being studied. The disappearance of the once diverse fauna may also play a role; MacNae & Kalk (1969) point out the close association between sesarmid crabs and the roots of *C. tagal* which enhance aeration of the root system.

Somewhat contradictory is the observation that, although *C. tagal* has the highest mortality rates, regeneration status is comparable with that of the Ponta Rasa mangrove on Inhaca Island (Fig. 4.b). This may be explained by *C. tagal* shedding its propagules at a time when conditions in the impounded mangrove were suitable for generation *viz.* sufficient water for rooting and germination. Propagules of *C. tagal* and *B. gymnorrhiza* on Inhaca Island begin germination in January and are shed in late January and February at about the same time as one of the last tidal floodings of the Portuguese Island mangrove (February and March 1989). Development of these seedlings is now being studied and the results will be presented in a later paper.

The future of the Portuguese Island mangrove will depend on the build-up of beach ridges around the mangrove. The rapid death of *C. tagal* may be followed by increased mortality of the co-dominant, *B. gymnorrhiza*. The newly-established seedlings of these two species will certainly require a regular exchange of sea water if they are to flourish. Ward *et al.* (1986) have shown that re-establishment of all mangrove species is possible even after mass mortality due to impoundment providing that physical factors preventing tidal recharge are removed (in this case a sand bar blocking the mouth of the Kosi Bay estuary south of Inhaca Island). If no natural tidal recharge of the Portuguese Island mangrove occurs in the near future and deleterious trends continue, then management options such as artificially creating a canal may have to be considered.

Acknowledgements

The authors gratefully acknowledge financial support by the Norwegian Organization for Research and Development for this project. We thank all members of staff of the Marine Biological Research Station, Inhaca Island for their logistic support throughout the period of this work.

References

Bennessalah, D., 1988. Manual on Mapping and Inventory of Mangroves. FAO, Rome.

Breen, C. M. & B. J. Hill, 1968. A mass mortality of mangroves in the Kosi Estuary. Trans. r. Soc. S. Afr., 38: 285–303.

Cott, H. B., 1929. Observation on the natural history of the land crabs *Sesarma meinerti*, with special reference to the theory of warning colours. Proc. Zool. Soc. Lond. 18: 679–692.

Chapman, V. J., 1976. Mangrove Vegetation. Cramer, Vaduz.

Hutchings, P. & P. Saenger, 1987. Mangrove Ecology. St. Lucia, Queensland: University of Queensland Press.

Jacques, J. M., 1975. Aperçu sur un biocerose cotiere; la mangrove. Nat. Belg. 56: 45–50.

Gordon, D. M., 1988. Disturbance to mangroves in tropical-arid Western Australia: hypersalinity and restricted tidal exchanges as factors leading to mortality. J. Arid Envir. 15: 117–145.

Kalk, M., 1959. The fauna of the shores of Northern Mocambique. Rev. Biol. Lisbon, 2, 1.

MacNae, W. & M. Kalk, 1962. The ecology of the mangrove swamps at Inhaca Island, Mocambique. J. Ecol. 50: 19–34.

MacNae, W. & M. Kalk, 1969. A Natural History of Inhaca Island, Mocambique. Witwatersrand University Press, Johannesburg.

Mogg, A., 1963. A preliminary investigation of the significance of salinity in the zonation of species in salt-marsh and mangrove swamp associations. S. Afr. J. Bot. 59: 81–86.

Sim, T. R., 1910. The flora of Portuguese East Africa. S. Afr. J. Sci. 6: 294–300.

Smith, T. J., 1988. Differential distribution between subspecies of the mangrove *Ceriops tagal*: competitive interactions along a salinity gradient. Aquat. Bot. 32: 78–89.

Tinley, K., 1985. Coastal Dunes of South Africa. South African National Scientific Programmes Report No 133. CSIR, Pretoria, 193 pp.

Ward, C. J., T. D. Steinke & M. C. Ward, 1986. Mangroves of the Kosi System, South Africa: Their re-establishment since mass mortality in 1965/66. S. Afr. J. Bot. 52: 501–512.

Hydrobiologia **247**: 59–75, 1992.
V. Jaccarini & E. Martens (eds), The Ecology of Mangrove and Related Ecosystems.

The seagrass and associated macroalgal vegetation of Gazi Bay (Kenya)

Eric Coppejans,[1] Hans Beeckman[2] & Mie De Wit
[1] *Laboratorium voor Morfologie, Systematiek & Ecologie van de Planten, R.U.G., K.L. Ledeganckstraat 35, B-9000 Gent, Belgium;* [2] *Koninklijk Museum voor Midden-Afrika, B-3080 Tervuren, Belgium*

Key words: seagrasses, macroalgae, Kenya

Abstract

The seagrass and macroalgal vegetation of Gazi Bay (at approximately 50 km south of Mombasa) have been studied by means of 88 relevés along 7 transects. Correlation between the distribution of the seagrasses and some abiotic factors (particle size fractions, chemical composition of the substrate) is not well marked. Nevertheless a general zonation and succession of seagrasses could be established:

1. a transition zone between the mangal and the seagrass beds is covered by *Boodleopsis pusilla*;
2. the pioneer association *Halophila ovalis* + *Halodule wrightii* forms low sandy bumps at the upper limit of the seagrass beds, but also occurs in the whole midlittoral where sandlayers have recently been accumulated (e.g. on coral platforms);
3. the climax vegetation of the intertidal zone seems to be *Thalassia hemprichii* which sometimes is associated with *Cymodocea rotundata* and *C. serrulata*, certainly in deeper pools and close to low water mark; *Halimeda opuntia*, *Gracilaria salicornia* and *G. corticata* are also frequent in this vegetation type;
4. from low water at neap tide downwards patches of monospecific *Enhalus acoroides* vegetation can also occur;
5. from mean low water down to approximately −1 m mixed meadows of *Thalassia*, *C. serrulata*, *C. rotundata* and *Halodule uninervis* are well developed; the seaweeds *Halimeda macroloba* and *Avrainvillea obscura* are also typical for this zone; locally patches of *Syringodium isoetifolium* grow on small bumps and *Halophila stipulacea* grows as a pioneer on bare sand;
6. from −1 m downwards the whole lagoon is covered by homogeneous, monospecific *Thalassodendron ciliatum* meadows, locally replaced by *E. acoroides*.

Introduction

The importance of the seagrass ecosystem has only been recognized recently. Now it is even supposed to be one of the most productive ecosystems (Mc. Roy & Mc Millan, 1977; Zieman & Wetzel, 1980). The ecological roles of seagrass beds are (den Hartog, 1967; Dawes, 1981: 481–484): sediment traps, stabilizing bottom sediments and improving water clarity; as primary producers they have high rates of production; they are a direct food source for many animals; they are important habitats and shelters for a number of animal species; they are important substrata for attachment of epiphytes (Harlin, 1980) which also provide important food sources for small fish and invertebrates; they are highly efficient in removing nutrients (nutrient stripping) from marine waters and surface sediments and are therefore critical plants in the control of water quality of shallow waters (Patriquin, 1972).

In Gazi Bay the large quantities of drifting seagrasses (especially of *Thalassodendron ciliatum*) and seaweeds (*Sargassum* sp.) brought into the mangrove at each high tide (certainly in some periods of the year) might account for a major import of organic material in that ecosystem.

Next to these ecological roles of the seagrass ecosystem, the economical uses of these marine angiosperms are emphasized by different authors: production of paper because of the high cellulose/low lignin content of the leaves; young rhizomes as well as fruits of some species are eaten in different parts of the world; dried seagrass leaves mixed with terrestrial plant material form an excellent fodder; seeds of *Zostera* are ground as flour in Sonora, Mexico; seagrasses like *Thalassia* are excellent for green manuring.

In Gazi Bay several biological research programs are carried out: Floristics, Faunistics and Ecology of Kenyan coastal biotopes (FKFO 2.0043.88: 1988–1991), Dynamics and assessment of Kenyan mangrove ecosystems (EEC project TS2-0240-C (GDF): 1989–1992), Systematics, Ecology and Biogeography of Marine Organisms in the Indian Ocean (FKFO 2.0009.92: 1992–1995).

The purpose of this paper is to give a description of the seagrass vegetation. It is hoped to be a main framework for more detailed studies of this ecosystem.

Material and methods

Description of the site

A map of Kenya with the localisation of Gazi is given in Coppejans & Beeckman (1989: 382); a general description and map of Gazi Bay (= Maftaha Bay), approximately 50 km south of Mombasa, in Coppejans & Gallin (1989: 48–49, Fig. 1); climatological data and mean salinity values are summarized in Gallin *et al.* (1989: 198).

Except for some restricted seagrass growth in mangal tide pools the seagrass beds are best developed from the lower margin of the mangal down to the subtidal zone. They cover the whole lower basin (lagoon) of Gazi Bay, stretching up to the fringing reef. Only a few sandy patches (Gazi Beach) or fossil coral outcrops do not have seagrasses. The complete area is subhorizontal, slightly sloping towards the tide channel and the shallow lagoon. At neap low tide most of the area remains inundated; at spring low tide vast areas (= the infralittoral fringe) are emergent except for the deepest part of the tide channel and the lagoon.

The substrate is very diverse because of sedimentation patterns (Coppejans & Gallin, 1989: 49; Onyango, 1989): locally it is rather coarse sand, on other places one sinks knee-deep in silty sand. Large areas are characterized by closely packed cone-like bumps, $\pm$ 0,5 m high, $\pm$ 1 m in diameter, enclosing small tide pools. The bumps are built up by still unknown animals which regularly extrude sediments from a central crater. The solid sandy bumps are devoid of seagrasses, the pools contain a mixed seagrass/seaweed vegetation. This diversity in environmental conditions results in a zonation and patchiness of the seagrass vegetation.

Sampling

Field work was concentrated in July and August 1987 (De Wit, 1988). Additional observations have been made in August 1989. Seven transects through the seagrass beds have been studied during spring low tides (Fig. 1). Five of them (3–7) are at right angles to the coastline, starting at the lowermost pneumatophores of *Sonneratia* (Coppejans & Gallin, 1989; Gallin *et al.*, 1989) going down to the rather homogeneous subtidal meadows. The vegetation of the tide channel was not studied here. A detailed study of it has been carried out by De Pauw (1990). Each transect was subdivided in homogeneous zones defined by the presence/absence of seagrass species, visible changes of substrate and/or relief. In total 88 relevés (1×1 m^2) were made in the different zones.

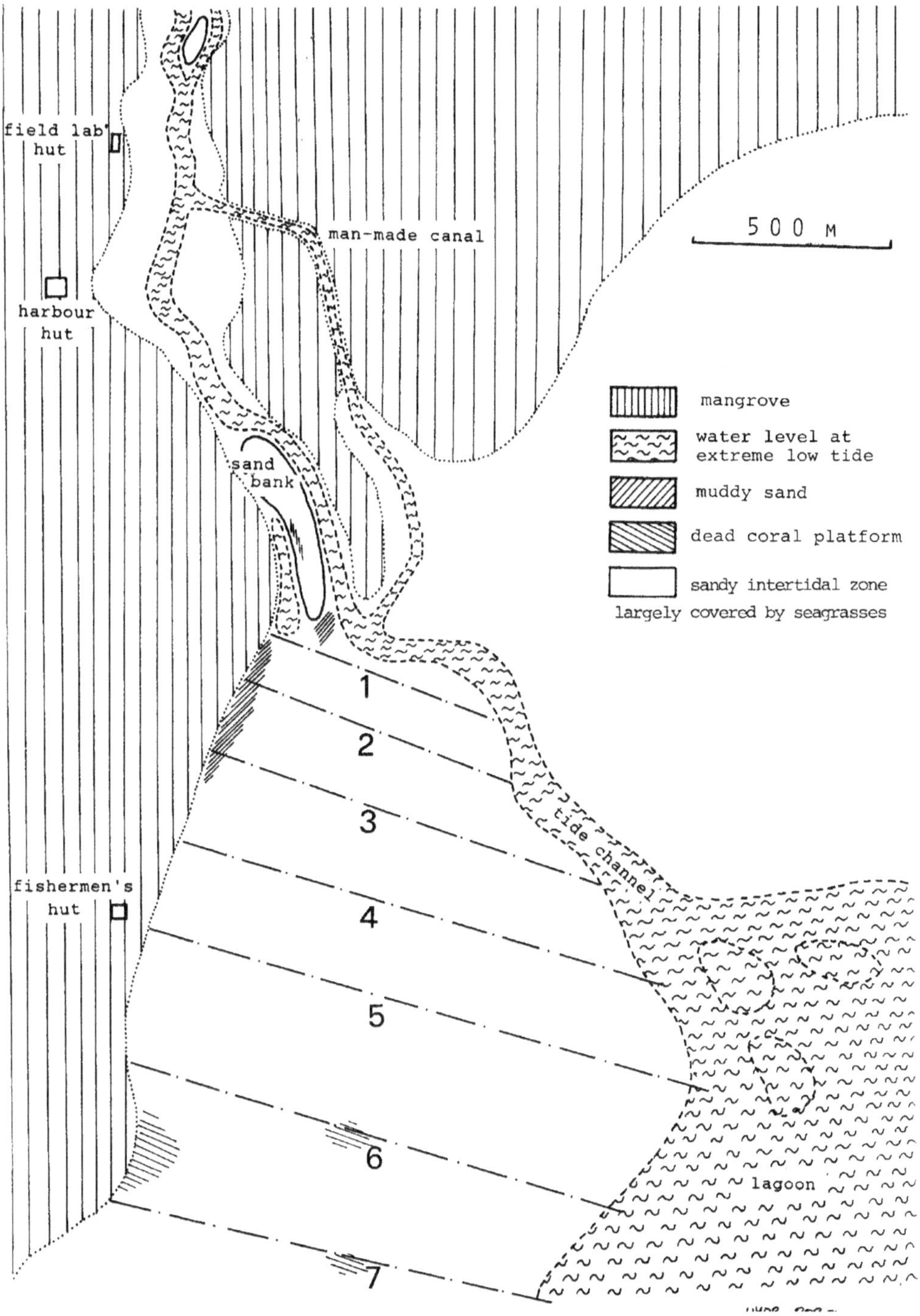

Fig. 1. Map of the study site at Gazi, with position of the transects 1–7.

The presence and cover of the seagrasses as well as of the macroalgae were recorded using the Londo-scale (Londo, 1984). Specimens which were unidentifiable on the field (mainly seaweeds) were collected and either pickled in 4% formalin/seawater or prepared as herbarium specimens which are deposited in the Herbarium Universitatis Gandavensis (GENT) under HEC-numbers (Herbarium Eric Coppejans); duplicates are present in the Kenya Marine and Fisheries Research Institute (K.M.F.R.I., Mombasa, Kenya).

Salinity and water temperature of the tidal pools along the transects 1, 3 and 4 were measured on the field with a combined conductivity/thermometer. Sediment samples were also taken along some of the transects.

Analysis of the samples

The unidentified seaweed specimens were analyzed in the laboratory for Plant Morphology, Systematics and Ecology. The chemical soil features were determined in the Laboratory for Analytical and Agrochemistry: the organic substance by the method of Walkley and Black (1934); the concentration of Ca, K and Na by flame photometry, the concentration of Mg by atomic absorption and the concentration of P by the colorimetric method of Scheel (Cottenie *et al.*, 1982: 43). The texture of the soil was analyzed by the pipette method in the Laboratory for Agronomic Soil Science (for more details: De Wit, 1988: 24–29).

Numerical analysis

The data on benthic plant species and on abiotic factors were submitted to multidimensional numerical analysis (Legendre P. & L., 1984a, b). The data were checked for possible clusterings between seagrasses and/or seaweeds and for correlations between biotic and abiotic components. Therefore a matrix of 37 species and 88 relevés was submitted to the FORTRAN 77 program TWINSPAN (Hill, 1979). The two-way table has been manually rearranged in order to show the main gradient in the vegetation. A part of the matrix, consisting of 32 taxa and 45 relevés, is compared to another matrix with information on 15 environmental variables for the same 45 sample units by the algorithm Canonical Correspondence Analysis (CCA) (CANOCO, ter Braak, 1988).

The involved environmental parameters are: the distance (m) of the plots to the starting point of each transect (= low water level), the concentrations of Na, K, Ca, Mg and P (mg/kg substrate), the quantity of organic material (%), the pH of the substrate, the proportion of seven particle size fractions (%).

Canonical Correspondence Analysis can be considered as a multidimensional direct gradient analysis into which regression and ordination have been integrated (ter Braak, 1986, 1987; Beeckman, 1989). Canonical or constrained ordination is based on Hutchinson's concept of the niche (Hutchinson, 1944, 1957, 1978; Schoener, 1989), which was first introduced in a footnote to a 1944 paper in Ecology on phytoplankton periodicity: 'the term niche is here defined as the sum of all the environmental factors acting on the organism; the niche thus defined is a region of an n-dimensional hyper-space, comparable to the phase-space of statistical mechanics' (Hutchinson, 1944: 20). Using this concept, Canonical Correspondence Analysis is an appropriate method to study ecological complexity and interactions (Beeckman & Lemeur, 1989). The general aim of CCA, when used in an exploratory way, is to turn out an ordination diagram of samples, species and environmental variables which optimally displays how community composition varies as a function of environment. Biplot scores of environmental variables give the coördinates of the heads of arrows in the ordination diagram. The rules for constructing and interpreting species-environment biplots are the same as those given in ter Braak (1987: 127–129) for PCA biplots: the direction of the arrow indicates the direction in which the amount of the corresponding environmental variable increases most; the length of the arrow equals the rate of change in that direction; the angle between arrows of a pair of variables (environmental, species or samples) provides an approximation of their pair-wise correlation. The biplot is constructed most easily by drawing separate plots of species, samples and environmental variables on transparant paper, each one with its own scaling. However, within each plot the scale units of the axes must have equal physical length. The biplot is obtained by superimposing the plots with the axes aligned and the origins of the coordinate systems coinciding (ter Braak, 1988).

Results

Species list of the plants

Species of plants listed here are exclusively from the transects through the intertidal and subtidal seagrass beds and partly sand-covered coral platform. Large intertidal pools with coral boulders and/or dead tree trunks as well as the tide channel have a much more diversified algal flora; these species are not included here.

Seagrasses (Helobiales)
Descriptions and illustrations are to be found in: Brouns & Heys (1986), den Hartog (1970), Isaac F. (1968), Phillips & Meñez (1988), Moorjani & Simpson (1988). They have been summarized in De Wit (1988: 46–55). Eleven of the twelve known seagrass species from Kenya (Isaac F., 1968) have been found in the study area: *Cymodocea rotundata* Asch. & Schweig. (HEC 8477), *C. serrulata* (Braun) Asch. & Magnus (HEC 8478), *Enhalus acoroides* (L.f.) Royle (HEC 8506), *Halodule uninervis* (Forssk.) Asch. (HEC 8476), *H. wrightii* Asch. (HEC 8475), *Halophila minor* (Zoll.) den Hartog (HEC 8471b), *H. ovalis* (Braun) Hooker (HEC 8503), *H. stipulacea* (Forssk.) Asch. (HEC 8472) *Syringodium isoetifolium* (Asch.) Dandy (HEC 8502), *Thalassia hemprichii* (Ehrenb.) Asch. (HEC 8479), *Thalassodendron ciliatum* (Forssk.) den Hartog (HEC 8480). Only *Zostera capensis* Setch., collected in Gazi Bay by Isaac (1968: 32) has not been observed again.

Following the morphological classification system of the seagrasses of den Hartog (1967, 1977: 92) the seagrasses from Gazi Bay can be classified in six categories of growth forms which can be linked with differences in ecology.

The parvozosterids (e.g. *Halodule*) with fine, linear leaves; they can be found in all kinds of habitats, from coarse sand to almost liquid mud, from the intertidal zone down to considerable depth, from open sea to the estuaries and mangrove swamps.

The magnozosterids (e.g. *Cymodocea*, *Thalassia*) with wide linear leaves; they also grow on various substrata, but they are typically subtidal. The syringodiids (e.g. *Syringodium*) with long subulate leaves; they do tolerate some emersion, but are usually restricted to the upper sublittoral.

The enhalids (e.g. *Enhalus*) with leathery strap-shaped leaves; they are restricted to the 'upper' part of the sublittoral and inhabit sandy and rocky substrate, being rarely found on mud.

The halophilids (e.g. *Halophila*) with delicate, fast growing rhizomes and elliptic, ovate or lanceolate delicate leaves; as with the parvozosterids they have a very wide ecological range.

The amphibolids (e.g. *Thalassodendron*) with sympodially branched vegetative system, the leaves being distichously arranged along the upright stems; as the enhalids they are strictly subtidal, euhaline, inhabiting sandy and rocky substrates (down to −30 m for *T. ciliatum*.

Seaweeds
Descriptions and illustrations of most of the species can be found in Jaasund (1976) and in Moorjani & Simpson (1988).

Chlorophyta
Avrainvillea obscura J. Ag. (HEC 6069), *Boodleopsis pusilla* (Coll.) Taylor (HEC 6741), *Bryopsis hypnoides* (HEC 6903), *Caulerpa sertularioides* (Gmel.) Howe (HEC 6076), *Chaetomorpha crassa* (Ag.) Kütz., *Cladophora patentiramea* (Mont.) Kütz. f. *longiarticulata* Reinb. (HEC 6840), *Enteromorpha clathrata*

(Roth) Ag., *E. flexuosa* (Wulf.) J. Ag. (HEC 5707), *E. kylinii* Bliding (HEC 6696), *E. ramulosa* (Smith) Hook., *Halimeda macroloba* Decsne. (HEC 5709, 6080), *H. opuntia* (L.) Lamour., *Nereodictyon imitans* Gerloff (HEC 6075), *Udotea orientalis* A. & E. S. Gepp (HEC 7377), *Ulva lactuca* L., *Ulva reticulata* Forssk.

Phaeophyta
Padina boryana Thivy (HEC 6907), *Rosenvingea orientalis* (Ag.) Børgs. (HEC 6695), *Sphacelaria rigidula* Kütz. (HEC 6705a), *S. tribuloides* Menegh. (HEC 6705b).

Rhodophyta
Acanthophora spicifera (Vahl) Børgs. (HEC 5711, 6700), *Amphiroa fragilissima* (L.) Lamour., *A. rigida* Lamour., *Ceramium strictum* (Kütz.) Harv. (HEC 5710), *Gracilaria corticata* Ag. (HEC 5705), *G. salicornia* (Ag.) Dawson (HEC 5704), *G. verrucosa* (Huds.) Papenf. (HEC 5703), *Hypnea* [identifications within this genus should be regarded with care] *H. cornuta* Lamour. ex Ag. (HEC 6073), *H. hamulosa* (Turn.) Mont. (HEC 5708), *H. nidifica* Ag., *H. valentiae* (Turn.) Mont., *Laurencia papillosa* (Forssk.) Grev.

These seaweeds can also been classified according to the substrate where they grow.

Most of the macroalgae are growing on hard substrate: on rock and coral substrate: *Cladophora*, *Ulva lactuca*, *Padina*, most of the red algae mentioned above; on shells and coral fragments: *Enteromorpha*, some *Hypnea*, *Gracilaria verrucosa*; epiphytic: *Bryopsis*, *Rosenvingea*, *Sphacelaria*, *Ceramium*, *Gracilaria corticata*.

Some only occur in or on loose substrate: the filamentous *Boodleopsis* largely grows in the silty-sandy substrate whereas the upright fronds of *Avrainvillea obscura* and *Halimeda macroloba* are fixed in the sand by a pseudobulbous holdfast; *Caulerpa sertularioides*, *Udotea orientalis* and *Halimeda opuntia* on the other hand develop numerous groups of rhizoids for their fixation.

A third group includes seaweeds which are not really attached to a substratum but are entangled to other algae or seagrasses: *Chaetomorpha crassa*, *Ulva reticulata*, *Nereodictyon*.

Abiotic variables

The salinity along transect 1 is significantly lower (variance analysis, unbalanced designs, STATGRAPHICS: De Wit, 1988: 15–18) than along transects 3 and 4, varying from 24 to 27‰ in 1, from 46 to 56‰ in 3 and from 47 to 52 in 4. The low salinities in 1 might be due to the nearby presence of the outlet of the Kidogoweni river combined with seepage which is clearly visible at some places. The high salinities in 3 and 4 are due to the evaporation from the pools at low tide.

The temperature of the tidal pools, measured along transects 1–4 from 8.45 to 12.40 h on a sunny day varies from 25,5 at 8.45 to 34,0 °C at 11.30 and 12.04. In a single shallow pool (covered with seagrasses) the temperature went up 5 °C in 1½ hours (from 10 to 11.30). In February 1988 a maximum temperature of 42 °C was measured at noon in one of these tidal pools.

Mean values of Na, K, Ca, Mg, P, organic material, pH are given in Table 1. According to Cox (1959) the relative amounts of some ions in seawater, compared to K are: Mg 3, Ca 1, Na 28. In the soil potassium is fixed to clay minerals and generally only represents less than 1% of the total exchangeable fraction. As clay minerals are absent from the soil samples from Gazi, the K

Table 1. Mean values, standard deviation and inflation factors of environmental variables.

Variables	Mean	Stand. dev.	Infl. fact.
Dist low	102.73	59.50	3.04
Na (mg/kg)	3122.58	1006.33	51.27
K (mg/kg)	244.74	129.47	91.30
Ca (mg/kg)	35901.72	17724.70	55.26
Mg (mg/kg)	1891.16	774.61	28.72
P (mg/kg)	22.01	6.68	5.69
Org mat (%)	1.39	1.45	16.35
pH	8.37	0.23	2.90
2000–1000 μm	0.34	0.48	17.61
1000–500 μm	3.09	3.83	55.81
500–200 μm	26.89	18.80	1088.62
200–100 μm	49.52	16.11	930.14
100–50 μm	20.29	13.92	636.84
50–2 μm	0.03	0.07	11.05
2–0 μm	0.08	0.09	17.93

present in them should be coming from the seawater. Based on the K/Mg relation it can be concluded (De Wit, 1988; Table 10 p. 34) that in transects 1 and 2 the quantity of Mg is also the same as in seawater. In transects 3, 4, 5 there is a marked enrichment, up to 5 times.

The amount of calcium is considerably higher than in the seawater (up to 275 times) in the samples of all transects. According to Patriquin (1975) the $CaCo_3$ fraction of seagrass substrate is from biotic origin, coming from the calcified *Halimeda* (Chlorophyta), the Corallinaceae (Rhodophyta), corals, foraminifera and gastropods.

The clay and loam fractions of the soil samples along the transects 1–5 are generally less than 0,5%. The amount of very coarse sand is also very limited. The proportion of coarse, medium coarse, medium fine and fine sand varies from one sampling spot to the other (De Wit, 1988, Table 8, p. 32), even within a transect. In transects 1–2 they mainly belong to the 500–200 and 200–100 μm fractions, in transects 3–4 to the (500–200), 200–100 and 100–50 μm fractions, in transect 5 to the 500–200, 200–100 and 100–50 μm fractions.

According to the texture triangle of soil taxonomy (FAO 1977) the samples can therefore be described as sand. Grady (1981) found 82–99% sand in seagrass flat sediments in Florida.

TWINSPAN classification

The 88 relevés are split up in 8 groups:

- TWIN-group 1 with *Boodleopsis pusilla* as indicator species, characteristic for the high intertidal zone (between the highest seagrasses and the lowermost mangroves);
- TWIN-group 2 with *Enteromorpha flexuosa* and *E. ramulosa* as indicator species, characteristic on muddy substrates in the intertidal zone;
- TWIN-group 3 with *Laurencia papillosa* as indicator species, characteristic for thinly sand-covered coral platform in the intertidal zone;

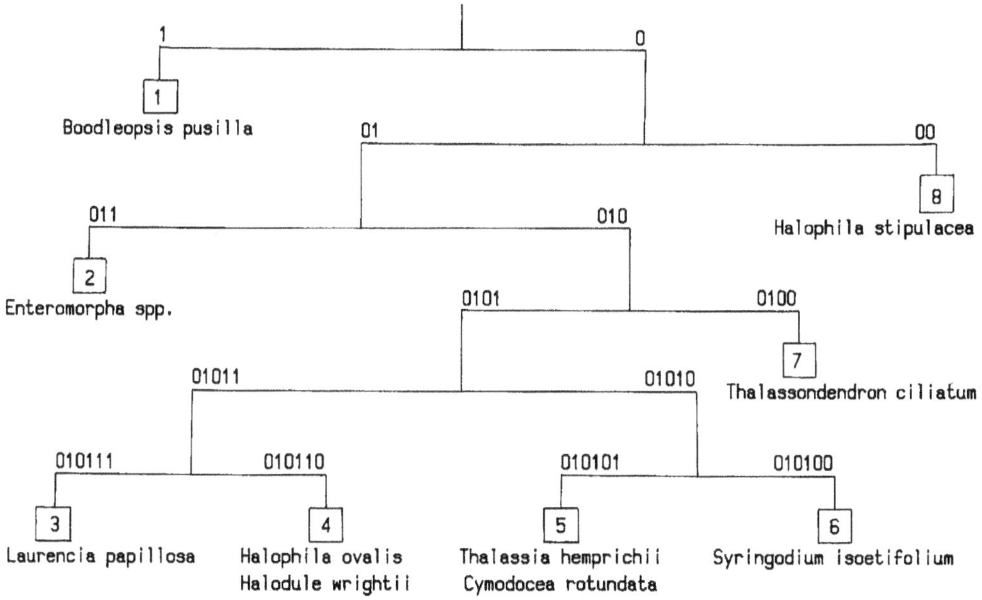

Fig. 2. TWINSPAN-dendrogram of the relevés indicating the subsequent splitting levels of the TWIN-groups and the indicator species.

Table 2. Two way ordered TWINSPAN-table after manual rearrangement. The significance of the name codes is given after table 7. [1 = <1%; 2 = 1–2%; 3 = 2–5%; 4 = 5–10%; 6 = 30–50%; 7 = 50–70%; 8 = 70–90%; 9 = >90%].

	19	20	1	2	3	70	59	68	69	71	72	73	75
BOO PUS	2	3	.	.	.	.	.	.	.	.	.	.	.
ENT FLE	.	1	.	6	7	3	.	.	.	.	.	.	.
ENT RAM	.	.	9	6	7	.	.	.	.	.	.	.	.
ENT CLA	.	.	.	.	.	3	.	.	.	.	.	.	.
ENT SPE	.	.	.	1	1	.	.	.	.	.	.	.	.
ULV LAC	.	.	.	.	.	3	.	.	.	.	.	.	.
LAU PAP	.	.	.	.	.	.	.	1	.	4	5	5	5
PAD BOR	.	.	.	.	.	.	.	5	5	.	2	.	.
ACA SPI	.	.	.	.	.	.	.	1	2	.	.	.	.
CHA CRA	.	.	.	.	.	.	.	.	1	.	1	.	1
GRA SAL	.	.	.	.	.	.	.	2	.	1	.	.	1
GRA EDU	.	.	.	.	.	.	.	.	.	.	.	.	1
HYP NID	.	.	.	.	.	.	.	.	.	.	.	.	1
CLA PAT	.	.	.	.	.	.	.	2	.	.	.	.	.
HAL OVA	.	.	.	.	.	.	.	.	.	.	.	.	.
HAL WRI	.	.	.	.	.	.	2	2	2	2	2	2	2
THA HEM	.	.	.	.	.	.	.	.	.	.	.	.	.
CYM ROT	.	.	.	.	.	.	.	.	.	.	.	.	.
HAL OPU	.	.	.	.	.	.	.	.	.	.	.	.	.
ROS ORI	.	.	.	.	.	.	.	.	.	.	.	.	.
HAL UNI	.	.	.	.	.	.	.	.	.	.	.	.	.
CER TRU	.	.	.	.	.	.	.	.	.	.	.	.	.
AMP FRA	.	.	.	.	.	.	.	.	.	.	.	.	.
HAL MIN	.	.	.	.	.	.	.	.	.	.	.	.	.
CYM SER	.	.	.	.	.	.	.	.	.	.	.	.	.
CAU SER	.	.	.	.	.	.	.	.	.	.	.	.	.
SYR ISO	.	.	.	.	.	.	.	.	.	.	.	.	.
ENH ACO	.	.	.	.	.	.	.	.	.	.	.	.	.
HAL MAC	.	.	.	.	.	.	.	.	.	.	.	.	.
HAL STI	.	.	.	.	.	.	.	.	.	.	.	.	.
THA CIL	.	.	.	.	.	.	.	.	.	.	.	.	.
ENT KYL	.	.	.	.	.	3	.	2	.	1	.	.	.
HYP COR	.	.	.	.	.	.	.	1	.	.	.	1	.
ULV RET	.	.	.	.	.	.	.	.	.	1	.	.	1
NER IMI	.	.	.	.	.	.	.	.	.	.	.	.	.
AMP RIG	.	.	.	.	.	.	.	.	.	.	.	.	.
HYP VAL	.	.	.	.	.	.	.	.	.	.	.	.	.
Group	1		2				3						

	7	10	13	14	16	17	21	22	25	29	33	40	41	42	43	44	45	48	57	58	60	63	66	67	74	76	77	78	80	81	82	83	84	85	86	87	88
BOO PUS	.	.	.	.	.	.	.	.	.	.	.	.	.	.	.	.	.	.	.	.	.	.	.	.	.	.	.	.	.	.	.	.	.	.	.	.	.
ENT FLE	.	.	.	.	.	.	.	.	.	.	.	.	.	.	.	.	.	.	.	.	.	.	.	.	.	.	.	.	.	.	.	.	.	.	.	.	.
ENT RAM	5	.	.	.	.	.	.	.	.	.	.	.	.	.	.	.	.	.	.	.	.	.	.	.	.	.	.	.	.	.	.	.	.	.	.	.	.
ENT CLA	.	.	.	.	.	.	.	.	.	.	.	.	.	.	.	.	.	.	.	.	.	.	.	.	.	.	.	.	.	.	.	.	.	.	.	.	.
ENT SPE	.	.	.	.	.	.	.	.	.	.	.	.	.	.	.	.	.	.	.	.	.	.	.	.	.	.	.	.	.	.	.	.	.	.	.	.	.
ULV LAC	.	.	.	.	.	.	.	.	.	.	.	.	.	.	.	.	.	.	.	.	.	.	.	.	.	.	.	.	.	.	.	.	.	.	.	.	.
LAU PAP	.	.	.	.	.	.	.	.	.	.	.	.	.	.	.	.	.	.	.	.	.	.	.	.	.	.	.	.	.	.	.	.	.	.	.	.	.
PAD BOR	.	.	.	.	.	.	.	.	.	.	.	.	.	.	.	.	.	.	.	.	.	.	.	.	.	.	.	.	.	.	.	.	.	.	.	.	.
ACA SPI	.	.	.	.	.	1	.	.	.	.	.	.	.	.	.	.	.	.	.	.	.	.	.	.	.	.	.	.	.	.	.	.	.	.	.	.	.
CHA CRA	.	.	.	1	.	.	.	.	.	.	.	.	.	.	.	.	.	.	.	.	.	.	.	1	.	.	.	.	.	.	.	.	.	.	.	.	.
GRA SAL	.	.	.	1	.	1	.	.	.	.	.	.	.	.	.	.	.	.	.	.	.	.	.	.	.	.	.	.	.	.	.	.	.	.	.	.	.
GRA EDU	.	.	.	.	.	.	.	.	.	.	.	.	.	.	.	.	.	.	.	.	.	.	.	.	.	.	.	.	.	.	.	.	.	.	.	.	.
HYP NID	.	.	.	.	.	.	.	.	.	.	.	.	.	.	.	.	.	.	.	.	.	.	.	.	.	.	.	.	.	.	.	.	.	.	.	.	.
CLA PAT	.	.	.	.	.	.	.	.	.	.	.	.	.	.	.	.	.	.	.	.	.	.	.	.	.	.	.	.	.	.	.	.	.	.	.	.	.
HAL OVA	2	.	3	.	2	2	2	2	3	2	2	3	.	2	3	5	2	.	2	.	2	2	2	.	2	3	5	5	.	.	2	2	5	.	2	5	5
HAL WRI	3	3	2	3	5	5	2	3	2	3	3	2	3	2	3	5	3	3	3	4	3	3	5	6	3	7	6	7	5	6	6	6	5	5	5	5	6
THA HEM	.	.	.	.	.	.	.	.	.	.	.	.	.	.	.	.	.	.	.	.	.	.	.	.	.	.	.	.	.	.	.	.	.	.	.	.	.
CYM ROT	.	.	.	.	.	.	.	.	.	.	.	.	.	.	.	.	.	.	.	.	.	.	.	.	.	.	.	.	.	.	.	.	.	.	.	.	.
HAL OPU	.	.	.	.	.	.	.	.	.	.	.	.	.	.	.	.	.	.	.	.	.	.	.	.	.	.	.	.	.	.	.	.	.	.	.	.	.
ROS ORI	.	.	.	.	.	.	.	.	.	.	.	.	.	.	.	.	.	.	.	.	.	.	.	.	.	.	.	.	.	.	.	.	.	.	.	.	.
HAL UNI	.	.	.	.	.	.	.	.	.	.	.	.	.	.	.	.	.	.	.	.	.	.	.	.	.	.	.	.	.	.	.	.	.	.	.	.	.
CER TRU	.	.	.	.	.	.	.	.	.	.	.	.	.	.	.	.	.	.	.	.	.	.	.	.	.	.	.	.	.	.	.	.	.	.	.	.	.
AMP FRA	.	.	.	.	.	.	.	.	.	.	.	.	.	.	.	.	.	.	.	.	.	.	.	.	.	.	.	.	.	.	.	.	.	.	.	.	.
HAL MIN	.	.	.	.	.	.	.	.	.	.	.	.	.	.	.	.	.	.	.	.	.	.	.	.	.	.	.	.	.	.	.	.	.	.	.	.	.
CYM SER	.	.	.	.	.	.	.	.	.	.	.	.	.	.	.	.	.	.	.	.	.	.	.	.	.	.	.	.	.	.	.	.	.	.	.	.	.
CAU SER	.	.	.	.	.	.	.	.	.	.	.	.	.	.	.	.	.	.	.	.	.	.	.	.	.	.	.	.	.	.	.	.	.	.	.	.	.
SYR ISO	.	.	.	.	.	.	.	.	.	.	.	.	.	.	.	.	.	.	.	.	.	.	.	.	.	.	.	.	.	.	.	.	.	.	.	.	.
ENH ACO	.	.	.	.	.	.	.	.	.	.	.	.	.	.	.	.	.	.	.	.	.	.	.	.	.	.	.	.	.	.	.	.	.	.	.	.	.
HAL MAC	.	.	.	.	.	.	.	.	.	.	.	.	.	.	.	.	.	.	.	.	.	.	.	.	.	.	.	.	.	.	.	.	.	.	.	.	.
HAL STI	.	.	.	.	.	.	.	.	.	.	.	.	.	.	.	.	.	.	.	.	.	.	.	.	.	.	.	.	.	.	.	.	.	.	.	.	.
THA CIL	.	.	.	.	.	.	.	.	.	.	.	.	.	.	.	.	.	.	.	.	.	.	.	.	.	.	.	.	.	.	.	.	.	.	.	.	.
ENT KYL	.	.	.	.	.	.	1	.	.	.	.	.	.	.	.	.	.	.	.	.	.	.	.	.	.	.	.	.	.	.	.	.	.	.	.	.	.
HYP COR	.	.	.	.	.	.	.	1	.	1	.	.	.	.	.	.	1	.	.	.	.	.	.	.	.	.	.	.	.	.	.	.	.	.	.	.	.
ULV RET	.	.	.	.	.	.	.	.	.	.	.	.	.	.	.	.	.	.	.	.	.	.	1	1	.	.	.	.	.	.	.	.	.	.	.	.	.
NER IMI	.	.	.	.	.	1	.	.	.	.	.	.	.	.	.	.	.	.	.	.	.	.	.	.	.	.	.	.	.	.	.	.	.	.	.	.	.
AMP RIG	.	.	.	.	.	.	.	.	.	.	.	.	.	.	.	.	.	.	.	.	.	.	.	.	.	.	.	.	.	.	.	.	.	.	.	.	.
HYP VAL	.	.	.	.	.	.	.	.	.	.	.	.	.	.	.	.	.	.	.	.	.	.	1	.	.	.	.	.	.	.	.	.	.	.	.	.	.
Group	4																																				

- TWIN-group 4 with *Halophila ovalis* and *Halodule wrightii* as indicator species, which can be considered as a pioneer community in the intertidal zone;
- TWIN-group 5 with *Thalassia hemprichii* and *Cymodocea rotundata* as indicator species, which can be considered as the climax community in shallow intertidal pools and flats;
- TWIN-group 6 with *Syringodium isoetifolium* as indicator species, which can be considered as the climax community between low water mark at neap tide and at spring tide; including the relevés with *Enhalus acoroides*;
- TWIN-group 7 with *Thalassodendron ciliatum* as indicator species which can be considered as the climax community of the subtidal zone;
- TWIN-group 8 with *Halophila stipulacea* as indicator species, which can be considered as an infralittoral pioneer community.

	4	5	6	8	9	12	15	18	26	27	28	30	31	32	34	39	46	47	49	50	61	62	64	65	79	35	36	37	38	51	52	55	56	23	24	53	54
BOO PUS	.	.	.	.	.	.	.	.	.	.	.	.	.	.	.	.	.	.	.	.	.	.	.	.	.	.	.	.	.	.	.	.	.	.	.	.	.
ENT FLE	.	.	.	.	.	.	.	.	1	.	.	.	.	.	.	.	.	.	.	.	.	.	.	.	.	.	.	.	.	.	.	.	.	.	.	.	.
ENT RAM	2	3	2	5	.	.	.	.	.	.	.	.	.	.	.	.	.	.	.	.	.	.	.	.	.	.	.	.	.	.	.	.	.	.	.	.	.
ENT CLA	.	.	.	.	.	.	.	.	.	.	.	.	.	.	.	.	.	.	.	.	.	.	.	.	.	.	.	.	.	.	.	.	.	.	.	.	.
ENT SPE	.	.	.	.	.	.	.	.	.	.	.	.	.	.	.	.	.	.	.	.	.	.	.	.	.	.	.	.	.	.	.	.	.	.	.	.	.
ULV LAC	.	.	.	.	.	.	.	.	.	.	.	.	.	.	.	.	.	.	.	.	.	.	.	.	.	.	.	.	.	.	.	.	.	.	.	.	.
LAU PAP	.	.	.	.	.	.	.	.	.	.	.	.	.	.	.	.	.	.	.	.	.	.	.	.	.	.	.	.	.	.	.	.	.	.	.	.	.
PAD BOR	.	.	.	.	.	.	.	.	.	.	.	.	.	.	.	.	.	.	.	.	.	.	.	.	.	.	.	.	.	.	.	.	.	.	.	.	.
ACA SPI	.	.	.	.	.	.	.	.	.	.	1	.	.	.	.	.	.	.	.	.	.	.	.	.	.	.	.	.	.	.	.	.	.	.	.	.	.
CHA CRA	2	.	.	.	.	.	.	2	.	.	.	.	.	.	.	.	.	.	.	.	.	.	.	.	.	.	.	.	.	.	.	.	.	.	.	.	.
GRA SAL	.	.	.	.	.	.	.	.	.	.	.	.	.	.	.	.	.	.	.	.	.	.	.	.	.	.	.	.	.	.	.	.	.	.	.	.	.
GRA EDU	.	.	.	.	.	.	.	.	.	.	.	.	.	.	.	.	.	.	.	.	.	.	.	.	.	.	.	.	.	.	.	.	.	.	.	.	.
HYP NID	.	.	.	.	.	.	.	.	.	.	.	.	.	.	.	.	.	.	.	.	.	.	.	.	.	.	.	.	.	.	.	.	.	.	.	.	.
CLA PAT	.	.	.	.	.	.	.	.	.	.	.	.	.	.	.	.	.	.	.	.	.	.	.	.	.	.	.	.	.	.	.	.	.	.	.	.	.
HAL OVA	.	2	.	1	.	.	.	.	.	3	.	.	.	.	.	3	.	.	.	.	.	.	.	.	.	.	.	.	.	.	.	.	.	.	.	.	.
HAL WRI	.	2	2	3	5	.	2	.	.	.	.	2	.	.	.	.	2	2	2	.	2	2	3	.	5	.	.	.	.	.	.	.	.	.	.	.	.
THA HEM	.	.	5	5	2	.	5	3	3	5	5	4	3	6	3	3	5	5.	.	5	2	5	4	5	3	2	5	5	6	6	5	6	3	.	.	.	.
CYM ROT	5	2	5	5	2	2	5	3	4	5	5	4	5	5	5	5	3	3	5	3	4	4	4	2	5	.	.	.	5	3	.	2	5	.	.	.	.
HAL OPU	2	.	3	2	3	.	.	3	.	.	.	.	1	.	.	.	.	3	.	2	.	.	1	2.	.	5	2	4	4	.	.	.	2	.	.	.	.
ROS ORI	.	.	.	.	.	.	.	.	.	.	1	.	2	.	2	.	.	.	1	1	1	.	.	.	.	.	.	.	.	1	1	.	.	.	.	.	.
HAL UNI	.	.	.	2	.	.	.	.	.	.	.	.	.	2	2	3	.	.	.	.	.	.	.	.	.	.	.	.	.	.	.	.	.	.	.	.	.
CER TRU	3	5	5	2	3	.	.	.	.	.	.	.	.	.	.	.	.	.	.	.	.	.	.	.	.	.	.	.	.	.	.	.	.	.	.	.	.
AMP FRA	.	.	.	2	2	.	.	.	.	.	1	.	.	.	.	.	.	.	.	.	.	.	.	.	.	.	.	.	.	.	.	.	.	.	.	.	.
HAL MIN	.	.	.	2	5	.	.	.	.	.	.	.	.	.	.	.	.	.	.	.	.	.	.	.	.	.	.	.	.	.	.	.	.	.	.	.	.
CYM SER	.	.	.	2	2	.	.	.	.	.	.	.	.	.	.	.	.	.	.	.	.	.	.	.	.	.	.	.	.	.	.	.	.	.	.	.	.
CAU SER	2	1	.	.	.	.	.	.	.	.	.	.	.	.	.	.	.	.	.	.	.	.	.	.	.	.	.	.	.	.	.	.	.	.	.	.	.
SYR ISO	.	.	.	.	.	.	.	.	.	.	.	.	.	.	.	.	.	.	.	.	.	.	.	.	.	5	2	4	3	5	5	5	3	.	.	.	.
ENH ACO	.	.	.	.	.	.	.	.	.	.	.	.	.	.	.	.	.	.	.	.	.	.	.	.	.	.	.	.	.	.	.	.	.	.	5	.	5
HAL MAC	.	.	.	.	.	.	.	.	.	.	.	.	.	.	.	.	.	.	.	.	.	.	.	.	.	.	.	.	.	.	.	.	1	.	.	.	.
HAL STI	.	.	.	.	.	.	.	.	.	.	.	.	.	.	.	.	.	.	.	.	.	.	.	.	.	.	.	.	.	.	.	.	.	3	3	.	.
THA CIL	.	.	.	.	.	.	.	.	.	.	.	.	.	.	.	.	.	.	.	.	.	.	.	.	.	.	.	.	.	.	.	.	.	.	.	8	9
ENT KYL	.	.	.	.	.	.	.	.	.	.	.	.	.	.	.	.	.	.	.	.	.	.	.	.	.	.	.	.	.	.	.	.	.	.	.	.	.
HYP COR	.	.	.	.	.	.	.	.	.	.	1	.	.	.	.	.	.	.	1	1	.	.	.	.	.	.	.	.	.	.	.	.	.	.	.	.	.
ULV RET	.	.	.	.	.	.	.	.	.	.	.	.	.	.	.	.	.	.	.	.	.	.	.	.	.	.	.	.	.	.	.	.	.	.	.	.	.
NER IMI	.	.	.	.	.	.	.	.	.	.	.	.	.	.	.	.	.	.	.	.	.	.	.	.	.	.	.	.	.	.	.	.	.	.	.	.	.
AMP RIG	.	.	.	.	.	.	.	2	.	.	.	.	.	.	.	.	.	.	.	.	.	.	.	.	.	.	.	.	.	.	.	.	.	.	.	.	.
HYP VAL	.	.	.	.	.	.	.	.	.	.	.	.	.	.	.	.	.	.	.	.	.	.	.	.	.	.	.	.	.	.	.	.	.	.	.	.	1
	5																									6								7		8	

Correlations between the vegetation and abiotic factors (CCA)

Intercorrelations between environmental factors
Some variables are strongly correlated with others (inflation factor > 20) and therefore do not have an univocal contribution to the explanation of the variation in vegetation. They are: Na, K, Ca, Mg, fractions 1000–500, 500–200, 200–100, 100–50 μm. Parameters which do have an univocal significance for explaining the variability are: distance to low water, P, organic material, pH, fractions 2000–1000, 50–2, 2–0 μm. Within the latter, the Canonical Correspondence Analysis was applied again.

The correlation matrix specifies the intercorrelations between some of the variables of the excluded group: Na: strong correlation with K and small particle fractions; K: strong correlation with organic material and small fractions; Ca: strong

Table 3. Correlation matrix.

Variables															
Dist low	1.00														
Na (mg/kg)	0.03	1.00													
K (mg/kg)	0.06	0.87	1.00												
Ca (mg/kg)	−0.24	0.12	−0.34	1.00											
Mg (mg/kg)	−0.25	0.51	0.10	0.87	1.00										
P (mg/kg)	−0.18	0.09	−0.13	0.63	0.67	1.00									
Org mat (%)	−0.07	0.65	0.90	−0.48	−0.08	−0.23	1.00								
pH	−0.29	−0.09	−0.29	0.29	0.24	0.07	−0.32	1.00							
2000–1000 μm	0.46	−0.22	−0.04	−0.53	−0.64	−0.63	−0.01	−0.31	1.00						
1000–500 μm	0.48	−0.23	0.06	−0.69	−0.70	−0.49	0.13	−0.22	0.86	1.00					
500–200 μm	0.27	−0.35	0.04	−0.86	−0.84	−0.46	0.17	−0.17	0.52	0.70	1.00				
200–100 μm	−0.67	0.07	−0.15	0.66	0.61	0.39	−0.10	0.25	−0.62	−0.73	−0.75	1.00			
100–50 μm	0.27	0.49	0.13	0.61	0.6	0.33	−0.13	0.01	−0.25	−0.39	−0.69	0.06	1.00		
50–2 μm	0.18	0.82	0.86	−0.24	0.18	0.03	0.69	−0.17	−0.02	0.12	0.03	−0.31	0.31	1.00	
2–0 μm	−0.06	0.86	0.86	−0.03	0.36	0.05	0.72	−0.06	−0.23	−0.08	−0.26	0.10	0.30	0.79	1.00

correlation with Mg and P, strong negative correlation with the coarse fractions; Mg: strong correlation with Ca and P, negative correlation with the coarse fractions; particle fraction 1000–500 μm: positive correlation with other coarse fractions, distance to low water, negative correlation with Ca, Mg, P and small fractions; fraction 500–200 μm: positive correlation with coarse fraction, negative with small fractions, Ca, Mg and P; fraction 200–100 μm: negative correlation with distance to low water, Ca, Mg, P and coarse fractions; fraction 100–50 μm: negative correlation with coarse fractions, positive correlation with Ca and Mg.

Canonical ordination axes

To evaluate the reaction of the species to the current environmental variables, a Monte Carlo permutation-test has been calculated. The first eigenvector calculated from the data – as observed – is larger than most of the test statistics calculated from the random data ($P = 0.13$), which indicates that the species do react to the current environmental variables.

For the evaluation of the contribution of the remaining variables in the variance of the data-set, the length of the vectors corresponding with each variable has been calculated in a four-dimensional space.

In decreasing importance the variables are: P, fraction coarse sand (1000–2000 μm), distance to low water, organic material, fraction 50–2 μm, fraction 2–0 μm.

Axes 1–4 explain 85.7% of the total variance of the data set.

The correlation of environmental variables with the axes is defined by Table 6.

Axis 1: most of the variables are positively correlated with this axis; only pH is negatively correlated. Axis 1 is mainly defined by the amount of organic material and the particle fraction 50–2 μm in positive direction and pH in negative direction (higher pH values corresponding with smaller coordinates).

Axis 2: the defining variables are the distance to low water mark in positive direction and P in negative direction. Coarse sediments are also

Table 4. Biplot scores of environmental variables. (Ax = Axis).

	Ax 1	Ax 2	Ax 3	Ax 4	Length vectors (4-dimensional)	Rank
Dist low	81	357	- 78	- 9	374	3
P (mg/kg)	86	- 260	- 308	- 85	421	1
Org mat (%)	277	- 91	93	167	349	4
pH	- 203	70	- 134	- 152	295	5
2000–1000 μm	60	218	290	- 164	403	2
50–2 μm	215	- 51	43	87	241	6
2–0 μm	54	- 190	- 21	116	230	7

Table 5. Percentage variance accounted for by first 4 axes.

Axis	Percentage	Eigenvalue
1	37.6	0.67
2	22.5	0.40
3	14.2	0.25
4	11.4	0.20

Table 6. Inter set correlations of environmental variables with axes. (Multiplier 100).

Variables	Axis 1	Axis 2	Axis 3	Axis 4
Dist low	157	518	- 128	- 15
P (mg/kg)	166	- 377	- 507	- 137
Org mat (%)	535	- 133	154	270
pH	- 391	102	- 221	- 247
2000–1000	115	316	478	- 266
50–2	414	- 74	72	141
2–0	104	- 276	- 35	189

positively correlated, small sediments rather negatively.

Axis 3: mainly defined by P in negative direction and the fraction very coarse sand in positive direction.

Axis 4: most positive correlation with the amount of organic material and most negative correlation with the fraction very coarse sand.

Ordination diagrams based on canonical correspondence analysis

The most evident correlations appear from the diagram of the first and second canonical ordination axes. Correlations amongst environmental factors are illustrated in diagram 1: positive correlation between distance to low water mark and the particle size fraction 2000–1000 μm, organic material and particle size fraction 50–2 μm, P and particle size fractions 2–0 μm; negative correlation between pH and organic material or 50–2 μm size fraction.

Well marked correlations between environmental factors and species (diagram 2) are not numerous: *Boodleopsis pusilla* (5), *Enteromorpha flexuosa* (17) and *E. ramulosa* (14) are positively correlated with high values of organic material and particle size fraction 50–2 μm and with low pH values. In the field these seaweed species are typically found (in and) on the mud flats devoid of seagrasses. *Boodleopsis pusilla* is a tiny filamentous algae, growing in dense mats, the vertically growing filaments fixing enormous amounts of extremely small particles.

Correlations between the TWINSPAN relevé clusters and environmental variables are expressed in diagram 3: the TWIN-groups 1, 2, 3, 4 + 5 are distributed along the axis distance to low water mark [from subtidal (1, 2) over infralittoral fringe (3) to midlittoral (4,5)]; the TWIN-groups 4 + 5, 6, 7, 8 (all from the midlittoral zone) are distributed along the axes organic material and particle size fraction 50–2 μm suggesting that the distribution of the seagrasses within the midlittoral zone is due to these environmental variables.

Conclusions and discussion

A zonation pattern of seagrasses and associated macroalgae can generally be recognized in Gazi Bay. This has been corroborated by the TWINSPAN and CCA-analysis.

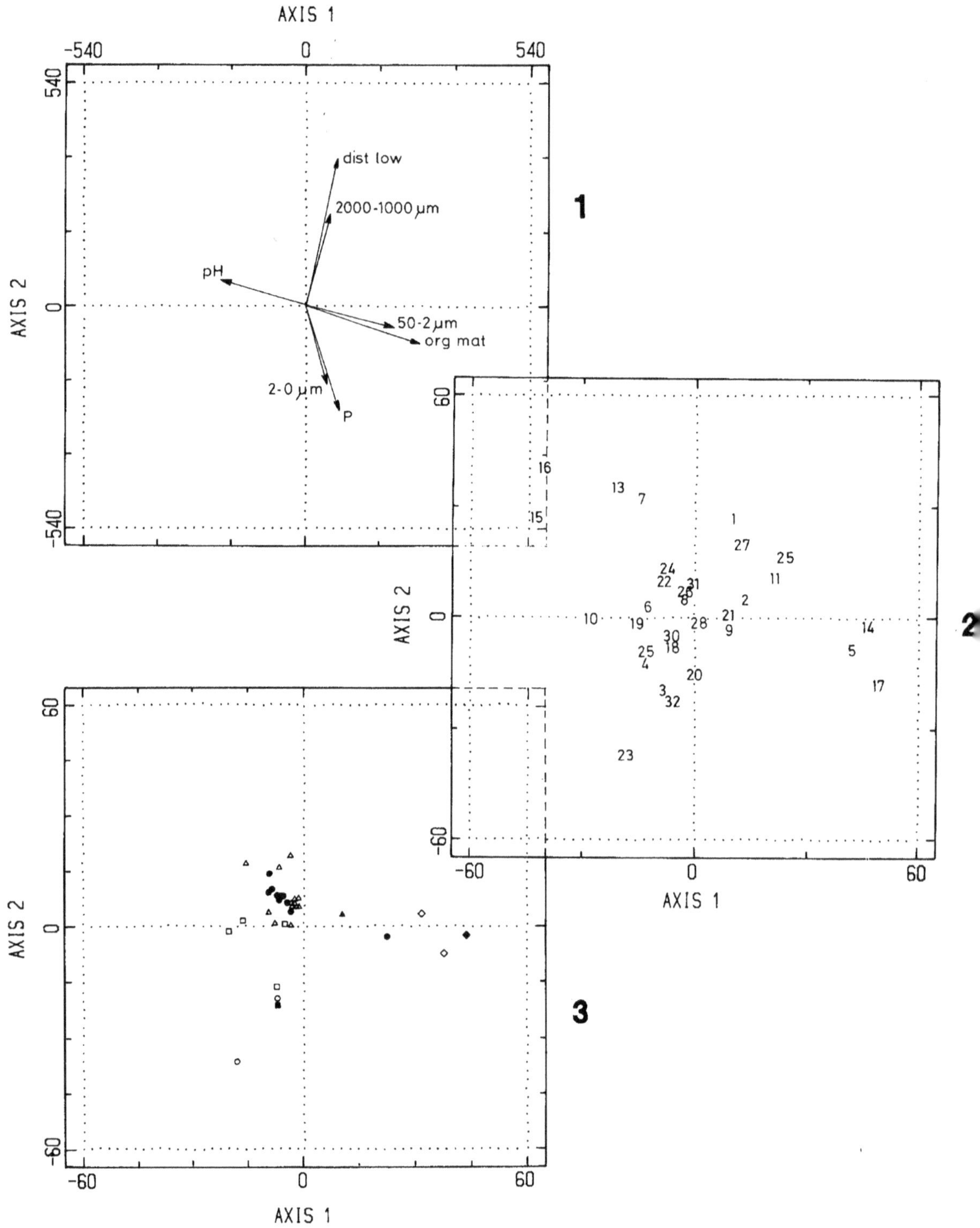
AXIS 1
-540
0
540
AXIS 2
dist low
2000-1000 μm
pH
50-2 μm
org mat
2-0 μm
P
1
AXIS 2
AXIS 1
-60
0
60
16
13
7
15
1
27
25
24
22
31
11
2
6
10
19
28
21
9
14
30
18
25
4
5
20
17
3
32
23
2
AXIS 2
AXIS 1
3

Table 7. Ranked species coördinates for the 4 canonical ordination axes.

Axis 1 Eig. val. = 0.669	Axis 2 Eig. val. = 0.401	Axis 3 Eig. val. = 0.253	Axis 4 Eig. val. = 0.204
ENT SP5 47	ENT SP3 39	HAL MIN 82	ENT SP2 106
ENT RAM 44	ENT KYL 34	CYM CER 58	ENT SP3 46
BOO PUS 41	CER cTR 31	AMP RIG 42	PAD SP1 41
SPH SPE 22	ACA SPI 26	AMP FRA 31	CER cTR 38
ENT cCR 20	ENT SP2 26	ENT KYL 29	CAU SER 35
AMP FRA 13	PAD SP1 19	HAL STI 26	AMP RIG 19
PAD SP1 11	SPH SPE 16	HAL WRI 24	SPH SPE 19
ACA SPI 10	HAL OVA 13	PAD SP1 22	GRA SAL 17
CYM SER 9	ENT cCR 11	GRA SAL 19	ENH ACO 15
HAL UNI 8	HAL WRI 10	CER cTR 18	HAL OPU 11
ROS ORI 0	THA HEM 9	CAU SER 16	HAL STI 9
HAL MAC -1	HYP COR 7	HAL OPU 9	HAL UNI 8
CYM ROT -2	CYM ROT 6	HAL UNI 9	ACA SPI 6
THA HEM -2	AMP FRA 5	SPH SPE 4	AMP FRA 6
HYP COR -3	CAU SER 3	ENT RAM 2	ENT RAM 6
GRA SAL -7	HAL UNI 1	THA CIL 0	THA CIL 3
SYR ISO -7	ENH ACO 0	HAL OVA -1	AMP SP1 1
THA CIL -7	HAL OPU -1	HAL MAC -3	CYM SER -2
AMP RIG -8	ROS ORI -1	ENT SP2 -4	HAL OVA -2
HAL OVA -8	ENT RAM -2	CYM ROT -5	CYM ROT -3
HAL WRI -9	CYM SER -3	THA HEM -5	BOO PUS -6
CAU SER -12	SYR ISO -4	ENT SP3 -6	THA HEM -7
AMP SP1 -13	GRA SAL -7	ROS ORI -8	SYR ISO -8
HAL MIN -13	BOO PUS -8	BPP PUS -9	ROS ORI -10
CER cTR -14	HAL MIN -9	SYR ISO -9	ENT cCR -14
HAL OPU -16	AMP SP1 -12	ENH ACO -11	ENT SP5 -15
HAL STI -18	HAL MAC -15	ENT cCR -13	HYP COR -15
ENT KYL -21	ENT SP5 -17	ENT SP5 -13	HAL WRI -17
ENH ACO -27	AMP RIG -19	AMP SP1 -20	ENT KYL -21
ENT SP3 -40	THA CIL -21	HYP COR -22	HAL MAC -23
ENT SP2 -42	HAL STI -36	ACA SPI -36	HAL MIN -33

The numbering of species used for the canonical ordination diagrams are added between [ ].
ACA SPI = Acanthophora spicifera [1], AMP FRA = Amphiroa fragilis [2], AMP RIG = A. rigida [3], AMP SPE = A. species [4], BOO PUS = Boodleopsis pusilla [5], CAU SER = Caulerpa serrulata [6], CER cTR = Ceramium cf. truncatum [7], CYM ROT = Cymodocea rotundata [8], CYM SER = C. serrulata [9], ENH ACO = Enhalus acoroides [10], ENT cCL = Enteromorpha cf. clathrata [11], ENT KYL = E. kylinii [13], ENT RAM = E. ramulosa [14], ENT SP2, ENT SP3, ENT SP5 = E. species 2 [15], 3 [16], 5 [17], GRA SAL = Gracilaria salicornia [18], HAL MAC = Halimeda macroloba [20], HAL MIN = Halophila minima [20], HAL OPU = Halimeda opuntia [19], HAL OVA = Halophila ovata [24], Halophila minima [20], HAL STI = H. stipulacea [23], HAL UNI = Halodule univervis [21], HAL WRI = H. wrightii [22], HYP COR = Hypnea cornuta [26], PAD SP1 = Padina species [27], ROS ORI = Rosenvingea orientalis [28], SPH SPE = Sphacelaria species [29], SYR ISO = Syringodium isoetifolium [30], THA HEM = Thalassia hemprichii [31], THA CIL = Thalassodendron ciliatum [32].

Fig. 3. Ordination diagrams based on canonical correspondance analysis: 1. Environmental factors (dist low = distance to low water mark, org mat = organic material, P = phosphorus, 2000–1000 μm, 50–2 μm, 2–0 μm = particle size fractions); 2. Species (numbers corresponding with those given after Table 7); 3. Relevé clusters belonging to the TWIN-groups 1: ◇, 2: ◆, 3: ▲, 4: ●, 5: △, 6: □, 7: ■, 8: ○.

The intertidal zone

The relevés with the green alga *Boodleopsis pusilla*, forming group 1 in the TWINSPAN-analysis, represent the intermediate zone between the seagrass beds and the mangal.

Extreme muddy substrates in the intertidal zone of Gazi Bay are generally devoid of seagrasses: they are covered by *Enteromorpha* spp., which form group 2 in the TWINSPAN analysis. At some places (fishermen's harbour) the pioneer seagrass vegetation *Halophila ovalis* + *Halodule wrightii* starts developing on the mud.

According to the CCA-analysis these TWIN-groups (1, 2) seem to be correlated to the small particle size fraction (50–2 μm) and to the amount of organic material.

The lower part of the intertidal coral platforms are generally covered by *Laurencia papillosa* (group 3 of the TWINSPAN analysis) combined with *Padina boryana*, *Hypnea cornuta* and *Acanthophora spicifera*. This vegetation enhances sand retention on which *Halophila ovalis* and *Halodule wrightii* develop as pioneer seagrass vegetation, possibly evolving into the climax *Thalassia* vegetation. This is a similar phenomenon to the one described by Molinier & Picard (1951, 1952) in the Mediterranean sea: seaweeds *Padina* and *Jania* on rocks, silting up, colonization by the seagrass *Cymodocea nodosa* (Ucria) Asch., development of the climax *Posidonia oceanica* (L.) Delile seagrass vegetation.

The combination of *Halophila ovalis* and *Halodule wrightii* (group 4 of the TWINSPAN-analysis) occurs in the whole seagrass covered part of the intertidal zone. They form the pioneer community in the intertidal zone, growing as a rather open cover on sandy patches within the *Thalassia* vegetation, as well in shallow pools as emerged at low tide. They frequently form low sandy bumps at the upper fringe of the seagrass beds.

Most of the intertidal sand flats from MLWN (mean low water at neap tide) downwards as well as the highest parts of the subtidal zone [down to −1 m under MLWS (mean low water at spring tide)] are covered by a mixed seagrass vegetation mainly composed of *Thalassia hemprichii* and *Cymodocea rotundata*. These magnozosterids which form group 5 in the TWINSPAN analysis can be considered as the climax vegetation of this zone. *Thalassia* especially is a slow growing species which is very firmly anchored in the substratum by well developed rhizomes. On subhorizontal stabilized surfaces of sand-covered coral rubble *Thalassia* forms an extremely dense, short, almost monospecific vegetation. After spring ebb tides a marked degree of 'burning' can be seen.

In slightly deeper pools with sandy substrate and on the banks of the tidal channel *Thalassia* is mixed with variable quantities of *Cymodocea rotundata* and/or *C. serrulata*. The latter spp. become dominant in pools where the substrate is regularly shifted (f. ex. the pools between the cone-like structures). On places with strong sedimentation *Halodule uninervis* develops as a secondary pioneer species and sediment accumator (*Halophila stipulacea* being the primary one). The relevés including this species do not appear as a separate group in the TWINSPAN analysis. On the banks of the tidal channel which are often eroded, the bare areas are first colonized by the quick growing *Halophila stipulacea*, followed by *Halodule uninervis*, restabilizing the bottom, giving the opportunity to *Cymodocea rotundata* and *C. serrulata* to develop. Finally *Thalassia* 'definitely' stabilizes the substrate. Newly formed sand bumps in the *Thalassia* bed will then be covered by *Syringodium*.

The subtidal zone

Bare surfaces such as 'potholes', originating either naturally (e.g. grazing, storm) or from human activities, and lower parts of tidal channel banks are colonized by *Halophila stipulacea*, a halophilid which forms group 8 in the TWINSPAN-analysis. *H. stipulacea* can therefore be considered as the sublittoral colonizing species.

The whole lagoon (from −1 m under MLWS) of Gazi Bay is covered by dense, homogeneous *Thalassodendron ciliatum* meadows. This euhaline amphibolid which forms group 7 in the TWINSPAN-analysis represents the climax vegetation of the subtidal zone, but this species can also develop in deep rocky intertidal pools. The leaves and stems of *Thalassodendron* are richly covered by epiphytic algae (which have also been studied in detail by De Wit, 1988). Some rhizophytic algae grow characteristically between the *T. ciliatum* stolons: *Udotea orientalis*, *Halimeda macroloba*. Both of these show a marked patchiness, being numerous on some surfaces and almost totally absent at other places. *Avrainvillea obscura* grows as isolated specimens all over the lagoon. *Halimeda opuntia* together with *Amphiroa fragilissima* form scattered brittle cushions on the bottom, but they also occur in the intertidal pools.

The loosened stolons of *Thalassodendron* at the margins of the 'potholes' are covered by the brown alga *Rosenvingea orientalis*. Low sandy bumps are generally covered by *Syringodium isoetifolium*, especially close to MLWS, emerging only at extreme spring tides. This syringodiid forms group 6 in the TWINSPAN-analysis and sometimes grows as an interrupted belt between the midlittoral magnozosterids and the subtidal amphibolids.

Enhalus acoroides locally grows as monospecific stands in the *Thalassodendron* beds. This submerged vegetation forms open communities, the leathery leaves being vertically directed, the eroded apices reaching the water surface at MLWS. Patches of *Enhalus* vegetation also occur in the lowest intertidal zone, between MLWN and MLWS. At low tide the leaves lie partly exposed, partly submerged on the bottom of shallow pools or in sandy grooves, all directed seaward by the flow off of the water.

Synthesis of the plant zonation in Gazi Bay

High midlittoral:
mangal:
- *Lumnitzera racemosa* Van Steenis
- *Avicennia marina* (Forsk.) Vierh.
- *Ceriops tagal* (Perr.) Rob. + *Bruguiera gymnorrhiza* Lamarck + *Xylocarpus granatum* Koenig
- *Sonneratia alba* J. Smith + *Rhizophora mucronata* Lamarck

Mid midlittoral:
silty substrate:
- *Boodleopsis pusilla*
- *Enteromorpha* spp.

locally evolving to *Halophila ovalis* + *Halodule wrightii*

coral platform:
- *Laurencia papillosa* + other algae sand covering--->
- *Halophila ovalis* + *Halodule wrightii*--->
- *Thalassia hemprichii* + *Cymodocea rotundata* + *Halimeda opuntia*, *Hypnea cornuta*, *Chaetomorpha crassa*

Deep (rock) pools with sandy bottom:
- *Thalassodendron ciliatum* on the bottom, *Cymodocea* on the pool margins

Infralittoral fringe (between MLWN and MLWS):
- *Halophila ovalis* + *Halodule wrightii*--->
- *Halodule uninervis*--->
- *Cymodocea rotundata* + *C. serrulata* + *Thalassia hemprichii*
- *Syringodium isoetifolium*
- *Enhalus acoroides*
- \+ *Halimeda macroloba*, *Avrainvillea obscura*, *Cladophoropsis vaucheriaeformis* (Aresch.) Cribb, *Gracilaria corticata*, *G. salicornia*, *Hypnea cornuta*

Infralittoral down to −1 m:
- *Halophila stipulacea*--->
- *Halodule uninervis*
- *Thalassia hemprichii* + *Cymodocea serrulata* (+ *C. rotundata*)
- *Syringodium isoetifolium*
- *Enhalus acoroides*
- \+ *Rosenvingea orientalis* (on seagrass stolons), *Halimeda macroloba*, *H. opuntia*, *Avrainvillea obscura*, *Udotea orientalis*, *Amphiroa fragilissima*, *Caulerpa sertularioides*

Infralittoral, deeper than −1 m:
- *Thalassodendron ciliatum* (+ numerous epiphytes) (very locally replaced by *Enhalus acoroides*).

According to our field observations this zonation scheme can be generalized for the Kenyan coast, except for *Enhalus* which seems to be a rather rare species in this region. It also corresponds well with the one described by Brouns & Heijs (1985) and Heijs (1987) for S. Papua New Guinea (Port Moresby area) except for the absence of *Thalassodendron ciliatum* in the subtidal zone around Port Moresby; it is there replaced by *Thalassia hemprichii* which either forms subtidal monospecific stands or is growing together with other species, forming extensive mixed seagrass meadows.

Acknowledgements

This study was sponsored by N.F.W.O. (Nationaal Fonds voor Wetenschappelijk Onderzoek), F.K.F.O. (Fonds voor Kollektief en Fundamenteel Onderzoek) and EEC (European Economic Community). We are also grateful to Ph. Polk for giving us the opportunity to do botanical research in Kenya within the framework of the Kenyan Belgian Research Project, to Mr. Allela for putting the infrastructure of the Kenya Marine and Fisheries Research Institute at our disposal. Our gratitude also goes to I. Wolfs, E. Martens and N. Revis for the local organization of and T. Beeckman for the help with the field work and laboratory research.

References

Beeckman, H., 1989. Mathematische verwerking van phytosociologische gegevens. Groene Band 76: 1–36.

Beeckman, H. & R. Lemeur, 1989. Ecology, science of complexity and interactions. FTI 1989 Project Sheet Interface RUG 9: 3 pp.

Brouns, J. & F. Heijs, 1985. Tropical seagrass ecosystems in Papua New Guinea. A general account of the environment, marine flora and fauna. Mar. Biol. 88: 13–50.

Brouns, J. & F. Heijs, 1986. Structural and functional aspects of seagrass communities and associated algae from the tropical West-Pacific. Ph.D. thesis Nijmegen (Netherlands). 431 pp.

Coppejans, E. & T. Beeckman, 1989. *Caulerpa* section Sedoideae (Chlorophyta, Caulerpales) from the Kenyan coast. Nova Hedwigia 49: 381–393.

Coppejans, E. & E. Gallin, 1989. Macroalgae associated with the mangrove vegetation of Gazi Bay (Kenya). Bull. Soc. r. Bot. Belg. 122: 46–60.

Cottenie, A., M. Verloo, L. Kiekens, G. Velghe & R. Camerlynck, 1982. Chemical analysis of Plants and Soils. I.W.O.N.L. Brussel. 63 pp.

Dawes, C. J., 1981. Marine Botany. John Wiley & Sons, New York.

den Hartog, C., 1967. The structural aspect in the ecology of sea-grass communities. Helgol. Wiss. Meeresunters. 15: 648–659.

den Hartog, C., 1970. The sea-grasses of the world. Verh. Kon. Ned. Ak. Wetensch., Natuurk. Reeks 2, 59: 1–275.

den Hartog, C., 1977. Structure, function, and classification in sea-grass communities. In: McRoy & Helfferich (eds), Seagrass Ecosystems: a scientific perspective. M. Dekker Inc., New York: 89–121.

De Pauw, K., 1990. De vegetatie van de getijdengeul van Gazi Bay (Kenia). M. Sc. thesis State University Ghent (Belgium): 142 + tab.

De Wit, M., 1988. Vegetatie-ecologische studie van de zeegrasgemeenschap (incl. wiercomponent) in Gazi Bay, Kenia. M. Sc. thesis State University Ghent (Belgium): 189 + 81 pp.

FAO, 1977. FAO-guidelines for soil profile description. Rome. 66 pp.

Gallin, E., E. Coppejans & H. Beeckman, 1989. The mangrove vegetation of Gazi Bay (Kenya). Bull. Soc. r. Bot. Belg. 122: 197–207.

Grady, J. R., 1981. Properties of sea grass and sand flat sediment from the intertidal zone of St. Andrew Bay, Florida. Estuaries 4: 335–344.

Harlin, M. M., 1980. Seagrass epiphytes. In Phillips R. C. & McRoy C. P. (eds), Handbook of seagrass biology. An ecosystem perspective. Garland STPM Press, New York: 147–152.

Heijs, F., 1987. Community structure and seasonality of macroalgae in some mixed seagrass meadows from Papua New Guinea. Aquat. Bot. 27: 139–158.

Hill, M. O., 1979. Twinspan – a FORTRAN program – for arranging multivariate data in an ordered two-way table by classification of the individuals and attribute: 90 pp. Cornell Univ. Ithaca, New York.

Hutchinson, G. E., 1944. Limnological studies in Connecticut VII. A critical examination of the supposed relationship between phytoplankton periodicity and chemical changes in lake waters. Ecology 25: 3–26.

Hutchinson, G. E., 1957. Concluding remarks. Cold Spring Harbor Symp. Quant. Biol. 22: 415–427.

Hutchinson, G. E., 1978. An introduction to population ecology. Yale Univ. Press, New Haven, Connecticut.

Isaac, F., 1968. Marine Botany of the Kenya coast. 4. Angiosperms. J. E. Afr. Nat. Hist. Soc. 27: 29–47.

Jaasund, E., 1976. Intertidal seaweeds in Tanzania. Univ. Tromsø. 160 pp.

Legendre, P. & L. Legendre, 1984a. Ecologie numérique. 1.

Le traitement multiple de données écologiques. Masson. Paris. 260 pp.

Legendre, P. & L. Legendre, 1984b. Ecologie numérique. 2. La structure des données écologiques. Masson. Paris. 335 pp.

Londo, G., 1984. The decimal scale for relevés of permanent quadrats. In R. Knapp (ed.), Handbook of vegetation science 4: 45–50.

McRoy, C. P. & C. McMillan, 1977. Production ecology and physiology of seagrasses. In: McRoy C. P. & Helfferich C. (eds), Seagrass ecosystems: a scientific perspective. M. Dekker Inc., New York: 53–88.

Molinier, R. & J. Picard, 1951. Biologie des herbiers de Zostéracées des côtes françaises de la Méditerranée. C. r. Acad. Sci., Paris 233: 1212–1214.

Molinier, R. & J. Picard, 1952. Recherches sur les herbiers de phanérogames marines du littoral Méditerranéen français. Ann. Inst. Océanogr. 27: 157–234.

Moorjani, S. & B. Simpson, 1988. Seaweeds of the Kenya coast. Oxford Univ. Press. Nairobi. 134 pp.

Onyango, H. B. A., 1989. Numerical modelling of Gazi Creek (South coast, Kenya): 1. Hydrodynamics model; 2. Descriptive ecological model. M. Sc. thesis FAME. VU Brussel.

Patriquin, D. G., 1972. The origin of nitrogen and phosphorus for growth of the marine angiosperm *Thalassia testudinum*. Mar. Biol. 15: 35–46.

Patriquin, D. G., 1975. 'Migration' of blowouts in seagrass beds at Barbados and Cariacon, West Indies and its ecological and geological implications. Aquat. Bot. 1: 163–189.

Phillips, R. C. & E. G. Meñez, 1988. Seagrasses. Smiths. Contrib. Mar. Sci. 34: 104 pp.

Schoener, T. W., 1989. The ecological niche. In: J. M. Cherrett (ed.), Ecological Concepts. Blackwell Scientific Publications, Oxford, 385 pp.

ter Braak, C. J. F., 1986. Canonical correspondance analysis: a new eigenvector technique for multivariate direct gradient analysis. Ecology 67: 1167–1179.

ter Braak, C. J. F., 1987. Ordination. In: Jongman R. H., C. J. F. ter Braak & O. F. R. Van Tongeren (eds), Data Analysis in community and landscape ecology. Pudoc Wageningen. 301 pp.

ter Braak, C. J. F., 1988. CANOCO – A FORTRAN program for canonical community ordination by partial detrended canonical correspondance analysis, principal components analysis and redundancy analysis (version 2.1.). Agric. Math. Group. Wageningen. 95 pp.

Walkley, A. & E. A. Black, 1934. An examination of the Degtjareff method for determining soil organic matter and proposed modification of the chromic acid titration method. Soil Sci. 37: 29.

Zieman, J. C. & R. G. Wetzel, 1980. Productivity in seagrasses: methods and rates. In: Phillips R. C. & McRoy C. P. (eds), Handbook of seagrass biology. An ecosystem perspective. Garland STPM Press, New York: 87–116.

Hydrobiologia **247**: 77–86, 1992.
V. Jaccarini & E. Martens (eds), The Ecology of Mangrove and Related Ecosystems.

Microphytoplankton of the Pitchavaram mangals, southeast coast of India: species composition and population density

L. Kannan & K. Vasantha
CAS in Marine Biology, Annamalai University, Parangipettai – 608 502, India

Key words: Microphytoplankton, Pitchavaram mangals, species composition, population density

Abstract

Species richness of phytoplankton of the Pitchavaram mangals was high. There were 82 species constituted by 67 species of diatoms, 12 species of dinoflagellates and 3 species of bluegreen algae. The diatoms formed the bulk with 72% followed by the dinoflagellates with 15%, in which there were autochthonous (temporary and permanent) and allochthonous forms. Phytoplankton population density exhibited a wide seasonal fluctuation with the minimum during monsoon and the maximum during summer, suggesting the possible differential influence of various environmental factors.

Introduction

The phytoplankton constitute 95% of the total marine production (Steemann Nielsen, 1975) so they form a vital source of energy at the first trophic tier and also serve as a direct source of food to many aquatic animals. Reports on the species composition, diversity, distribution, succession, seasonal, tidal and diurnal variations and decomposition of phytoplankton are available from various parts of the world.

In India, studies on marine phytoplankton systematics, species composition, community structure, distribution and productivity in relation to environmental parameters were carried out extensively from both the east and west coasts (eg. Qasim *et al.*, 1974; Gopinathan, 1975; Vijayalakshmi & Venugopalan, 1975; Santhanam, 1976; Sundararaj, 1978; Abidi *et al.*, 1983; Chandran, 1985; Jayalakshmi *et al.*, 1986; Raman and Phani Prakash, 1989; Sumitra Vijayaraghavan & Krishnakumari, 1989). The present investigation was carried out to focus attention on the species composition and population density of phytoplankton of the Pitchavaram mangals and to establish the relationship between phytoplankton and various environmental parameters including particulate organic carbon (POC) and microbiota.

Materials and methods

The study area (Pitchavaram mangals) is shown in Fig. 1 along with the sampling stations. The study area is located 10 km south of Porto Novo (Lat. 11° 29′ N; Long. 79° 46′ E) in the Vellar-Coleroon estuarine complex. Station 1 is located in the Pitchavaram mangal region where a luxuriant mangrove vegetation exists with the dominance of *Avicennia marina* and *Rhizophora apiculata*. Station 2 is situated at a distance of 2 km north of first station and this area is characterized by the rich growth of the seaweeds and seagrasses.

Samples were collected at fortnightly intervals from the sampling stations during January to De-

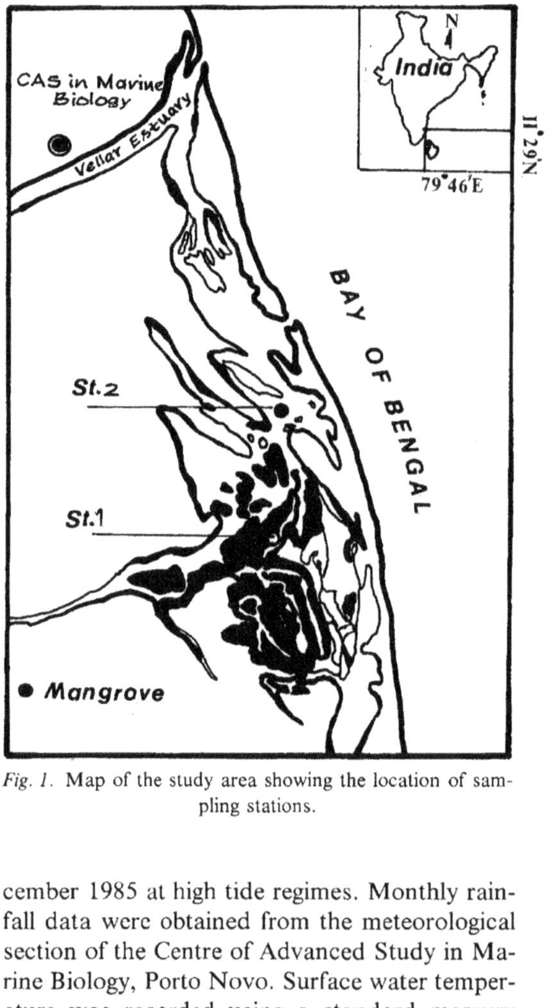

Fig. 1. Map of the study area showing the location of sampling stations.

cember 1985 at high tide regimes. Monthly rainfall data were obtained from the meteorological section of the Centre of Advanced Study in Marine Biology, Porto Novo. Surface water temperature was recorded using a standard mercury thermometer of ± 0.5 °C accuracy. Light penetration was measured with a Secchi disc and the light extinction coefficient (K) was calculated using Pool & Atkins (1929) formula. Salinity and dissolved oxygen were measured following the methods described by Strickland & Parsons (1972). pH was measured with a pH meter (Elico-Model LI-10). Dissolved fractions of the nutrients and POC were measured adopting the procedures described by Parsons *et al.* (1984). Total heterotrophic bacteria were estimated adopting serial dilution plating technique using ZoBell's 2216 e agar medium (Aaronson, 1970).

Phytoplankton was collected only from the surface waters from both the stations. For qualitative analysis, a plankton net, No. 30 (Bolting silk: mesh size – 48 μm) was used for the collection. For assessing the phytoplankton density, surface water samples were collected in 1 litre polythene bottles to which a few drops of 5% neutralized formalin were added as preservative. The samples were allowed to settle for 48 hrs. The supernatant was discarded and the sedimented portion was used for the quantitative analysis. Numerical plankton studies were carried out using Utermöhl's inverted plankton microscope. Whenever necessary, sufficient dilutions of the samples were made and the number of dilutions in all the instances were noted and the values multiplied accordingly. For each sample, countings were made five times and the mean was taken (expressed as cells l^{-1}) as the final value. In the present investigation, only larger phytoplankton was considered and this study provides only a partial contribution to the total phytoplankton biomass and primary production of the Pitchavaram mangals.

For the sake of convenience and interpretation, the calendar year was divided into four seasons based on the northeast monsoon which alone brings bulk rainfall to the southeast coast of India where the present study area is situated. The seasons are: Monsoon (October–December), Postmonsoon (January–March), Summer (April–June) and Premonsoon (July–September).

Results and discussion

Environmental parameters

Monthly variations of different meteorological, physico-chemical and biological parameters recorded at stations 1 and 2 are shown in Figs 2 to 8.

Species composition

Species composition of phytoplankton revealed a total number of 82 species from both the stations. A checklist of phytoplankton recorded from both

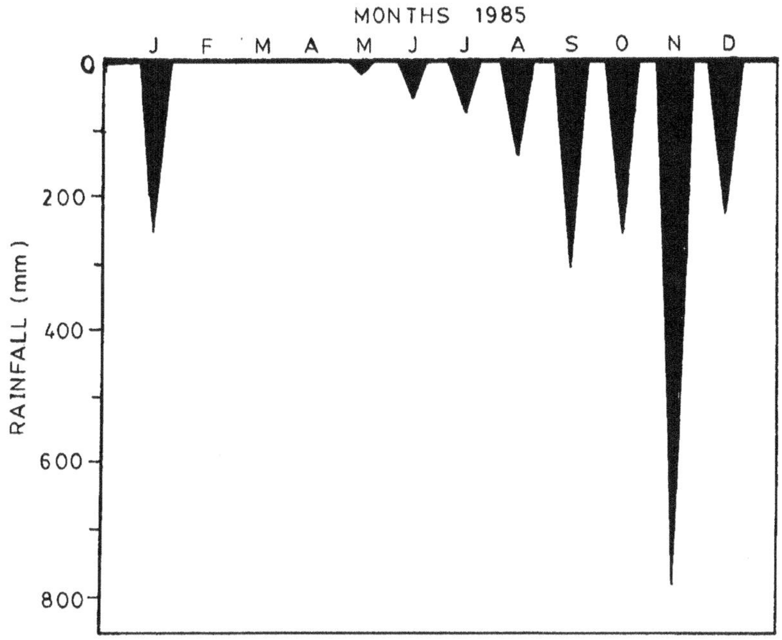

Fig. 2. Monthly mean of rainfall recorded at Porto Novo (common to stations 1 & 2).

the stations during the study period is given in Table 1.

Population density

The fluctuations in phytoplankton population density were wide in different months of the present study period (Fig. 9). In general, the maximum density was recorded during summer (May) and the minimum, during monsoon and postmonsoon seasons. At station 1, the phytoplankton density ranged from 360 (December) to 685 260 cells l^{-1} (May) and at station 2, the density varied between 420 (January) and 784 320 cells l^{-1} (May).

The phytoplankton comprised 67 species of diatoms (Bacillariophyceae), 12 species of dinoflagellates (Dinophyceae) and 3 species of bluegreens (Cyanophyceae). A total number of 76 species of phytoplankton was recorded from station 1 which included 84.21% of diatoms, 13.15% of dinoflagellates and 2.64% of bluegreens. At station 2, a total number of 80 species of phytoplankton was recorded which included 83.75% of diatoms, 13.75% of dinoflagellates and 2.50% of bluegreens. Hence diatoms were the dominant phytoplankton group in the study areas. Khalil (1990) also reported that diatoms were the dominant group comprising 52 to 90% in lake Manzala (Egypt) owing to their relative tolerance to salinity.

Generally, the diatoms ranked first in species composition followed by dinoflagellates and bluegreens. This tendency was observed throughout the study period at both the stations except during the freshwater flow in monsoon season when the dominance of freshwater phytoplankton constituting green algae such as *Pediastrum simplex* was observed.

The permanent autochthonous diatom species described from here by Sivakumar (1982) viz. *Asterionella glacialis*, *Bacillaria paradoxa*, *Chaetoceros affinis*, *Coscinodiscus marginatus*, *Odontella mobiliensis*, *Thalassiothrix frauenfeldii* and the dinoflagellates, *Ceratium furca* and *Protoperidinium* sp. were recorded from station 1 dur-

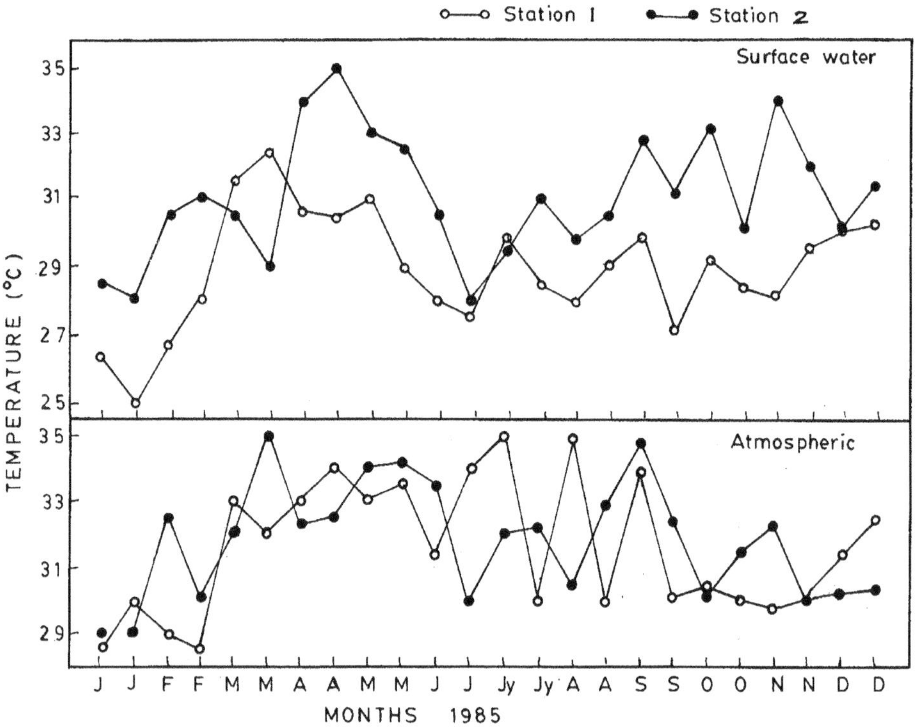

Fig. 3. Monthly variations in atmospheric and surface water temperatures.

Table 1. Checklist of microphytoplankton recorded from stations 1 and 2.

Serial no.	Species
	BACILARIOPHYCEAE
1	*Actinoptychus undulatus* (Bailey) Ralfs
2	*Amphiprora gigantea* Grun.
3	*Asterionella glacialis* Castracane
4	*Bacillaria paradoxa* Gmelin
5	*Bacteriastrum comosum* Pavill.
6	*B. hyalinum* Lauder
7	*Chaetoceros affinis* Lauder
8	*C. curvisetus* Cl.
9	*C. lascinosus* Schutt non Gran
10	*C. lorenzianus* Grun.
11	*C. messanensis* Castr.
12	*Climacosphenia moniligera* Ehr.
13	*Coscinodiscus centralis* Ehr.
14	*C. eccentricus* Ehr. (= *Thalassiosira eccentrica*)
15	*C. gigas* Ehr.
16	*C. lineatus* Ehr.
17	*C. marginatus* Ehr.

Table 1. (Continued)

Serial no.	Species
18	*C. oculus-iridis* Ehr.
19	*C. rothii* (Ehr.) Grun.
20	*C. sub-lineatus* (Grun.) Ratt.
21	*C. thorii* Pavill.
22	*Diploneis* sp.
23	*Ditylum brightwellii* (West) Grun.
24	*D. sol* (Grun.) De Toni
25	*Eucampia cornuta* (Cl.) Grun.
26	*Fragilaria intermedia* (Grun.) Grun.
27	*Guinardia flaccida* (Castr.) Perag. H.
28	*Gyrosigma balticum* (Ehr.) Rabenh.
29	*Hemidiscus hardmannianus* (Grev.) Mann.
30	*Lauderia annulata* (Grev.) Cl.
31	*Leptocylindrus danicus* Cl.
32	*L. minimus* Gran.
33	*Lithodesmium undulatum* Ehr.
34	*Melosira granulata* (Ehr.) Ralfs
35	*M. sulcata* (Ehr.) Kutz.
36	*Navicula* sp.

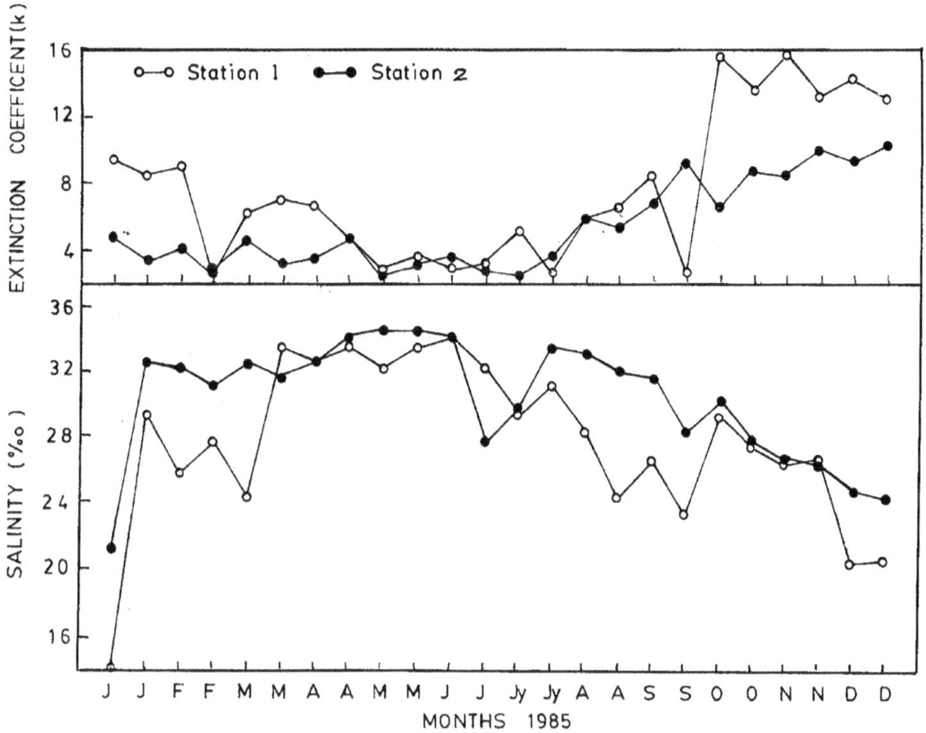

Fig. 4. Monthly variations in light extinction coefficient and salinity.

Table 1. (Continued)

Serial no.	Species
37	*Navicula* sp.
38	*Nitzschia closterium* (Ehr.) Sm.
39	*N. longissima* (Breb.) Ralfs
40	*N. seriata* Cl.
41	*N. sigma* (kutz.) W. Sm.
42	*N. vitrea* Norman
43	*Odontella mobiliensis* (Bail.) Grun.
44	*O. rhombus* (Ehr.) Kutz.
45	*O. sinensis* (Grev.) Grun.
46	*Planktoniella sol* (Wallich) Schutt
47	*Pleurosigma aestuarii* Breb. W. Sm.
48	*P. directum* Grun.
49	*P. elongatum* W. Sm.
50	*P. normanii* Ralfs
51	*Rhizosolenia alata* Btw.
52	*R. imbricata* Btw.
53	*R. robusta* Norman
54	*R. setigera* Btw.
55	*R. stolterforthii* Perag. H.

Table 1. (Continued)

Serial no.	Species
56	*R. styliformis* Btw.
57	*Skeletonema costatum* (Grev.) Cl.
58	*Stephanopyxis palmeriana* (Grev.) Grun.
59	*S. turris* (Grev.) Ralfs
60	*Synedra formosa* Hantzsch
61	*S. ulna* (Kutz.) Ehr.
62	*Thalassionema nitzschioides* Grun.
63	*Thalassiosira subtilis* (Ostenf.) Gran.
64	*Thalassiothrix frauenfeldii* Grun. non Cl.
65	*T. longissima* Cl. & Grun.
66	*Triceratium favus* Ehr.
67	*T. reticultum* Ehr.
	DINOPHYCEAE
68	*Ceratium furca* (Ehr.)/Clap. & Lachm.
69	*C. fusus* (Ehr.) Dujardin
70	*C. massiliense* (Gourret) Karsten
71	*C. trichoceros* (Ehr.) Kofoid
72	*C. tripos* (O.F. Mull.) Nitzsch

Table 1. (Continued)

Serial no.	Species
73	*Dinophysis caudata* Sav.
74	*Gymnodinium* sp.
75	*Prorocentrum micans* Ehr.
76	*Protoperidinium* sp.
77	*P. depressum* (Bail.) Bal.
78	*P. obtusum* (Karsten) Parke & Dodge
79	*P. oceanicum* (Van.) Bal.
	CYANOPHYCEAE
80	*Anabaena ambigua* Rao, C.B.
81	*Oscillatoria subbrevis* Schmidle
82	*Trichodesmium erythraeum* Ehrenberg ex Gomont

ing the present study period. At station 2, the diatoms, *Bacteriastrum comosum*, *Odontella sinensis*, *Melosira granulata*, *Rhizosolenia alata* and the dinoflagellates, *Ceratium trichoceros* and *C. tripos* were recorded as permanent autochthonous species.

The temporary autochthonous diatom species *Bacteriastrum comosum*, *Fragilaria intermedia*, *Planktoniella sol* and *Stephanopyxis palmeriana* and the dinoflagellate species *Dinophysis caudata* and *Gymnodinium* sp. were recorded from station 1. From station 2, the temporary autochthonous diatom species *Guinardia flaccida*, *Lauderia annulata*, *Lithodesmium undulatum* and *Synedra formosa* and the dinoflagellate species *Prorocentrum micans* were recorded.

At both stations, the important allochthonous constituents were *Amphiprora gigantea*, *Ditylum brightwellii* and *Rhizosolenia setigera* among diatoms and *Ceratium fusus* and *C. massiliense*, among dinoflagellates. The common freshwater forms recorded from the two stations particularly during periods of low salinity were *Anabaena ambigua* and *Pediastrum simplex*.

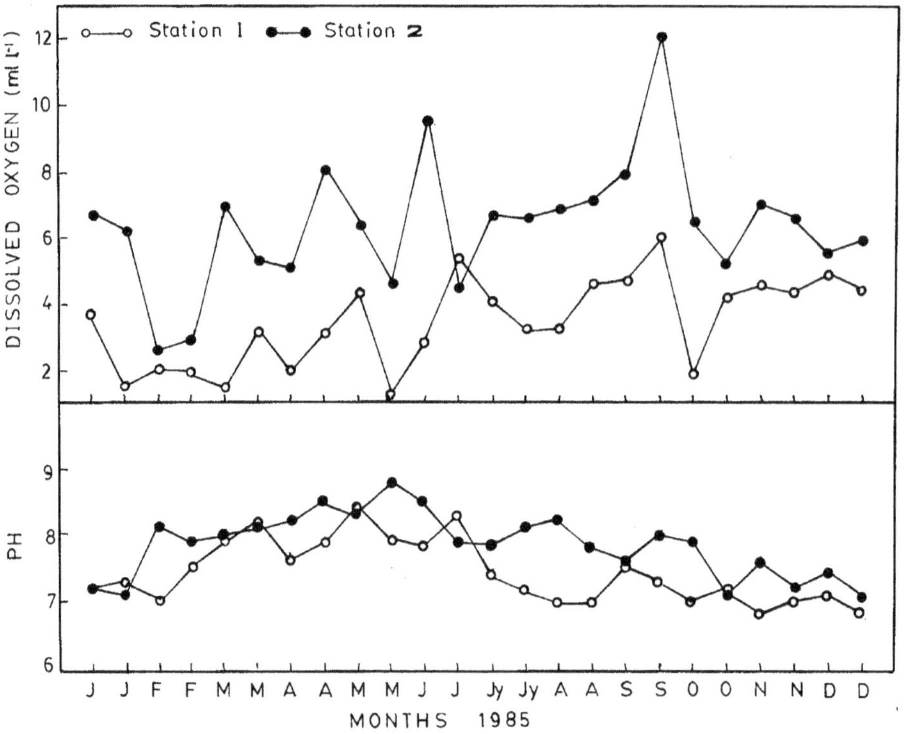

Fig. 5. Monthly variations in dissolved oxygen and hydrogen-ion concentrations.

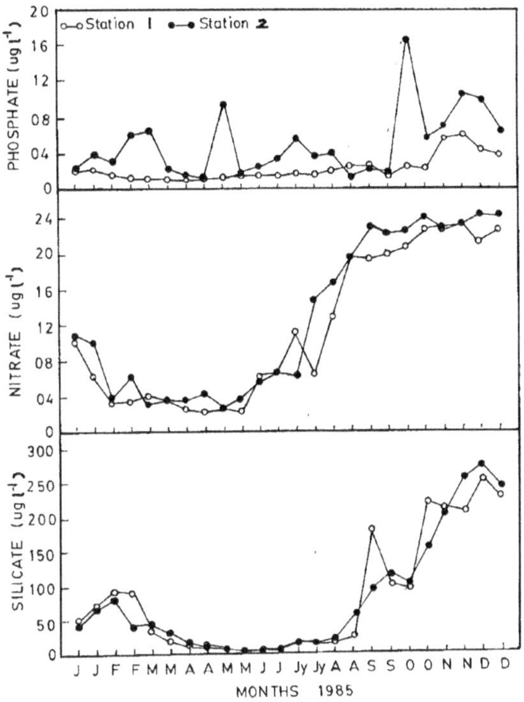

Fig. 6. Monthly variations in phosphate, nitrate and silicate concentrations.

The importance of temperature as a factor in controlling the growth of phytoplankton was emphasized by many workers (Williams & Murdoch, 1976; Goldman, 1977). In the present study, maximum biomass of phytoplankton was observed when the temperature of surface water was high. The peaks of phytoplankton density (685260 cells l^{-1} at station 1 and 784320 cells l^{-1} at station 2) were observed when the surface water temperature (31.0 °C at station 1 and 33.1 °C at station 2) was high. However, a low phytoplankton density (360 cells l^{-1}) was also observed in December when the surface water temperature was high (30.2 °C) at station 1. Sivakumar (1982) also noticed the coincidence of low temperature with low phytoplankton density in the backwaters of Porto Novo. Hence influence of surface water temperature on phytoplankton density is rather obscure in tropical marine environs.

Salinity acts as a major ecological factor controlling the distribution and seasonal succession of phytoplankton (Ketchum, 1964; Krishnamurthy *et al.*, 1974; Raman *et al.*, 1990). In the present study, a low density of phytoplankton (360 cells l^{-1} at station 1 and 384 cells l^{-1} at station 2) was noticed during late monsoon and early postmonsoon seasons respectively when the salinity was low (20.23‰ at station 1 and 21.21‰ at station 2) due to the inflow of freshwater from the upper reaches causing greater turbidity. Recording of a lower density of phytoplankton during monsoon months was due to the heavy freshwater inflow which flushed most of the native organisms including phytoplankton from study areas into the sea. Cadee (1984) also reported that excessive run-off might lead to dilution of the phytoplankton and produce 'washout'.

Subsequently the population density of phytoplankton gradually increased during postmonsoon season (8490 cells l^{-1} at station 1 and 19450 cells l^{-1} at station 2) and reached the peak (685260 cells l^{-1} at station 1 and 784320 cells l^{-1} at station 2) during summer and continued till early premonsoon season (102970 cells l^{-1} at station 1 and 127450 cells l^{-1} at station 2). The summer peak was aided by the inflow of clear neritic waters coupled with penetration of more light into the water column and flourishing growth of mostly the net phytoplankton (*Bacillaria paradoxa*, *Chaetoceros affinis*, *Odontella mobiliensis*, *Rhizosolenia alata* and *Thalassiothrix frauenfeldii*), tolerating a higher salinity (32.32‰ at station 1 and 34.85‰ at station 2). Such a bimodal oscillation of phytoplankton recording high density during summer and premonsoon was observed by Ramadhas (1977) in the Vellar estuary.

In general, low concentrations of nitrate and silicate recorded at stations 1 and 2 during postmonsoon to premonsoon seasons suggest the possible utilization of these nutrients in large amount by the abundantly occurring phytoplankton. However, these two nutrients were not limiting as revealed by the good population density of phytoplankton especially diatoms though silicate could become a limiting nutrient for diatoms (Wetsteyn *et al.*, 1990).

High standing crop of phytoplankton (685260 cells l^{-1} at station 1 and 784320 cells l^{-1} at sta-

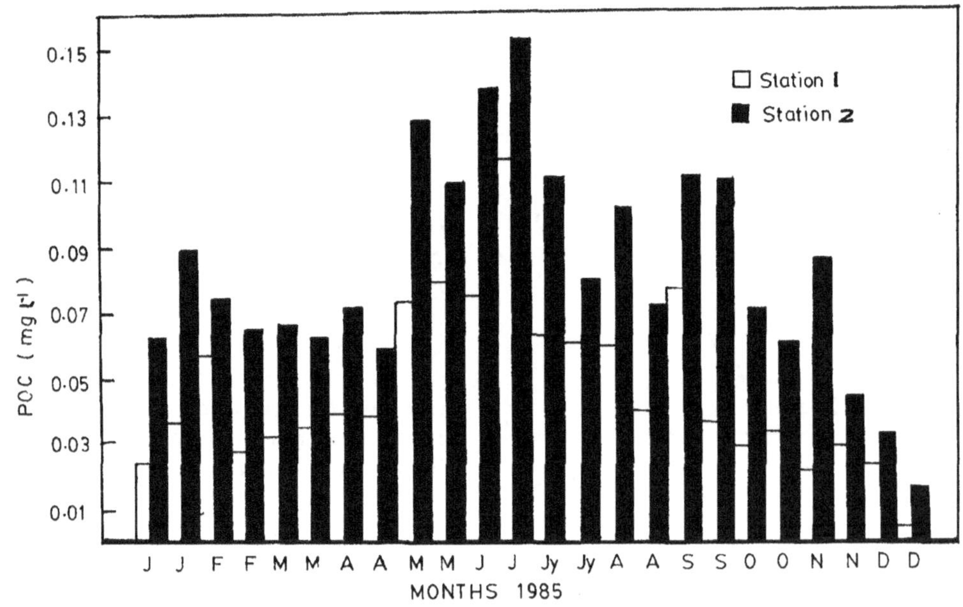

Fig. 7. Monthly variations in particulate organic carbon.

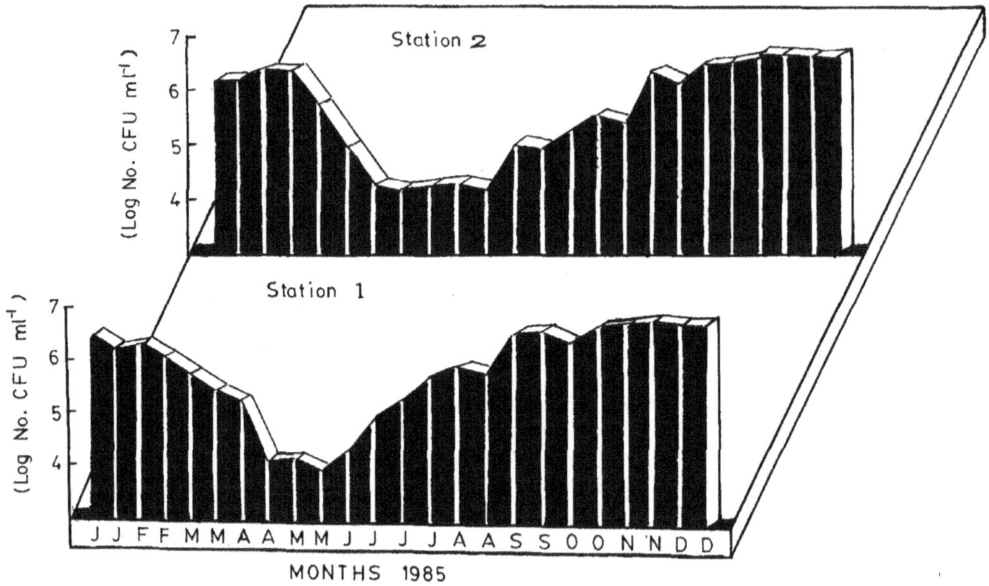

Fig. 8. Monthly distribution of heterotrophic bacterial population in water.

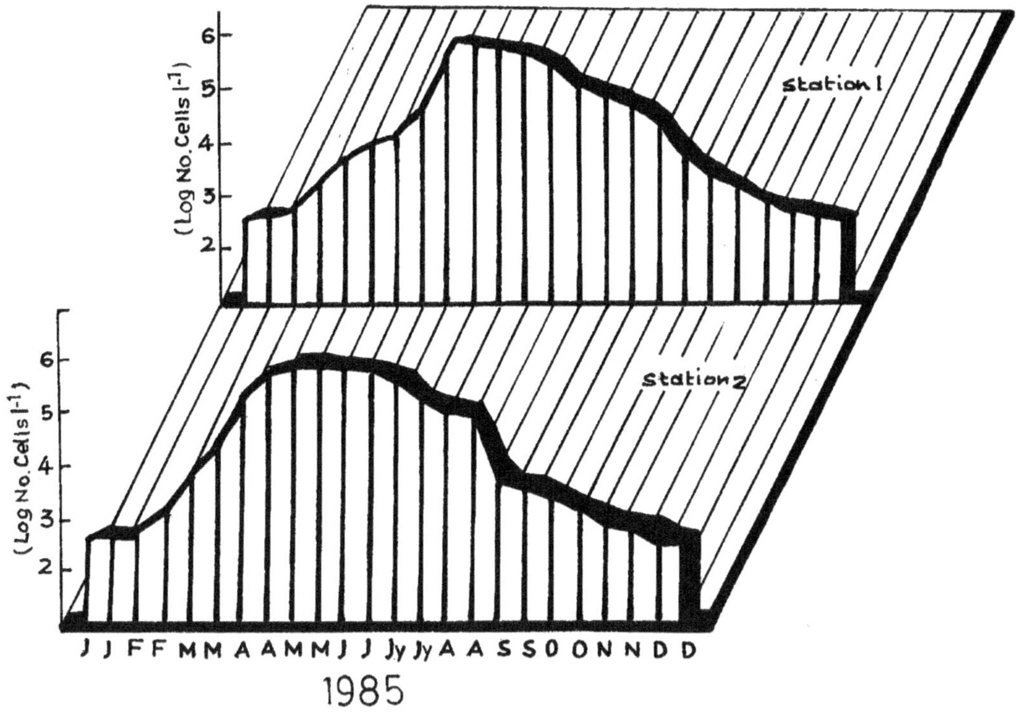

Fig. 9. Monthly variations in phytoplankton population density.

tion 2) dominated by large diatoms were recorded during summer in which season high contents (0.115 mg C l^{-1} at station 1 and 0.148 mg C l^{-1} at station 2) of POC and low bacterial counts (1.213×10^4 CFU ml^{-1} at station 1 and 1.643×10^4 CFU ml^{-1} at station 2) were recorded. This revealed the importance of phytoplankton for the production of POM (Sarma & Nageswara Rao, 1989) and retardation of heterotrophic bacterial growth owing to their extracellular release of antibacterial substances (Kogure *et al.*, 1979; Sieburth, 1979).

Summary

Pitchavaram mangals are rich in microphytoplankton resources with diatoms, dinoflagellates and blue greens. Both autochthonous and allochthonous forms are common. Phytoplankton population density varies between 360 and 784 320 cells l^{-1} recording low density during monsoon and high density during summer. Summer crop of microphytoplankton contributes to the production of POM and retardation of heterotrophic bacterial growth.

Acknowledgements

We thank the Director of the CAS in Marine Biology and the authorities of the Annamalai University for facilities and encouragement. Financial help by the UGC, New Delhi, is gratefully acknowledged.

References

Aaronson, S., 1970. Experimental microbial ecology. Academic press, New York. 236 pp.

Abidi, S. A. H., B. N. Desai & M. Jiyalal Ram, 1983. Studies on the hydrography and plankton of waters off Akarpati near Navapur, west coast of India during February–March. Mahasagar-Bull. natn. Inst. Oceanogr. 16: 91–94.

Cadee, G. C., 1984. Particulate and dissolved organic carbon and chlorophyll *a* in the Zaire river estuary and Plume. Neth. J. Sea Res. 17: 426–440.

Chandran, R., 1985. Seasonal and tidal variations of phytoplankton in the gradient zone of Vellar estuary. Mahasagar-Bull. natn. Inst. Oceanogr. 18: 37–48.

Goldman, J. C., 1977. Temperature effects of phytoplankton growth in continuous culture. Limnol. Oceanogr. 22: 932–936.

Gopinathan, C. P., 1975. Studies on the estuarine diatoms of India. Bull. Dept. mar. Sci. Univ. Cochin. 7: 995–1000.

Jayalakshmi, K. V., S. Kumaran & M. Vijayan, 1986. Phytoplankton distribution in Cochin backwaters – a seasonal study. Mahasagar-Bull. natn. Inst. Oceanogr. 19: 29–38.

Ketchum, B. H., 1964. Relation between circulation and planktonic populations in estuaries. Ecology 35: 191–200.

Khalil, M. T., 1990. Plankton and primary productivity in lake Manzala, Egypt. Hydrobiologia 196: 201–207.

Kogure, K., U. Simidu & N. Taga, 1979. Effect of *Skeletonema costatum* (Grev.) Cleve on the growth of marine bacteria. J. exp. mar. Biol. Ecol. 36: 201–215.

Krishnamurthy, K., R. Santhanam & V. Sundararaj, 1974. Species distribution in aquatic environment. Indian J. mar. Sci. 3: 135–138.

Parsons, T. R., Y. Maita & C. M. Lalli, 1984. A manual of chemical and biological methods for seawater analysis. Pergamon press, Oxford. 173 pp.

Pool, H. H. & W. R. G. Atkins, 1929. Photoelectric measurements of submarine illumination throughout the year. J. mar. biol. Ass. U.K. 16: 297–324.

Qasim, S. Z., Sumitra Vijayaraghavan, K. J. Joseph & V. K. Balachandran, 1974. Contribution of microplankton and nannoplankton in the waters of a tropical estuary. Indian J. mar. Sci. 3: 146–149.

Ramadhas, V., 1977. Studies of phytoplankton, nutrients and some trace elements in Porto Novo waters. Ph.D. thesis, Dept. Mar. Biol. Annamalai University. 235 pp.

Raman, A. V. & K. Phani Prakash, 1989. Phytoplankton in relation to pollution in Visakhapatnam harbour, east coast of India. Indian J. mar. Sci. 18: 33–36.

Raman, A. V., Ch. Satyanarayana, K. Adiseshasai & K. Phani Prakash, 1990. Phytoplankton characteristics of Chilka lake, a brackish water lagoon along east coast of India. Indian J. mar. Sci. 19: 274–277.

Santhanam, R., 1976. Hydrobiological studies in Porto Novo waters – Ecology of phytoplankton in the different biotopes. Ph.D. thesis, Dept. Mar. Biol. Annamalai University. 101 pp.

Sarma, N. S. & I. Nageswara Rao, 1989. Distribution of chlorophyll *a*, organic carbon, nitrogen and carbohydrate of particulate matter of Visakhapatnam harbour waters, east coast of India. Indian J. mar. Sci. 18: 24–29.

Sieburth, J., 1979. Sea microbes. Oxford University Press, New York. 491 pp.

Sivakumar, V., 1982. An environmental inventory of the tidal zone of the Vellar estuary. Ph.D. thesis, Dept. Mar. Biol. Annamalai University. 178 pp.

Steemann Nielsen, E., 1975. Marine photosynthesis with special emphasis on ecological aspects. Elsevier Oceanography series 13, Amsterdam.

Strickland, J. D. H. & T. R. Parsons, 1972. A practical handbook of seawater analysis. Fish. Res. Bd. Can. Bull. 167: 1–311.

Sumitra Vijayaraghavan & L. Krishna Kumari, 1989. Primary production in the southern Arabian Sea during southwest monsoon. Indian J. mar. Sci. 18: 30–32.

Sundararaj, V., 1978. Hydrobiological studies in the Vellar-Coleroon estuarine system. Ph.D. thesis, Dept. Mar. Biol., Annamalai University. 111 pp.

Vijayalakshmi, G. S. & V. K. Venugopalan, 1975. Studies on diversity of plankton species, pigments and succession with a note on primary production at a tidal zone in the Vellar estuary. In R. Natarajan (ed.), Recent researches in estuarine biology. Hindustan Publishing Corporation, Delhi: 237–246.

Wetsteyn, L. P. M. J., J. C. H. Peters, R. N. M. Duin, F. Vegter & P. R. M. de Visscher, 1990. Phytoplankton primary production and nutrients in the Oosterschelde (The Netherlands) during the pre barrier period 1980–1984. Hydrobiologia 195: 163–167.

Williams, R. B. & M. B. Murdoch, 1976. Phytoplankton production and chlorophyll concentrations in the Beaufort channel, North Carolina. Limnol. Oceanogr. 11: 73–82.

Hydrobiologia **247**: 87–98, 1992.
V. Jaccarini & E. Martens (eds), The Ecology of Mangrove and Related Ecosystems.

The ecology of Belizean mangrove-root fouling communities: patterns of epibiont distribution and abundance, and effects on root growth

Aaron M. Ellison & Elizabeth J. Farnsworth[1]
Department of Biological Sciences, Clapp Laboratory, Mount Holyoke College, South Hadley, Massachusetts 01075-1484, USA; Field Naturalist Program, Department of Botany, 120 Marsh Life Sciences Building, University of Vermont, Burlington, Vermont 05405-0086, USA; [1] *Current address: Department of Organismic and Evolutionary Biology, Biological Laboratories, Harvard University, 16 Divinity Ave., Cambridge, MA 02138, USA*

Key words: Belize, epibionts, isopods, mangroves, *Melongena*, *Rhizophora mangle*

Abstract

The aerial prop roots of the neotropical red mangrove, *Rhizophora mangle* L., begin growing well above highest high water (HHW) and often extend well below lowest low water (LLW) before rooting in the benthic substratum. In Belize, Central America, prop roots growing below LLW are colonized by diverse assemblages of organisms, including macroalgae, hydrozoans, ascidians, sponges, anemones, hard corals, and isopod crustaceans. Mangroves, root-fouling epibionts, root herbivores, and benthic predators engage in complex interactions that are major determinants of mangrove growth and production. Species richness of root epibionts increases with distance from the mainland and with proximity to the barrier reef. Species richness decreases with variability in water temperature and salinity. Ascidians and sponges transplanted from Lark Cay into the coastal Placencia Lagoon failed to survive, but anemones from Lark Cay survived in Placencia Lagoon. Reciprocal transplants survived off-shore. The gastropod predator, *Melongena melongena* L., present only in mainland estuaries, reduced local barnacle abundance and epibiont species richness in Placencia Lagoon. Isopod species richness also increases with distance from shore, but the number of roots bored by these species decreases. These isopods can reduce root relative growth rate (RGR_{root}) by 55%. On off-shore cays, sponges and ascidians ameliorate negative effects of isopods. In mainland estuaries where epibionts are less common, isopod damage to roots is more severe. Experimental studies in mangrove swamps throughout the world would clarify the importance of plant-animal interactions in these widespread tropical ecosystems.

Introduction

Epibionts are organisms that colonize and grow on virtually any living, solid, exposed surface in the marine environment (reviewed by Wahl, 1989). In mangal, littoral mangrove roots are often the only local substratum not subject to heavy sedimentation and provide ideal habitat for algal and invertebrate epibionts (Rützler, 1969). Although epibionts may significantly impact their host (and *vice versa*; Wahl, 1989), mangrove epibionts have received only brief mention in even the most comprehensive reviews (Macnae, 1968; Chapman, 1976; Tomlinson, 1986; Wahl, 1989; but see Rützler, 1969). Except for species inventories (*e.g.* Mattox, 1949; Macnae & Kalk, 1962;

Macnae, 1968; Rützler, 1969; Hutchings & Recher, 1974; Sasekumar, 1974; Pinto & Wignarajah, 1980; Sutherland, 1980; Alvarez I., 1989), investigators have ignored epibionts (and mobile animals) because they have been thought to contribute insignificantly to direct matter and energy flow in mangrove ecosystems (*e.g.* Golley *et al.*, 1962; Odum & Heald, 1972; Lugo & Snedaker, 1974), and because the mangrove fauna is not considered host-specific (*e.g.* Macnae, 1968; Tomlinson, 1986).

Throughout the neotropics, prop roots of *R. mangle* extending below LLW are colonized by numerous algal and invertebrate epibionts (Rützler, 1969; Sutherland, 1980; Taylor *et al.*, 1986; Alvarez I., 1989; Ellison & Farnsworth, 1990). Intertidal portions of roots are often colonized by barnacles and red algae (Rützler, 1969; M. M. Littler *et al.*, 1985; Perry, 1988; D. S. Littler *et al.*, 1989). Both subtidal and intertidal roots are bored by isopod crustaceans (Sphaeromatidae and Limnoriidae: Rehm & Humm, 1973; Snedaker, 1973; Estevez & Simon, 1975; Simberloff *et al.*, 1978; Ribi, 1982; Perry, 1988; Kensley & Schotte, 1989; Ellison & Farnsworth, 1990).

Recent experimental studies have shown that epibionts directly and indirectly affect root growth and production (Perry, 1988; Ellison & Farnsworth, 1990) and total energy flow (Wada & Wowor, 1989; Rodriguez & Stoner, 1990), and that herbivorous crabs can influence mangrove zonation (Smith, 1987; Smith *et al.*, 1989). Here, we describe patterns of epibiont distribution and abundance on mangrove roots in Belize, Central America; discuss some of the apparent causes of the observed patterns; document interactions between various epibionts; and summarize the known effects of these epibionts on root and tree growth.

Study sites

Belize, located in northeastern Central America, includes the largest continuous barrier reef in the western hemisphere (Rützler & Macintyre, 1982). Tidal amplitude throughout Belize is ≈30 cm (Kjerfve *et al.*, 1982). The mangroves *Rhizophora mangle* L., *Avicennia germinans* (L.) Stearn, *Laguncularia racemosa* (L.) Gaertn. f., and *Conocarpus erectus* L. (taxonomy follows Tomlinson, 1986) are found along rivers, in estuaries, and on numerous mangrove cays in the lagoon shoreward of the barrier reef (Hartshorn *et al.*, 1984; Stoddart *et al.*, 1982). All four species co-occur from HHW to mean water (MW). Below MW, *R. mangle* forms monospecific stands extending seaward to below LLW.

We studied mangrove-root epibionts at six locations in southern Belize (Fig. 1). These sites are representative of mainland estuaries (Placencia Lagoon and Big Creek), near-shore cays (Lark and Bugle Cays), and cays close to the barrier reef (Northeast Cay and Twin Cays). Detailed descriptions of these study sites are given in Ellison & Farnsworth (1990) and Farnsworth & Ellison (1991).

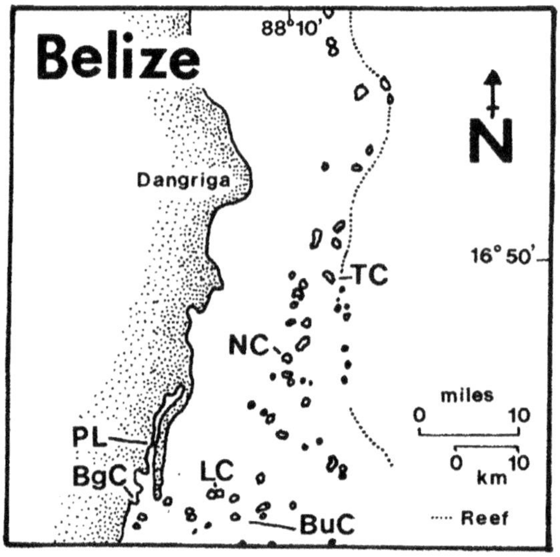

Fig. 1. Map of southern Belize, showing the six study sites. *PL*: Placencia Lagoon; *BgC*: Big Creek; *LC*: Lark Cay; *BuC*: Bugle Cay; *NC*: Northeast Cay; *TC*: Twin Cays. Names and locations follow British Survey Sheets of Belize, nos. 32 (1966), 36 (1976), and 40 (1982).

Patterns of epibiont distribution

Methods

At all six sites (Fig. 1), we sampled a minimum of 100 roots for epibionts. Roots were selected haphazardly along a 100 m transect parallel to shore. We distinguished roots as *aerial* or *ground* (rooted in the substratum). We catalogued all epibiont species encountered. At Placencia Lagoon, Lark Clay, and Twin Cays (hereafter referred to as Placencia, Lark, and Twin, respectively), we measured percent cover and vertical zonation of epibionts on roots. Percent cover was assessed by laying a transparent 5 × 20 cm sheet of acetate divided into grids of 100 1 × 1 cm squares over the seaward-facing side of the root. The number of squares under which each epibiont occurred was used as an estimate of percent cover (after correction for actual root diameter). We used additional grids to measure epibiont cover over the entire root. To assess epibiont zonation, we recorded presence of each species in 2 cm intervals along each root from 1 cm above the root cap to the upper limit of epibiont occurrence. Species occurring at the root tip (up to 1 cm above the root cap) were recorded separately.

Results

Species encountered at the six sites are listed in Table 1. Throughout, we will refer to all these species generically, except where such reference would be ambiguous. For all major taxa, species richness increased from the mainland towards the barrier reef (Table 1). Ascidians and cnidarians were absent from the Placencia estuary, and only one species in each of these two taxa was present in Big Creek. Twin and Northeast Cay, each only 4 km from the barrier reef, had the highest species richness of algae, sponges, crustaceans, and ascidians.

Epibiont species richness at Lark was much higher than at Bugle Cay, although both islands had fewer species than Twin and Northeast Cay (Table 1). Lark and Bugle Cay are both 7 km off shore; Bugle Cay is 2 km SE of Lark. However, Lark is protected from extreme wave action by a series of cays to the east, while Bugle Cay is more wave-swept. We suspect that the low number of species present at Bugle Cay reflects increased wave exposure there. Differences in root fauna observed between Lark and Bugle Cay are comparable to differences observed between exposed and sheltered roots at Twin (A. M. Ellison & E. J. Farnsworth, pers. obs.), and we are currently exploring relationships between epibiont patterns and wave exposure.

Species composition and percent cover differed between aerial and ground roots (Fig. 2), and these patterns varied with location. Across all sites filamentous algae, fleshy algae, and balanoid barnacles were generally more abundant on aerial roots (Fig. 2B, 2C). Crustose coralline algae and chthamaloid barnacles were more abundant on ground roots (Fig. 2A). At Lark, ascidians other than *Diplosoma* were more abundant on ground roots (Fig. 2B). At Twin, however, only *Didemnum conchyliatum* was found on both aerial and ground roots; all other ascidians were found only on aerial roots (Fig. 2C). Sponges were found on both aerial and ground roots at all sites. At Lark, all sponges other than *Haliclona curacaoensis* were more abundant on ground roots (Fig. 2B), but at Twin, sponges were generally equally abundant on aerial and ground roots.

At all sites, isopods attacked a much higher percentage of aerial roots than ground roots (Fig. 3). We encountered only *Sphaeroma* at Placencia and only *Phycolimnoria* at Lark. The latter species accounted for 95% of the isopod individuals found at Twin (Ellison & Farnsworth, 1990), where we also found two other sphaeromatids (Table 1).

Taken over all roots surveyed, epibiont species richness tended to be higher on aerial roots than on ground roots. At Placencia, all five epibiont species occurred on both aerial and ground roots. At Lark, 33 species were found on aerial roots, and 26 occurred on ground roots. At Twin, 47 species were found on aerial roots, while only 28 were found on ground roots. Ephemeral cnidaria (*Halocordyle*, *Halecium*, and *Sertularia*) occurred

Table 1. Taxa encountered on mangrove roots in various locations of coastal Belize (Site abbreviations as in Fig. 1).

Taxa	PL	BgC	LC	BuC	NC	TC
Cyanobacteria[1]						
Scytonema polycystum Barnet & Flahault					√	√
Chlorophyta[2]						
Acetabularia crenulata Lamour.		√	√	√		√
Caulerpa mexicana (Sond.) J. Ag.					√	√
Caulerpa racemosa (Forssk.) J. Ag.			√		√	√
Caulerpa sertulariodes (S. G. Gmelin) Howe						√
Caulerpa verticillata J. Ag.	√	√	√			√
Derbesia osterhoutii (L. & A. H. Blinks) Page			√		√	
Halimeda spp.			√	√	√	√
Valonia macrophysa Kütz.			√			√
Ventricaria ventricosa J. Ag.			√			√
Phaeophyta[2]						
Dictyota spp.		√	√	√	√	√
Lobophora variegata (Lamour.) Worm.			√	√	√	√
Padina sanctae-crucis Børg.		√	√	√	√	√
Rhodophyta[2]						
Amphiroa rigida var. *antillana* Børg.						√
Bostrychia spp.	√	√	√	√	√	√
Corallinaceae			√	√	√	√
Wrangelia argus (Montagne) Montagne			√		√	√
Porifera[3]						
Amphimedon viridis (D. & M.)		√	√	√	√	√
Chondrilla nucula Schmidt				√		
Halichondria melanadocia de Laubenfels						√
Haliclona implexiformis (Hechtel)		√	√		√	√
Haliclona curacaoensis (van Soest)			√		√	√
Haliclona pseudomolitiba de Weerdt, Rützler, and Smith						√
Lissodendoryx isodictyalis (Carter)			√		√	√
Mycale microsigmatosa Arndt						√
Mycale sp. 2 *fide* Rützler	√	√			√	
Pellina carbonaria (Lamarck)	√	√				
Tedania ignis (D. & M.)		√	√	√	√	√
Ulosa ruetzleri Wiedenmayer			√	√	√	√
Cnidaria[1]						
Aiptasia pallida (Verrill)			√	√	√	√
Bartholomea annulata (Lesueur)					√	√
Condylactis gigantea (Weinland)			√		√	√
Halocordyle disticha (Goldfuss)		√	√		√	√
Halecium spp.			√		√	√
Millepora alicornis L.			√	√		
Palythoa caribaea D. & M.			√	√	√	
Sertularia turbinata (Lamour.)			√		√	√
Annelida						
Sabellidae			√	√	√	√
Serpulidae			√	√	√	√

Table 1. (Continued)

Taxa	PL	BgC	LC	BuC	NC	TC
Crustacea (Cirripedia)[1]						
Balanus eburneus Gould	√	√	√	√	√	√
Chthamalus angustitergum thompsoni (Henry)	√	√				
Crustacea (Isopoda)[4]						
Dynamenella cf. *angulata* (Richardson)						√
Paracerceis caudata Say						√
Phycolimnoria clarkae Kensley & Schotte			√			√
Sphaeroma terebrans Bate	√	√				
Mollusca (Bivalvia)[5]						
Crassostrea rhizophorae Guilding	√				√	√
Isognomon alatus Gmelin	√	√	√		√	√
Chordata (Tunicata)[6]						
Ascidia nigra (Savigny)			√		√	
Botrylloides nigrum Herdman			√		√	√
Botrylloides cf. *perspicuum* Herdman			√		√	√
Clavelina picta (Verrill)					√	
Didemnum conchyliatum (Sluiter)			√		√	√
Didemnum psammathodes Sluiter						√
Diplosoma glandulosum (Milne Edwards)			√		√	√
Distaplia corolla Monniot			√		√	√
Ecteinascidia minuta (Berrill)						√
Ecteinascidia turbinata Herdman			√		√	√
Eudistoma olivaceum (van Name)		√	√		√	√
Eudistoma obscuratum (van Name)						√
Perophora formosana (Oka)						√
Perophora regina Goodbody & Cole						√
Perophora viridis (Verrill)			√		√	√
Styela partita (Stimpson)			√		√	

[1] Nomenclature follows Sterrer (1986).
[2] Nomenclature follows Littler *et al.* (1989); algae identified by D.S. Littler, Smithsonian Institution.
[3] Sponges identified by K. Rützler, Smithsonian Institution. *Haliclona* nomenclature follows de Weerdt *et al.* (1991).
[4] Nomenclature follows Kensley & Schotte (1989); isopods identified by B. Kensley, Smithsonian Insitution.
[5] Nomenclature follows McLean (1951).
[6] Tunicates identified by I. Goodbody, University of the West Indies.

only on aerial roots, while other cnidarians (*Palythoa*, *Millepora*, and *Condylactis*) occurred only on ground roots. Ascidian species richness also was much higher on aerial roots at these two sites (Lark: 5 species only on aerial roots; Twin: 11 species only on aerial roots). At Lark and Twin, isopods only attacked aerial roots.

Few clear patterns in vertical zonation of epibionts on roots were observed. Isopods were found most commonly in the root tip ($P<0.01$, G-test), and barnacles only occurred on the intertidal portions of the roots ($P<0.001$, G-test). *Bostrychia* was most common at the low water line ($P<0.05$, G-test). All of the other species occurred only below LLW, and were found with equal likelihood at all locations along the subtidal portion of the root ($P>0.50$, G-test).

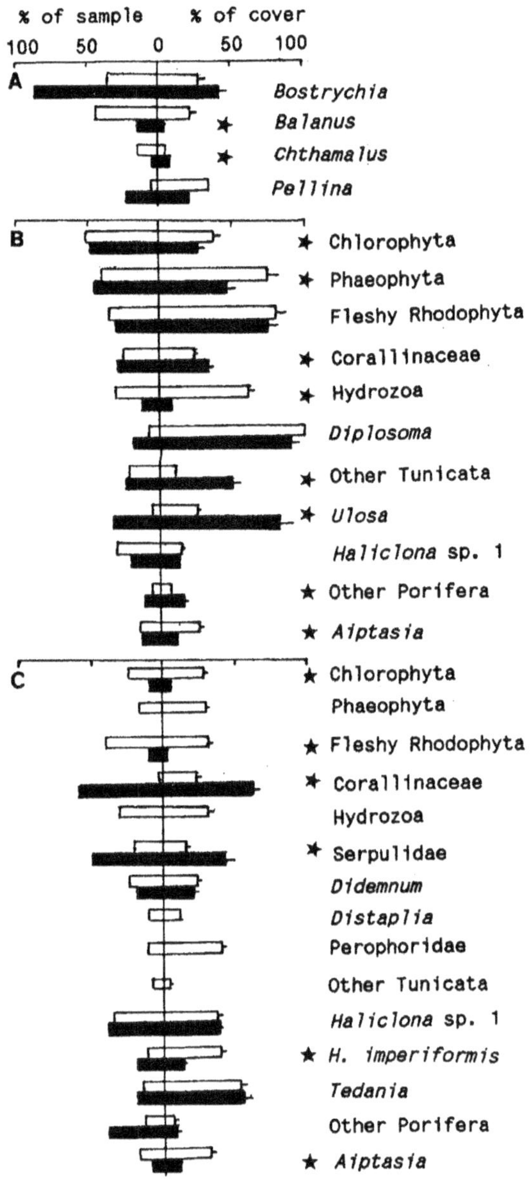

Fig. 2. Distribution and abundance of common taxa at Placencia (A), Lark (B), and Twin (C). Only taxa that occurred on at least 10% of the roots sampled (% of sample) are presented. Mean % cover (± 1 SE) of each taxa for roots on which it occurs is given for aerial (open bars) and ground (solid bars) roots. Stars indicate significant differences in % cover between aerial and ground roots ($P < 0.05$, 2-tailed t-test on angularly transformed data [Sokal & Rohlf, 1981]).

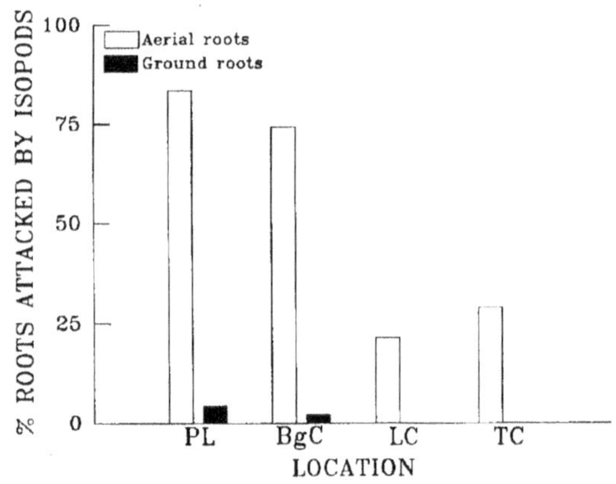

Fig. 3. Percent of aerial and ground roots attacked by isopods. Site names as in Fig. 1.

Factors affecting epibiont distribution

Methods – effects of salinity and water temperature on epibionts

Our survey revealed that several epibionts common on the cays, including the colonial ascidian *Diplosoma*, the sponge *Amphimedon*, and the anemone *Aiptasia* were absent from Placencia Lagoon. To determine if site-specific salinity and water temperature affected epibiont distribution, we reciprocally transplanted several epibionts between Lark and Placencia. On 24 November 1988, colonies of *Diplosoma*, *Amphimedon*, and *Aiptasia* were transplanted from Lark into Placencia. *Pellina*, a sponge only found in the mainland estuaries, was transplanted from Placencia to Lark. For each species, 20 replicate aerial roots with at least 50% cover of the epibiont were collected at each site. We transplanted ten of the roots *in situ* (transplant controls), and transplanted 10 to the other site. We measured percent cover of epibionts on the roots, then tied the transplants onto intact hanging roots with plastic cable ties. We transported roots in covered buckets between sites, and all roots were transplanted within 1 h of collection. Transplants were examined every 2 d for 28 d. Salinity (with a refractometer) and temperature were measured concurrently. After 1 month, transplanted roots had begun to decay and the experiment was terminated.

Results – effects of salinity and water temperature on epibionts

Diplosoma and *Amphimedon* died within 4 d of transplanting into Placencia. At Lark, after 28 d, 50% and 65% respectively of the *Diplosoma* and *Amphimedon* transplant controls remained (Fig. 4). *Aiptasia* from Lark survived equally well *in situ* and when transplanted into Placencia (Fig. 4). *Pellina* transplanted to Lark survived unchanged through 4 wks (Fig. 4). During the first 4 d, salinity at Placencia fluctuated between 18 and 26‰, and water temperature increased by 2 °C. Over the 4 wks of the experiment, water temperature averaged 27 °C at Lark and 28 °C at Placencia with low variance. Salinity at Lark was a constant 34‰, but at Placencia ranged from 18‰ to 31‰ (Fig. 5). Salinity at Placencia changed as much as 6‰ over 48 hr, with common diurnal variation of 1–3‰ (Fig. 5).

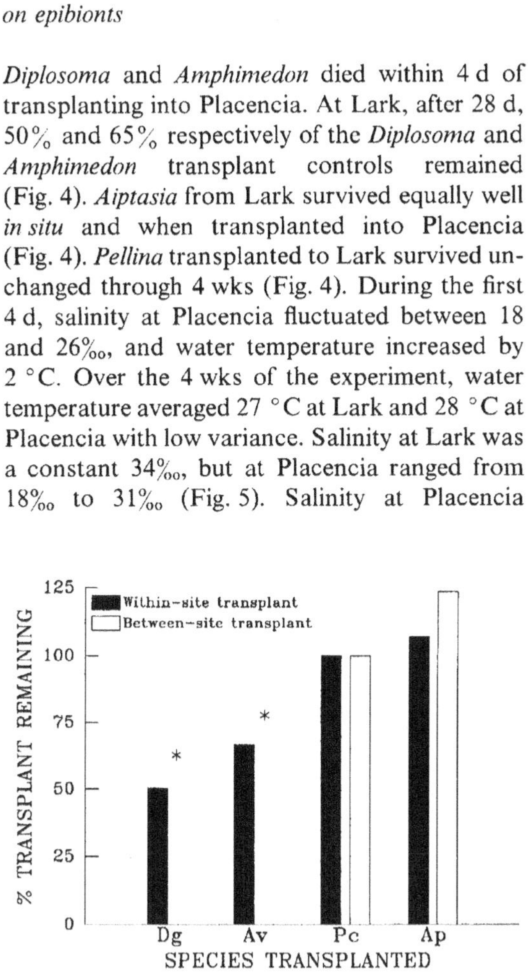

Fig. 4. Success of epibiont transplants between Placencia and Lark. *Diplosoma* (Dg), *Amphimedon* (Av), and *Aiptasia* (Ap) were transplanted from Lark to Placencia (and *in situ* at Lark). *Pellina* (Pc) was transplanted from Placencia to Lark (and *in situ* at Placencia). Bars show the mean % (cover) of the transplant remaining after 28 d when transplanted *in situ* (solid bars) or between sites (open bars). A * above a pair of bars indicates a significant difference between *in situ* and between-site transplants ($P<0.05$, 2-tailed t-test on angularly transformed data).

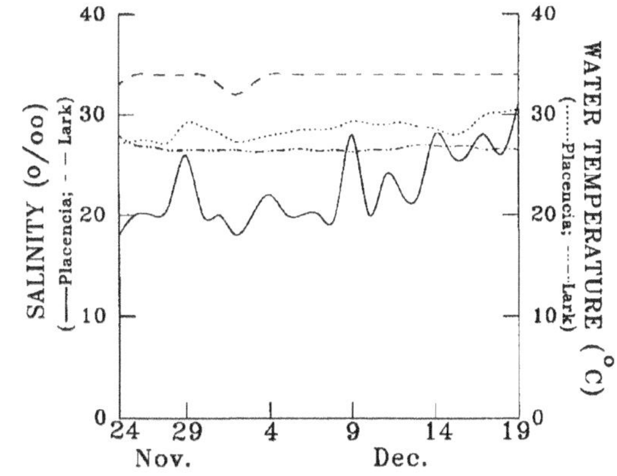

Fig. 5. Salinity (‰) and water temperature (°C) at Placencia and Lark during the transplant experiment.

Methods – effects of predation on epibionts

We also studied the effects of the predaceous gastropod *Melongena melongena* on epibionts at Placencia. At this site, *Melongena* is the most common benthic predator, and we have observed it preying on barnacles on aerial and ground roots at Placencia. Sea urchins, thaid snails, and hermit crabs, which affect epibiont distributions in Belize and Costa Rica (Taylor *et al.*, 1986; Perry, 1988), are absent at Placencia. We randomly chose 16 small (1.0–1.5 m ht) *R. mangle* trees growing between LW and LLW for manipulation. On 1 February 1988, we constructed snail exclosure fences (1.5 × 1.5 × 1.0 m, plastic mesh cages sunk 30 cm into the mud) around 6 of these trees. Of the remaining trees, 4 were designated cage controls (cages with only two parallel sides) and six were unmanipulated. We sampled percent cover of all epibionts on all roots of each tree (> 300 roots/treatment) on 2 February, 22 March, 6 November, 26 November, and 14 December 1988. At each sampling date, we counted the number of *Melongena* in a 100 × 20 m transect that included these trees within it. We simultaneously measured shell

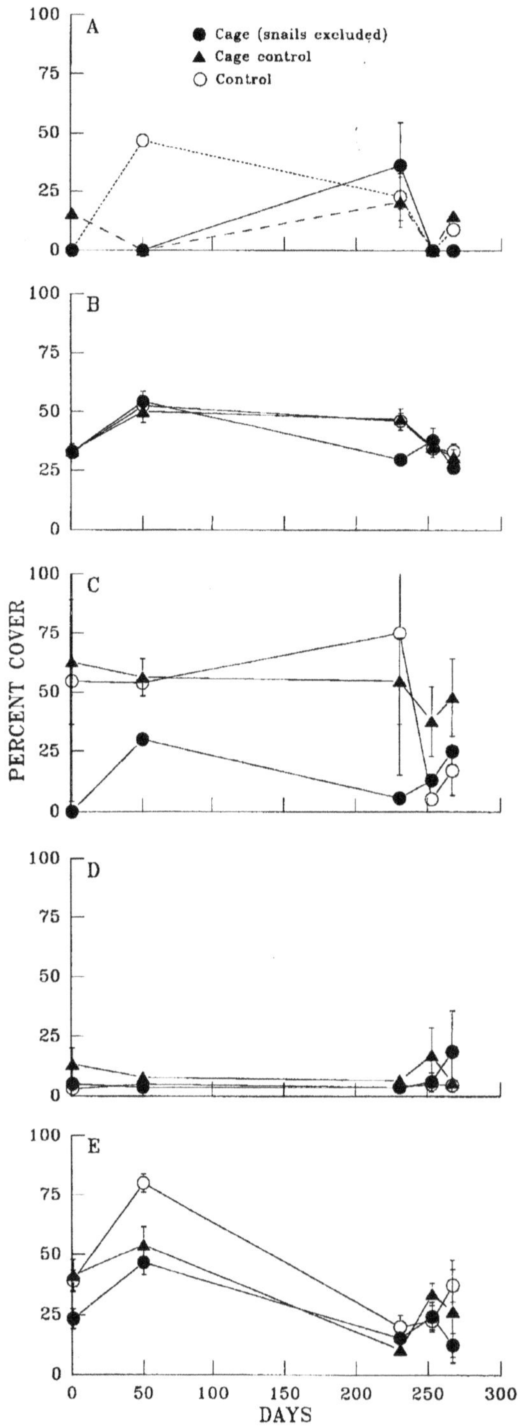

length and width (± 1 mm) of the snails with vernier calipers.

Results – effects of predation on epibionts

Abundance of algae (*Bostrychia* spp. and *Caulerpa verticillata*) fluctuated throughout the year on both aerial and ground roots, but there was no treatment effect on these species (Figs. 6A, 6B). Barnacles (*Balanus* and *Chthamalus*) were more abundant on aerial roots (Fig. 6C) than on ground roots (Fig. 6D) and barnacle abundance remained relatively constant until November. At the beginning of November, snail abundance increased dramatically (Fig. 7) and barnacle abundance declined on unprotected aerial roots (Fig. 6C). Simultaneously, barnacle abundance increased on ground roots of caged plants and decreased in the caged controls (Fig. 6D). The barnacle abundance data (Figs. 6C, 6D) and the snail size data (Fig. 7) both indicate possible recruitment of these species in late November. However, we have no quantitative data on recruitment in these species. Snails had no effects on *Pellina* abundance on roots in any treatment (Fig. 6E).

Effects of epibionts on mangroves

Interactions with Melongena

To determine the effects of these faunal interactions on the trees themselves, we measured tree height and stem diameter ('dbh': above the highest prop root) in the *Melongena* experiment (above). There was a trend towards increased height and 'dbh' in control and cage-control trees where snails foraged on barnacles, relative to trees

Fig. 6. Results of the snail exclosure experiment on epibiont % cover (x ± 1 SE) at Placencia. *A*: Algae (*Bostrychia* spp. and *Caulerpa verticillata*) on aerial roots; *B*: Algae on ground roots; *C*: Barnacles (*Balanus* and *Chthamalus*) on aerial roots; *D*: Barnacles on ground roots; *E*: Sponges (*Pellina*) on ground roots. Day 0 is 1 February 1988.

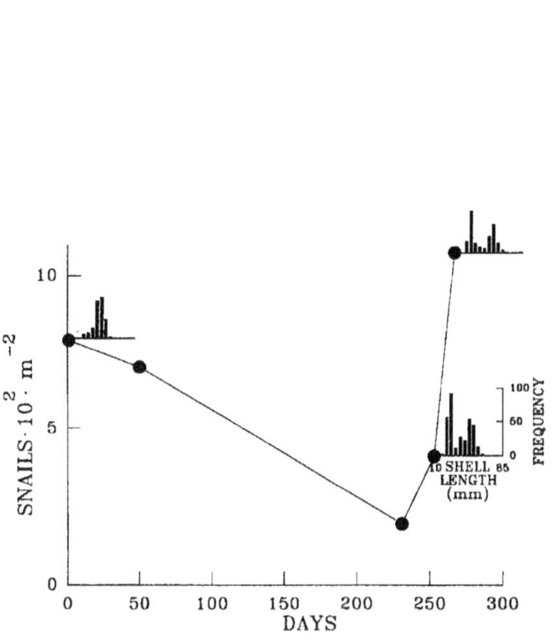

Fig. 7. Density (m^{-2}) and size distributions of *Melongena* during the snail exclosure experiment. Day 0 is 1 February 1988. The inset histograms show snail shell length distributions on 2 February, 26 November, and 14 December 1988. Shell length and width are allometrically correlated (width = $0.43 \cdot \text{length}^{1.22}$ $r^2 = 0.90$).

in snail exclosures ($P = 0.07$, Kruskal-Wallis non-parametric ANOVA). This result indicates some negative effects on tree growth by barnacles and possible amelioration of these effects by snails.

Interactions with isopods

Isopods (*S. terebrans*) were rare in these trees, but in another experiment at Placencia (Ellison & Farnsworth, in prep.), we found that *S. terebrans* reduced RGR_{root} by 60% relative to unattacked roots. At Twin Cays, Ellison & Farnsworth (1990) found that the root-boring isopod *Phycolimnoria* reduced RGR_{root} ($cm\ cm^{-1}\ d^{-1}$) of *R. mangle* by 55%. In that same study, we then asked if sponges (*Haliclona curacaoensis* and *Tedania*), ascidians (*Perophora formosana* and *Didemnum conchyliatum*), anemones (*Aiptasia*) and cyanobacteria affect isopod colonization of *R. mangle* roots. The sponges and ascidians inhibited isopod colonization, thus indirectly facilitating root growth (Ellison & Farnsworth, 1990). RGR_{root} of sponge-covered roots grew 83% faster than roots where we removed sponges *in situ* at Twin (Ellison & Farnsworth, 1990) and 50% faster than roots from which we excluded all fauna (Ellison & Farnsworth, in prep.).

Discussion

Together, water conditions, wave exposure, root status (aerial or ground), proximity of roots to the barrier reef, and predation pressure influence epibiont distribution and diversity on Belizean mangrove roots. Epibiont species richness increases with increasing distance from the mainland (Table 1), and decreasing wave exposure and variability in water salinity and temperature (Figs. 4, 5; Bacon, 1971).

Epibiont species composition differs between aerial and ground roots. Species richness is higher on aerial roots than on ground roots. Although one might expect that ground roots would be more species-rich because of their greater age and size, we observed the opposite patterns. Ascidians and ephemeral hydrozoans are more speciose on aerial roots. These species may be outcompeted by those sponges and colonial anemones (as they are in Venezuela [Alvarez, I., 1989]) that are more abundant on older ground roots (Rützler, 1969; Sutherland, 1980). Young aerial roots may therefore serve as refugia for these weakly competitive ascidians and hydrozoans. We also observed a decline in fleshly algae abundance on ground roots relative to aerial roots. Aerial roots may provide fleshy algae an escape from herbivorous sea urchins (Taylor *et al.*, 1986). Other differences in species composition between aerial and ground roots appear to be site-specific (Fig. 2), and may reflect local variability in recruitment dynamics or epibiont succession (Sutherland, 1980; Rützler, 1987; Alvarez I., 1989).

Many root epibionts affect root growth and production directly (Perry, 1988; Ellison & Farnsworth, in prep.) and indirectly by ameliorating the negative effects of root-boring isopods (Ellison & Farnsworth, 1990). Isopods colonize only young, aerial, uncorticated roots (Rehm & Humm, 1973; Perry, 1988; Ellison & Farnsworth, 1990). At Twin, isopods significantly reduce RGR_{root}, but the presence of dense epibiont cover prevents isopod colonization of roots (Ellison & Farnsworth, 1990). In Costa Rica, isopods (*Sphaeroma peruvianum* Richardson) also reduced root growth rate by 50% and net root production by 62% (Perry, 1988). Perry's experimental results also indicated some reduction in the effects of isopods in roots exposed to water column predators, but she was unable to identify the isopod predator.

In Belize, predation may also ameliorate potentially negative effects of epibionts on tree growth. In the mainland estuary at Placencia, *Melongena* predation reduces barnacle cover on both aerial and ground roots (Fig. 6). Those trees on which *Melongena* reduced barnacle cover grew more rapidly than trees with heavy barnacle cover. In the only other study of the effects of barnacles on mangrove roots, Perry (1988) found that *Balanus* spp. reduced *R. mangle* root growth rate (cm $month^{-1}$) by 30% and net root production (g $root^{-1}$ $month^{-1}$) by 52% along the Pacific coast of Costa Rica. Perry (1988) found that barnacles were preyed upon by snails (*Thais kiosquiformis* Duclus and *Morula lugubris* Adams) and hermit crabs (*Clibanarius panamensis* Stimpson) and proposed that predators indirectly facilitate tree growth.

Epibionts are found world-wide on mangrove roots (Mattox, 1949; Macnae & Kalk, 1962; Macnae, 1968; Rützler, 1969; Hutchings & Recher, 1974; Sasekumar, 1974; Pinto & Wignarajah, 1980; Sutherland, 1980; Alvarez I., 1989), but studies of their roles in the mangal have only just begun (Perry, 1988; Wada & Wowor, 1989; Ellison & Farnsworth, 1990; Rodriguez & Stoner, 1990). Historically, ecological studies of mangroves have emphasized floristic inventories (reviewed in Chapman, 1976; Tomlinson, 1986) and descriptions of energy flow and nutrient cycling (Golley *et al.*, 1962; Odum & Heald, 1972; Lugo & Snedaker, 1974). Although an oft-stated goal of faunistic studies in mangal is the determination of the trophic structure and energy budget of a given mangrove swamp (Sasekumar, 1984), this goal can not be achieved without detailed experimental studies. Intensive manipulative experiments in the marine rocky intertidal have revealed trophic complexity and determinants of community structure not predicted by observation alone (*e.g.* Connell, 1961; Connell, 1975; Paine, 1966; Paine, 1988; Menge *et al.*, 1986; Menge & Sutherland, 1987). Similar studies in temperature zone salt marshes have documented major effects of benthic invertebrates on plant production (Bertness, 1984; Bertness, 1985). In mangrove swamps, recent studies have uncovered unexpected interactions between herbivorous craps and plant zonation (Smith, 1987; Smith *et al.*, 1989), and between epibionts and root production (Perry, 1988; Ellison & Farnsworth, 1990; Rodriguez & Stoner, 1990). Long-term experimental studies on plant-animal interactions in mangroves throughout the world would likely yield new insights into community structure, functioning, and complexity.

Acknowledgements

EARTHWATCH (Center for Field Research), the National Geographic Society, and the Smithsonian Institution NMNH (with additional support from EXXON) supported this research. We thank six teams of dedicated EARTHWATCH volunteers for field assistance, and Chris, Skip, Jessica, and Meredith White for friendship and logistical support at Lark and Placencia. Meredith White collected the salinity data at Placencia. This is contribution no. 303, Caribbean Coral Reef Ecosystems (CCRE) Program, Smithsonian Institution, and paper 2 in the series 'The ecology of Belizean mangrove-root fouling communities'.

References

Alvarez, I. A., 1989. Establecimiento, desarrollo y mantenimiento de una comunidad epibentonica tropical. Ph.D. Thesis, Universidad Central de Venezuela, Facultad de Ciencias. 150 pp.

Bacon, P. R., 1971. The maintenance of a resident population of *Balanus eburneus* (Gould) in relation to salinity fluctuations in a Trinidad mangrove swamp. J. exp., mar. Biol. Ecol. 6: 187–198.

Bertness, M. D., 1984. Ribbed mussels and *Spartina*

alterniflora production in a New England salt marsh. Ecology 65: 1794–1807.

Bertness, M. D., 1985. Fiddler crab regulation of *Spartina alterniflora* production on a New England salt marsh. Ecology 66: 1042–1055.

Chapman, V. J., 1976. Mangrove vegetation. J. Cramer, Vaduz, Lichtenstein, 447 pp.

Connell, J. H., 1961. Effects of competition, predation by *Thais lapillus* and other factors on natural populations of the barnacle *Balanus balanoides*. Ecol. Monogr. 31: 61–104.

Connell, J. H., 1975. Some mechanisms producing structure in natural communities: a model and evidence from field experiments. In M. L. Cody & J. Diamond (eds), Ecology and evolution of communities. Belknap, Cambridge, MA, USA: 460–490.

Ellison, A. M. & E. J. Farnsworth, 1990. The ecology of Belizean mangrove-root fouling communities: I. Epibenthic fauna are barriers to isopod attach of red mangrove roots. J. exp. mar. Biol. Ecol. 142: 91–104.

Estevez, E. D. & J. L. Simon, 1975. Systematics and ecology of *Sphaeroma* (Crustacea: Isopoda) in the mangrove habitats of Florida. In C. Walsh, S. Snedaker & H. Teas (eds), Proceedings of international symposium on biology and management of mangroves. University of Florida, Gainesville, FL: 286–304.

Farnsworth, E. J. & A. M. Ellison, 1991. Patterns of herbivory in Belizean mangrove swamps. Biotropica 23: 555–567.

Golley, F., H. T. Odum & R. F. Wilson, 1962. The structure and metabolism of a Puerto Rican red mangrove forest in May. Ecology 43: 9–19.

Hartshorn, G., L. Nicolait, L. Hartshorn, G. Bevier, R. Brightman, J. Cal, A. Cawich, W. Davidson, R. DuBois, C. Dyer, J. Gibson, W. Hawley, J. Leonard, R. Nicolait, D. Weyer, H. White & C. Wright, 1984. Belize country environmental profile. Trejos Hnos Sucs. S.A., San Jose, Costa Rica, 152 pp.

Hutchings, P. A. & H. F. Recher, 1974. The fauna of Careel Bay with comments on the ecology of mangrove and seagrass communities. Aust. Zool. 18: 99–128.

Kensley, B. & M. Schotte, 1989. Guide to the marine isopod crustaceans of the Caribbean. Smithsonian Institution Press, Washington, D.C., 308 pp.

Kjerfve, B., K. Rützler & G. H. Kierspe, 1982. Tides at Carrie Bow Cay, Belize. In K. Rützler & I. G. Macintyre (eds), The Atlantic barrier reef ecosystem at Carrie Bow Cay, Belize, I. Structure and communities. Smithsonian Inst. Press, Washington, D.C.: 47–51.

Littler, D. S., M. M. Littler, K. E. Bucher & J. E. Norris, 1989. Marine plants of the Caribbean: a field guide from Florida to Brazil. Smithsonian Inst. Press, Washington, D.C., 263 pp.

Littler, M. M., P. R. Taylor, D. S. Littler, R. H. Sims & J. N. Norris, 1985. The distribution, abundance, and primary productivity of submerged macrophytes in a Belize barrier-reef mangrove system. Atoll Res. Bull. 282: 1–18.

Lugo, A. E. & S. C. Snedaker, 1974. The ecology of mangroves. Ann. Rev. Ecol. Syst. 5: 39–64.

Macnae, W., 1968. A general account of the fauna and flora of mangrove swamps and forests in the Indo-West-Pacific region. Adv. mar. Biol. 6: 73–270.

Macnae, W. & M. Kalk, 1962. The ecology of the mangrove swamps at Inhaca island, Mozambique. J. Ecol. 50: 19–34.

Mattox, N. T., 1949. Studies on the biology of the edible oyster, *Ostrea rhizophorae* Guilding in Puerto Rico. Ecol. Mono. 19: 339–356.

McLean, R. A., 1951. The Pelecypoda or bivalve mollusks of Porto Rico and the Virgin Islands. In Scientific survey of Porto Rico and the Virgin Islands. N.Y. Acad. Sci. 17: 1–183.

Menge, B. A., J. Lubchenco, S. D. Gaines & L. R. Ashkenas, 1986. A test of the Menge-Sutherland model of community organization in a tropical rocky intertidal food web. Oecologia (Berl.) 71: 75–89.

Menge, B. A. & J. P. Sutherland, 1987. Community regulation: variation in disturbance, competition, and predation in relation to environmental stress and recruitment. Am. Nat. 130: 730–757.

Odum, W. E. & E. J. Held, 1972. Trophic analyses of an estuarine mangrove community. Bull. mar. Sci. 22: 671–738.

Paine, R. T., 1966. Food web complexity and species diversity. Am. Nat. 100: 65–75.

Paine, R. T., 1988. Food webs: road maps of interactions or grist for theoretical development. Ecology 69: 1648–1654.

Perry, D. M., 1988. Effects of associated fauna on growth and productivity in the red mangrove. Ecology 69: 1064–1075.

Pinto, L. & S. Wignarajah, 1980. Some ecological aspects of the edible oyster *Crassostrea cucullata* (Born) occurring in association with mangroves in Negombo lagoon, Sri Lanka. Hydrobiologia 69: 11–19.

Rehm, A. & H. J. Humm, 1973. *Sphaeroma terebrans*: a threat to the mangroves of southwestern Florida. Science 182: 173–174.

Ribi, G., 1982. Does the wood-boring isopod *Sphaeroma terebrans* benefit red mangroves (*Rhizophora mangle*)? Bull. mar. Sci. 31: 925–928.

Rodriguez, C. & A. W. Stoner, 1990. The epiphyte community of mangrove roots in a tropical estuary: distribution and biomass. Aquat. Bot. 36: 117–126.

Rützler, K., 1969. The mangrove community, aspects of its structure, faunistics and ecology. In Lagunas Costeras, un Simposio. UNAM-UNESCO, Mexico: 515–535.

Rützler, K., 1987. Recruitment of sponges and their space competitors to settling plates in Twin Cay mangrove habitats. In K. Rützler (ed.), Caribbean Coral Reef Ecosystems: progress report 1987. National Museum of Natural History (Dept. of Echinoderms & Lower Invertebrates), Smithsonian Inst., Washington, D.C.: 23–24.

Rützler, K. & I. G. Macintyre (eds), 1982. The Atlantic barrier reef ecosystem at Carrie Bow Cay, Belize, I. Structure and communities. Smithsonian Institution Press, Washington, D.C., 539 pp.

Sasekumar, A., 1974. Distribution of macrofauna on a Malayan mangrove shore. J. anim. Ecol. 43: 51–69.

Sasekumar, A., 1984. Methods for the study of mangrove fauna. In S. C. Snedaker & J. G. Snedaker (eds), The mangrove ecosystem: research methods. UNESCO, Paris, France: 145–161.

Simberloff, D., B. J. Brown & S. Lowrie, 1978. Isopod and insect root borers may benefit Florida mangroves. Science 201: 630–632.

Smith, T. J. III, 1987. Seed predation in relation to tree dominance and distribution in mangrove forests. Ecology 68: 266–273.

Smith, T. J. III, H. T. Chan, C. C. McIvor & M. B. Robblee, 1989. Comparisons of seed predation in tropical tidal forests from three continents. Ecology 70: 146–151.

Snedaker, S. C., 1973. Mangroves, isopods, and the ecosystem. Science 183: 1036–1037.

Sokal, R. R. & F. J. Rohlf, 1981. Biometry, second edition. W. H. Freeman & Co., New York, N.Y., 859 pp.

Sterrer, W. (ed.), 1986. Marine fauna and flora of Bermuda. John Wiley & Sons, New York, N.Y., 742 pp.

Stoddart, D. R., F. R. Fosberg & D. L. Spellman, 1982. Cays of the Belize barrier reef and lagoon. Atoll Res. Bull. 256: 1–76.

Sutherland, J. P., 1980. Dynamics of the epibenthic community on roots of the mangrove *Rhizophora mangle* at Bahia de Buche, Venezuela. Mar. Biol. 58: 75–84.

Taylor, P. R., M. M. Littler & D. S. Littler, 1986. Escapes from herbivory in relation to the structure of mangrove island macroalgal communities. Oecologia (Berl.) 69: 481–490.

Tomlinson, P. B., 1986. The botany of mangroves. Cambridge University Press, Cambridge, U.K., 413 pp.

Wada, K. & D. Wowor, 1989. Foraging on mangrove pneumatophores by ocypodid crabs. J. exp. mar. Biol. Ecol. 134: 89–100.

Wahl, M., 1989. Marine epibiosis. I. Fouling and antifouling: some basic aspects. Mar. Ecol. Prog. Ser. 58: 175–189.

Weerdt, W. H. de, K. Rützler & K. P; Smith, 1991. The Chalinidae (Porifera) of Twin Cays, Belize, and adjacent waters. Proc. Biol. Soc. Wash. 104: 189–205.

Hydrobiologia **247**: 99–108, 1992.
V. Jaccarini & E. Martens (eds), The Ecology of Mangrove and Related Ecosystems.

The meiobenthos of five mangrove vegetation types in Gazi Bay, Kenya

S. Vanhove, M. Vincx, D. Van Gansbeke, W. Gijselinck & D. Schram
University of Ghent, Zoology Institute, Marine Biology Section, K.L. Ledeganckstraat 35, B-9000 Gent, Belgium

Key words: meiobenthos, mangroves, distribution, Kenya

Abstract

The vertical distribution of meiofauna in the sediments of *Avicennia marina*, *Bruguiera gymnorrhiza*, *Ceriops tagal*, *Rhizophora mucronata* and *Sonneratia alba* at Gazi Bay (Kenya), is described. Seventeen taxa were observed, with highest densities in the sediments of *Bruguiera* (6707 ind. 10 cm^{-2}), followed by *Rhizophora* (3998 ind. 10 cm^{-2}), *Avicennia* (3442 ind. 10 cm^{-2}), *Sonneratia* (2889 ind. 10 cm^{-2}) and *Ceriops* (1976 ind. 10 cm^{-2}). Nematodes accounted for up to 95% of total densities; other common taxa were copepods, turbellarians, oligochaetes, polychaetes, ostracods and rotifers. High densities occurred to about 20 cm depth in the sediment. Especially *Ceriops* sediments show still high densities of nematodes (342 ind. 10 cm^{-2}) and copepods (11 ind. 10 cm^{-2}) in the deepest layer (15–22 cm). Particle size and oxygen conditions were major factors influencing meiobenthic distribution; *Uca* burrows had a major impact on distribution and abundance of meiofauna.

Introduction

Most publications on the distribution of meiobenthos in mangrove sediments focus on the upper layers of the sediment, while studies on vertical distribution are scarce (Gerlach, 1958; Decraemer & Coomans, 1978; Hodda & Nicholas, 1985; Dye & Lasiak, 1986; Lalana-Rueda & Gosselck, 1986; Alongi, 1987a, b). Only two studies examine the meiobenthos densities in deeper layers (Dye, 1983a; Nicholas *et al.*, 1991).

The bulk of mangrove work concerns primary production and decomposition of litter (Odum, 1988; Boaden & Seed, 1985; Lugo & Snedaker, 1974; Boto *et al.*, 1984; Twilley *et al.*, 1986; Andersen & Hargrave, 1984; Benner & Hodson, 1985; Cundell *et al.*, 1979; Alongi *et al.*, 1989; Woodroffe, 1987; Steinke & Ward, 1988). This pilot study of meiobenthos in an east African mangrove swamp, deals with vertical distribution within the sediment with emphasis on the role of sediment structure on meiofauna densities.

Description of sites

Studies were carried out at Gazi Bay circa 50 km south of Mombasa (Fig. 1). The mixed mangrove, described by Gallin *et al.* (1989) was chosen, because it is undisturbed and relatively easy to reach. Moreover, studies on vegetation patterns and on the ecology of the oyster *Crassostrea cucullata* of that area are available (Gallin *et al.*, 1989; Okemwa *et al.*, 1986; Ruwa & Polk, 1986; Ruwa, 1990).

Samples were collected from the sediment

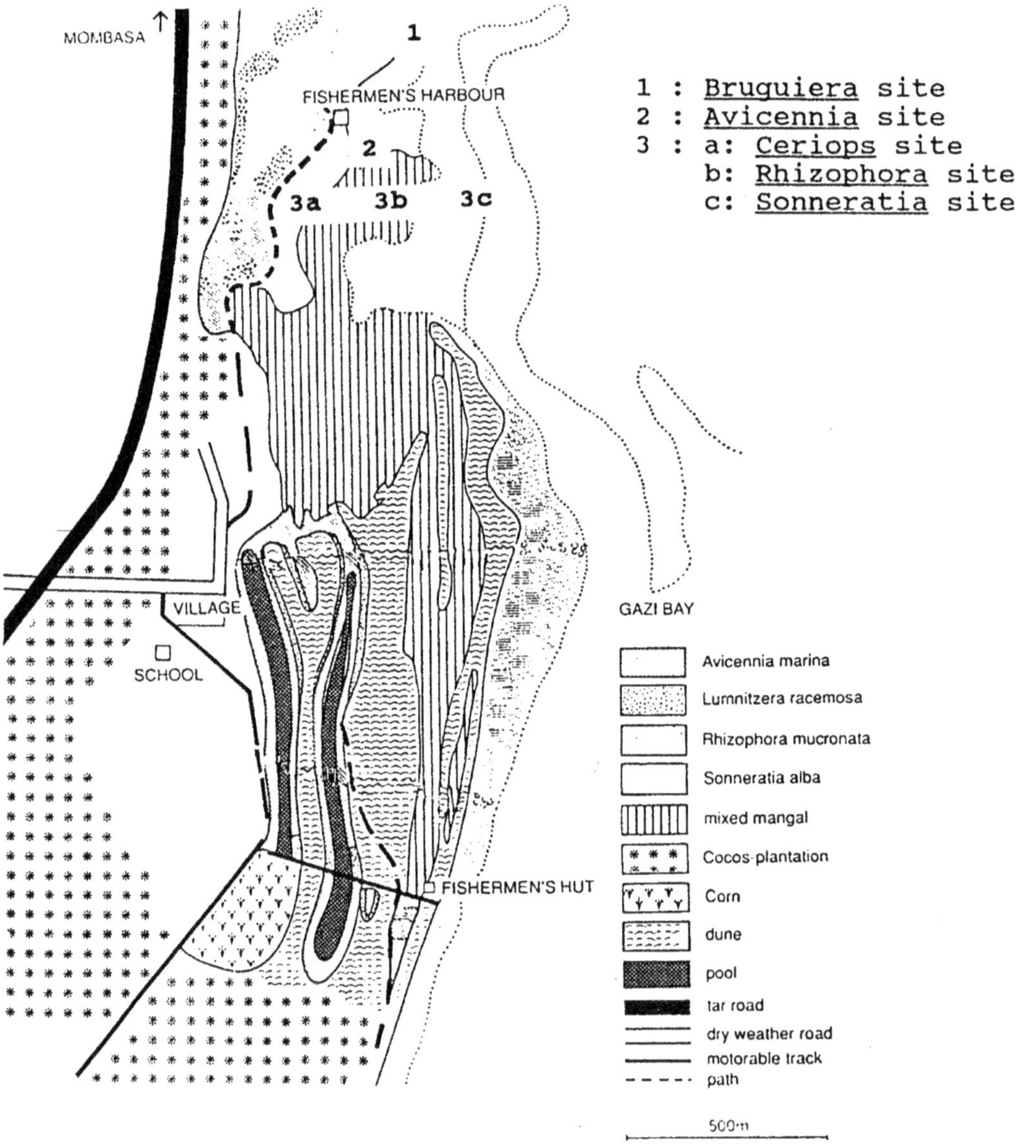

Fig. 1. Study area: vegetation map of the mangrove in Gazi Bay (after Gallin *et al.*, 1989).

around five mangrove species (*Avicennia marina*, *Bruguiera gymnorrhiza*, *Ceriops tagal*, *Rhizophora mucronata*, *Sonneratia alba*) during spring tide, in July 1989.

Material and methods

Meiofauna

Handcores (3.6 cm diameter, 30 cm length) were taken to a depth of 20 cm and subdivided into 1 cm portions for the upper 5 cm, 2.5 and 5 cm intervals for deeper layers. Each core was preserved in a hot (60 °C) 4% formaldehyde solution. Laboratory treatment consisted of extracting the meiofauna by a combination of sedimentation, sieving (38 μm), centrifugation, and staining with Rose Bengal. Roots and organic detritus were not separated, what made counting of the major taxa very time consuming.

Sediment analysis

One core was used for granulometric analysis. Preliminary drying at 100 °C was necessary to remove peat. A mechanical shaker was used to pass the sand fraction through a graded series of standard sieves (Buchanan, 1984).

Sediment temperature was measured by pushing a mercury thermometer into the sediment. Salinity was measured with a refractometer.

Results

Sediment temperature reached 24–25 °C and salinity 32–35‰; a small input of fresh water by the small rivers Kidogoweni and Mkurumu had no measurable influence on the mangrove. pH was 7.15–7.85.

Figures 2–6 show for each station, (a) the meiofauna taxa in the whole core, (b) the meiofauna taxa, excluding nematodes, (c) depth distributions of nematodes and copepods, (d) depth profiles of median grain size and (e) of mud content, (f) depth distribution of total meiofaunal density.

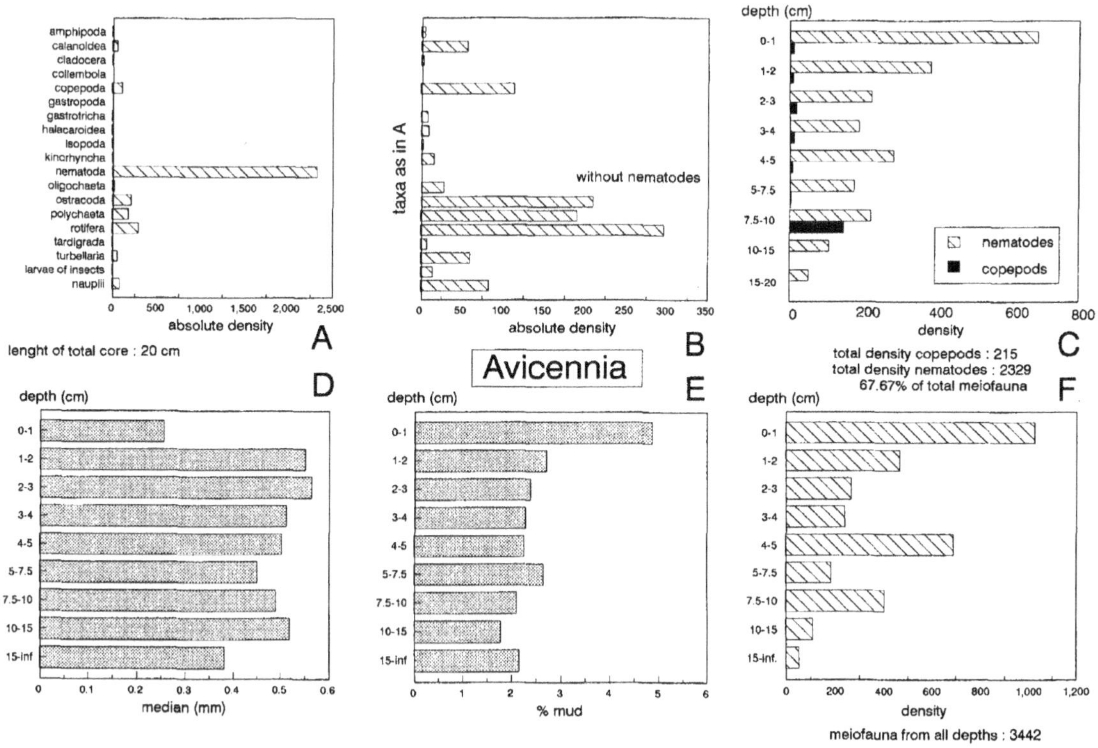

Fig. 2. Avicennia site (a): density of all meiofauna taxa in the whole core (ind./10 cm^2). (b): density of the meiofauna, excluding the nematodes (ind./10 cm^2) (c): depth distribution patterns of densities of nematodes and copepods (d): depth profiles of median grain size of the sand fraction (in mm) (e): depth profile of the percentage of mud (%) (f): depth distribution of the total densities of meiofauna.

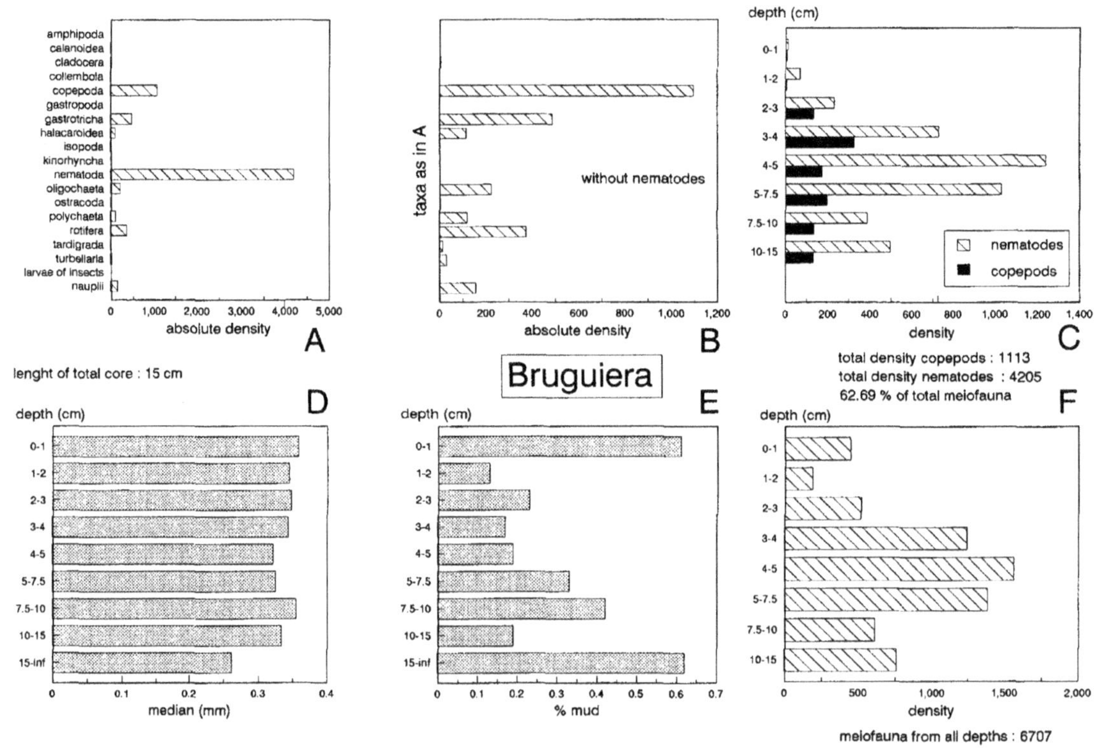

Fig. 3. Bruguiera site (a): density of all meiofauna taxa in the whole core (ind./10 cm^2). (b): density of the meiofauna, excluding the nematodes (ind./10 cm^2) (c): depth distribution patterns of densities of nematodes and copepods (d): depth profiles of median grain size of the sand fraction (in mm) (e): depth profile of the percentage of mud (%) (f): depth distribution of the total densities of meiofauna.

Meiofauna

Total densities

Total densities are high in all five stations: 6707 ind. 10 cm^{-2} in *Bruguiera*, 3998 ind. 10 cm^{-2} for *Rhizophora*, 3442 ind. 10 cm^{-2} for *Avicennia*, 2889 ind. 10 cm^{-2} in *Sonneratia* and 1976 ind. 10 cm^{-2} in *Ceriops*.

Except for *Bruguiera*, the density of meiofauna is highest near the sediment surface, and especially in *Sonneratia* where the upper centimeter contains 50% of the total (Fig. 6F). In *Ceriops* (Fig. 4F) and *Rhizophora* (Fig. 5F), total meiofauna fluctuates with depth, and in *Bruguiera* (Fig. 3F) a maximum is reached at 4–5 cm. In *Avicennia* (Fig. 2F), again, higher densities occur in the top centimeter.

Taxa

Seventeen meiofaunal taxa are recorded. Nematodes dominate: 77.8% (*Avicennia*), 53.5% (*Bruguiera*), 85.5% (*Ceriops*), 89.0% (*Rhizophora*) and 95.1% (*Sonneratia*). Copepod densities, particularly harpacticoids, are low in comparison with nematodes (less than 1% of total density). Other common taxa are turbellarians, oligochaetes, polychaetes, ostracods and rotifers which occur in high numbers in some areas. Few gastrotrichs, kinorhynchs and tardigrades are recorded.

In Figs 2–6 (A, B) no distinction is made between adults and larvae of polychaetes, though high numbers were found for the larvae in *Avicennia* and *Bruguiera*: 93% and 78% respectively. The other sediments show remarkable lower data

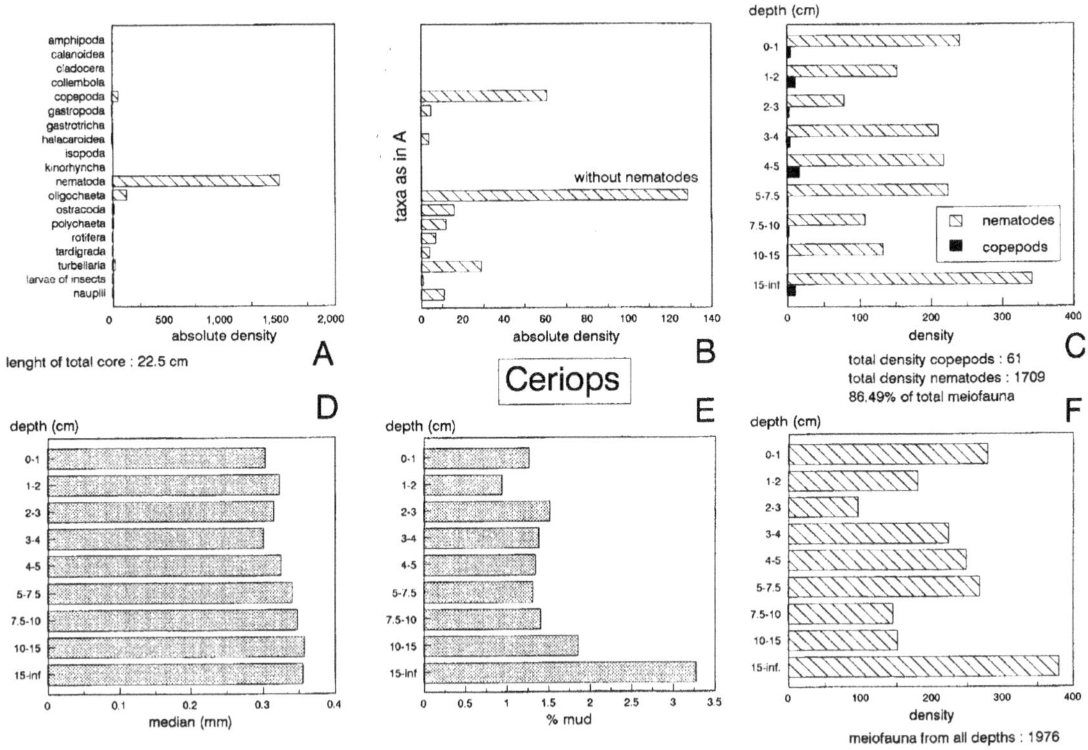

Fig. 4. *Ceriops* site (a): density of all meiofauna taxa in the whole core (ind./10 cm^2). (b): density of the meiofauna, excluding the nematodes (ind./10 cm^2) (c): depth distribution patterns of densities of nematodes and copepods (d): depth profiles of median grain size of the sand fraction (in mm) (e): depth profile of the percentage of mud (%) (f): depth distribution of the total densities of meiofauna.

(as an extreme 2 of the 85 polychaetes of *Sonneratia* are larvae).

Although no attempt is made to detail the different taxa, a lot of epsilonematid nematodes are counted, especially in *Bruguiera* sediments where they make up about 70% of nematode numbers.

Depth profiles

Avicennia (Figs 2 C, F)
The upper layer (0–1 cm) shows a high abundance of nematodes (682 ind.; 67% of total meiofauna at this interval). Copepods cover only 1% of total meiofauna, with nauplii four times more abundant than adults. This is in contrast with the subsequent depths where adult copepods keep almost the same numbers, but numbers of larvae drop. From 2–3 cm on, only nematodes and copepods remain abundant, respectively 83 and 7%. At 7.5–10 cm copepods reach a maximum of 150 ind. (abundance: 37%), while nematodes follow a decreasing trend (223 ind.; 55%). Densities of total meiofauna quickly drop with depth (Fig. 2f). Nevertheless there are still 161 nematodes in the sediment beneath 10 cm.

Bruguiera (Figs 3 C, F)
Abundances of nematodes are high in comparison with other taxa (maximum value of 1241 ind. at the 4–5 cm interval); only the first cm shows a low percentage (3%), mainly due to high amounts of gastrotrichs (87%). Copepod data are very low in the first 2 cm (6 and 9 ind.), but increase deeper in the sediment, reaching a max-

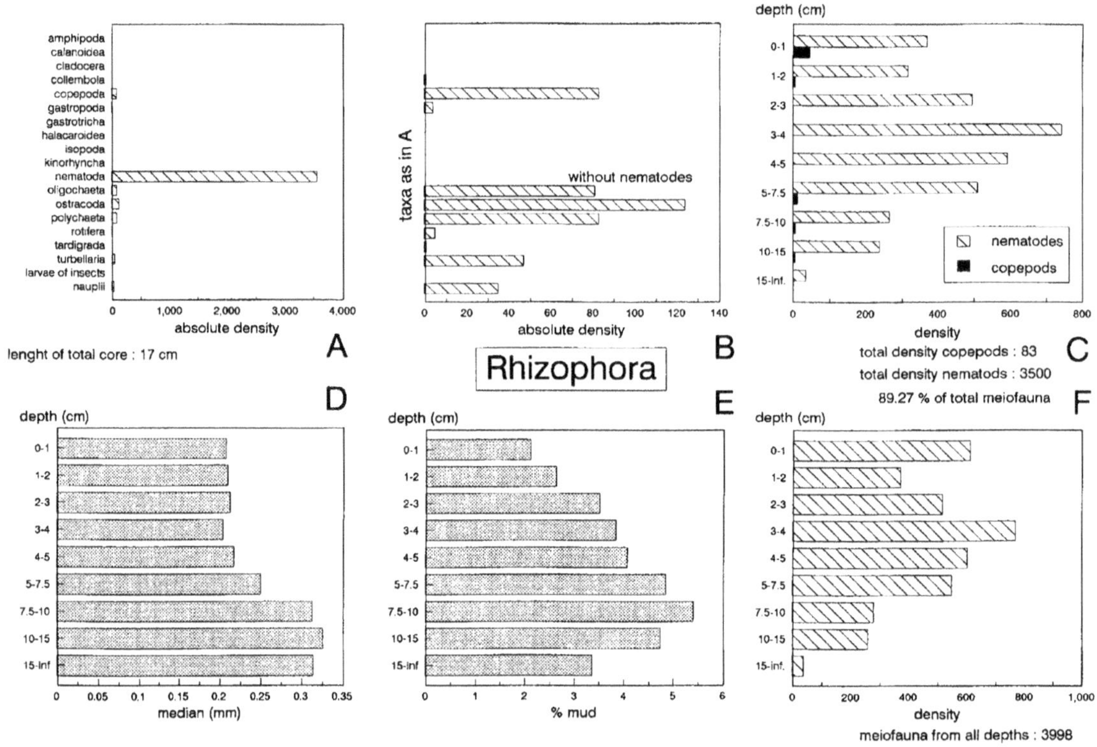

Fig. 5. Rhizophora site (a): density of all meiofauna taxa in the whole core (ind./10 cm^2). (b): density of the meiofauna, excluding the nematodes (ind./10 cm^2) (c): depth distribution patterns of densities of nematodes and copepods (d): depth profiles of median grain size of the send fraction (in mm) (e): depth profile of the percentage of mud (%) (f): depth distribution of the total densities of meiofauna.

imum of 325 ind. at 3–4 cm. Nematodes and copepods are still present in high numbers in the deepest layer (342 and 11 ind.), though other taxa (oligochaetes, polychaetes and rotifers) also remain present and account for 17% at this depth where a total of 759 individuals is reported.

Ceriops (Fig. 4 C, F)

The entire core is characterized by lower numbers. Nematodes dominate and are most frequent in the uppermost and deepest core layers; 342 nematodes are still present at a depth of 15–22.5 cm. Copepods only reach 4% of nematode numbers. Sampling deeper than the present depth would give a better idea of how density evolves.

Rhizophora (Figs 5 C, F)

The general aspect of high nematode abundances is also present (mean value: 89%). Copepods are only abundant in the first sediment layer (47 adults and 20 nauplii), but this number is low in comparison with the 371 nematodes at this depth. The remaining 32% of the total meiofaunal density is mainly made up by ostracods (19%). In sediment layers deeper than 2 cm, nematodes constitute more than 90% of the total meiofauna in each interval, and even 98% at 4–5 cm. In the deepest layer total density drops a minimum of 37 individuals.

Sonneratia (Figs 6 C, F)

A typical profile of total density, which is again similar to the figure of nematode distribution,

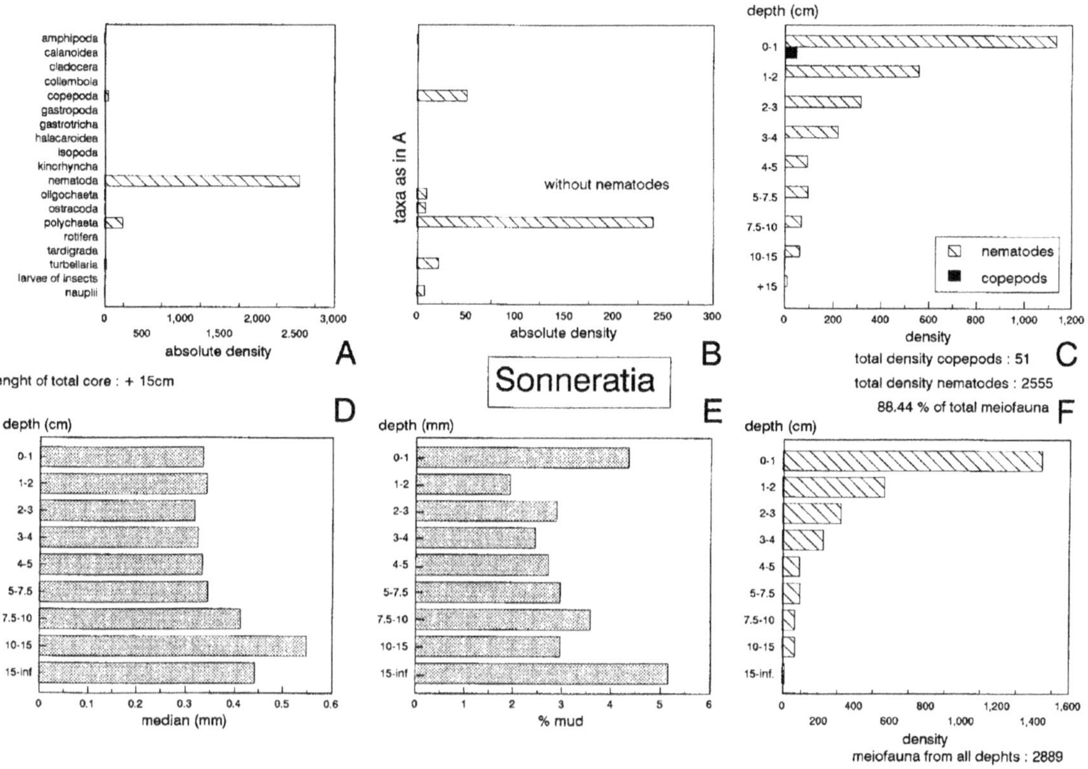

Fig. 6. Sonneratia site (a): density of all meiofauna taxa in the whole core (ind./10 cm^2). (b): density of the meiofauna, excluding the nematodes (ind./10 cm^2) (c): depth distribution patterns of densities of nematodes and copepods (d): depth profiles of median grain size of the sand fraction (in mm) (e): depth profile of the percentage of mud (%) (f): depth distribution of the total densities of meiofauna.

shows high densities in the upper layer (1443 ind.; 50% of total density), but a very quick drop deeper in the sediment. Copepods only occur in the first layer (48 ind.).

In conclusion. Although densities tended to be higher in the upper layers, several taxa were distributed in high densities in deeper sediment intervals as well. Regarding the deepest interval *Sonneratia* has the lowest amount of individuals (2%), followed by *Avicennia* (3.2%), *Rhizophora* (6.5%), *Ceriops* (8%) and *Bruguiera* (11%). Both, nematodes and copepods live at sediment depths deeper than the one examined. Especially *Ceriops* shows a high nematode and copepod number in the deepest layer (15–22.5 cm: 342 nematodes, 11 copepods).

Sediment

Figures 2–6 (D, E) show the vertical profile of median sediment grain size and mud content. The mean median particle size of each core fluctuates between 0.266 mm (*Rhizophora*) and 0.468 mm (*Avicennia*); consequently all sediments are classified as medium sand. A constant median grain size vertical profile is shown for *Avicennia* sediments (except for the uppermost and deepest intervals), *Bruguiera* and *Ceriops*; *Rhizophora* and *Sonneratia* have an increasing trend in median grain size of the sand fraction with depth. The mud percentage of *Avicennia* and *Sonneratia* is higher in the top centimeter in comparison with deeper intervals. Unlike the stable median grain size, mud content shows larger variation, with highest values at 0–1 cm and 15-inf. cm (*Sonneratia*). Higher mud percentages are also reported in the deepest interval of *Ceriops*. A clear increase up to a depth of 7.5–10 cm (with 5.41% mud), is followed by a decrease of the mud content in deeper layers for *Rhizophora* mud content.

Distribution of meiofauna with regard to sediment type

The taxon composition of the meiofauna in the different vegetation sites is mostly related with sediment grain size, and probably to some extent with oxygen conditions. Sediments with higher mud content are mainly inhabited by burrowers such as oligochaetes, nematodes and turbellarians. The sediment of *Bruguiera* contains a low mud percentage, and consequently shows a higher diversity of taxa. *Rhizophora* and *Sonneratia*-samples, with finer sediments, are mainly composed of nematodes. Small meiofauna such as rotifers, gastrotrichs, nauplii and larvae of polychaetes are not linked with sediment characteristics. Apparently the activity of currents is responsible for the distribution of these animals. Copepods and epsilonematid nematodes do not withstand high silt fractions.

With the exception of *Bruguiera* no evident relationship with depth distribution is detected. The decreasing density with depth in *Sonneratia* is probably governed by low oxygen conditions.

Discussion

Sheltered regions with muddy sediments rich in detritus generally are characterised by low diversity, but high densities (Heip *et al.*, 1985; Coull, 1988; Giere *et al.*, 1988). In the present study mud percentages reached 5% and total densities up to 6707 ind. are recorded, though a high number of taxa (17) is found. Table 1 gives a summary of densities recorded in several studies on the meiobenthos of mangroves (range is between 36 and 12000 ind. 10 cm^{-2}); values of Gazi are in between the recorded values.

Table 1. Densities of meiofauna in mangroves (ind./10 cm^2).

	Density	Author
S. Cuba	36–245	Lalana-Rueda & Gosselck, 1980
S.E. Cuba	341	Herrera & del Valle, 1980
N.E. Australia	57–2,454	Alongi, 1987 (a + b)
S.E. Australia	64–12,058	Hodda & Nicholas, 1985
S. Africa	840–5,300	Dye, 1983 (b)
E. Africa	1,976–6,707	This study

Nematodes were also the most abundant taxon in the surveys of Australian mangroves (Hodda and Nicholas, 1985; Nicholas *et al.*, 1991) and South African mangroves (Dye, 1983b). Nematodes seem to be less important in Cuban mangroves (Lalana-Rueda & Gosselck, 1986: 35.3 and 61.1%).

Alongi (1987a, b), Hodda & Nicholas (1985) and Dye (1983a, b) noted a low copepod density (respectively 1–22 ind. 10 cm^{-2}, under 10% and 30–80 ind. 10 cm^{-2}). Lalana-Rueda & Gosselck (1986) in contrast found the copepods to be the second most common taxon in terms of individual dominance.

Unlike most other studies, turbellarians were the most abundant meiobenthic group in the data of Alongi (1987a), probably due to the used extraction technique. Hodda & Nicholas (1985) only found the oligochaetes as abundant group next to nematodes and copepods, while Dye (1983a, b) recorded 6.4% ciliates, followed by oligochaetes (4.5%), turbellarians (3%) and kinorchychs (2.7%). Copepods, polychaetes, gastrotrichs, larvae of crustaceans and insects only appeared sporadically. The five most important taxa in the Toledo Lagoon were nematodes, copepods, Tanaidacea, oligochaetes and polychaetes and in the Basto Lagoon nematodes, copepods, ostracods, amphipods and polychaetes (Lalana-Rueda & Gosselck, 1986).

Among the different sampling sites, *Rhizophora* and *Sonneratia* sediments contained the highest abundance (respectively 89 and 95%) of nematodes, but also had the highest mud content (respectively 4 and 3%). Low copepod percentages, both in Gazi and the study areas cited above, suggest that this taxon is more related to coarser grain texture, and that nematodes show a broad tolerance to diverse sediments. Interstitial groups, e.g., Gastrotricha, Tardigrada are typically excluded from the muddy sediment. Sediments with a median particle size smaller than 125 mm are usually dominated by burrowers (Coull, 1988).

In muddy sediments, almost no life persist under a depth of 8 cm (McIntyre & Warwick, 1984). The present study, however, shows that high densities, especially of nematodes, occur beneath this level.

In a study of vertical distribution of meiofauna in mangrove sediments in Transkei, Dye (1983a) found also that meiofauna occurs to considerable depths, although highest densities occur in the upper 10 cm of the sediment. He calculated that when one restricts meiofauna sampling to the top 5 cm of the sediment only 40% of the total density can be obtained.

A recent study by Nicholas *et al.* (1991) confirms those findings: about 85% of the nematodes occur in the top cm of soft mud, but 5–7 species inhabit the deeper anoxic mud down to 10 cm.

Low oxygen conditions are supposed to be an important regulator for the low densities versus depth of *Sonneratia*. Coull (1988) asserts that vertical distribution patterns typically are controlled by the depth of the redox potential discontinuity (RPD) level, i.e., the boundary between aerobic and anaerobic sediments, which is primary determined by the 'super variable' oxygen. In this context Wieser *et al.* (1974) claim harpacticoid copepods are the most sensitive meiobenthic taxon to decreased oxygen (this in combination with the previously discussed sediment factor). Nematodes seem to be more tolerant to reducing circumstances (Boaden, 1975 in Hodda & Nicholas, 1985).

Fiddler crab *Uca* spp. are a prevalent aspect of the Kenyan mangroves (personal observation of the authors). DePatra & Levin (1989) found a fundamental influence of the *Uca* burrows on the abundance and distribution of several meiofaunal taxa in a North Carolina salt marsh. Both by passive deposition and enhancement of the micro environment, *e.g.*, food resources, stagnation of water (which provides suitable habitats for the development of young stages of several taxa), densities are higher in these 'biogenetic depressions'. On the other hand abundances tend to be lower due to the accumulation of toxic components and the fouraging activities of the fiddler crabs themselves. A study on their influence on energy flow in the mangrove ecosystem is not yet analysed, but a similar phenomenon is present in the mangroves of northeastern Australia, where the sesarmid crab *Sesarma messa* has a high impact on the weight loss from leaves by their leaf burying and consuming activities (Robertson, 1986), and in the mangroves of Queensland, where the crab *Sesarma erythrodactyla* enhances the decomposition of mangrove litter (Camilleri, 1989).

This pilot study of meiobenthic composition in the mangroves of Gazi Bay will be followed by a more intensive study investigating the quantity (in the light of tidal regime and seasons) and quality (identification of the most important taxa and trophic groups) of some ecological concepts like predation, consumption and interactions between macrofauna and meiobenthos.

Acknowledgements

The authors acknowledge the Kenyan Marine Fisheries and Research Institute in Mombasa (Kenya) for providing logistic facilities for the field work. Dr A. Vanreusel and Dr P. Aerts are thanked for the sampling assistance. This study is financed by the Belgian National Science Foundation (FKFO 2.0043.88) and the EEC project TS2-0240-C (GDF).

References

Alongi, D. M., 1987a. Intertidal zonation and seasonality of meiobenthos in tropical mangrove estuaries. Mar. Biol. 95: 447–458.

Alongi, D. M., 1987b. Inter-estuary variations and intertidal zonation of free-living nematode communities in tropical mangrove systems. Mar. Ecol. Prog. Ser. 40: 103–114.

Alongi, D. M., K. G. Boto & F. Tirendi, 1989. Effects of exported mangrove litter on bacterial productivity and dissolved organic carbon fluxes in adjacent tropical nearshore sediments. Mar. Ecol. Prog. Ser. 56: 133–144.

Andersen, F. O. & B. T. Hargrave, 1984. Effects of *Spartina* detritus enrichment on aerobic/anaerobic benthic metabolism in an intertidal sediment. Mar. Ecol. Prog. Ser. 16: 161–171.

Benner, R. & R. E. Hodson, 1985. Microbial degradation of the leachable and lignocellulosic components of leaves and wood from *Rhizophora mangle* in a tropical mangrove swamp. Mar. Ecol. Prog. Ser. 23: 221–230.

Boaden, P. J. S., 1975. Anaerobiosis, meiofauna and early metazoan evolution. Zool. Scr. 4: 21–4.

Boaden, P. J. S. & R. Seed, 1985. An introduction to Coastal Ecology, Chapter 7, Blackie USA: Chapman & Hall, N.Y., 218 pp.

Boto, K. G., J. S. Bunt & J. T. Wellington, 1984. Variations in mangrove forest productivity in northern Australia and Papua New Guinea. Estuar. coast. Shelf Sci. 19: 321–329.

Buchanan, J. B., 1984. Sediment analysis. In N. A. Holme & McIntyre (eds), Methods for the Study of Marine Benthos. IBP No 16, 2nd edn. Blackwell Scientific Publications, 387 pp.

Camilleri, J., 1989. Leaf choice by crustaceans in a mangrove forest in Queensland. Mar. Biol. 102: 453–459.

Coull, B. C., 1988. Ecology of the marine meiofauna. In R. P. Higgins & H. Thiel (eds), Introduction to the study of meiofauna: 18–38.

Cundell, A. M., M. S. Brown, R. Stanford & R. Mitchell, 1979. Microbial degradation of *Rhizophora mangle* leaves immersed in the sea. Estuar. coast. mar. Sci. 9: 281–286.

Decraemer, W. & A. Coomans, 1978. Scientific report on the belgian expedition to the Great Barrier Reef in 1967. Nematodes XII. Ecological Notes on the Nematode Fauna in and around Lizard Island. Aust. J. mar. Freshwat. Res. 29: 497–508.

DePatra, K. D. & L. A. Levin, 1989. Evidence of the passive deposition of meiofauna into fiddler crab burrows. J. exp. mar. Biol. Ecol. 125: 173–192.

Dye, A. H., 1983a. Vertical and horizontal distribution of meiofauna in mangrove sediments in Transkei, Southern Africa. Estuar. coast. Shelf Sci. 16: 591–598.

Dye, A. H., 1983b. Composition and seasonal fluctuations on meiofauna in a southern african mangrove estuary. Mar. Biol. 73: 165–177.

Dye, A. H. & T. A. Lasiak, 1986. Microbenthos, meiobenthos and fiddler crabs: trophic interactions in a typical mangrove sediment. Mar. Ecol. Prog. Ser. 32: 259–264.

Gallin, E., E. Coppejans & H. Beeckman, 1989. The mangrove vegetation of Gazi Bay (Kenya). Bull. Soc. r. Bot. Belg. 122: 197–207.

Gerlach, S. A., 1958. Die Mangroveregion tropischer Küsten als Lebensraum. Z. Morph. Ökol. Tiere 46: 636–730.

Giere, O., A. Eleftheriou & D. J. Murison, 1988. Abiotic factors. In R. P. Higgins & H. Thiel (eds), Introduction to the Study of Meiofauna: 61–78.

Heip, C., M. Vincx & G. Vranken, 1985. The ecology of marine nematodes. Oceanogr. Mar. Biol. ann. Rev. 23: 399–489.

Herrera, A. & R. del Valle, 1980. Caracteristicas de la meiofauna bentósica en la laguna y zona costera de Guayabal en relación con el grado de contaminacion. Ciencias Biológicas 5: 29–45.

Hodda, M. & W. L. Nicholas, 1985. Meiofauna associated with mangroves in the Hunter River Estuary and Fullerton Cove, south-eastern Australia. Aust. J. mar. Freshwat. Res. 36: 41–50.

Lalana-Rueda, R. & F. Gosselck, 1986. Investigations of the Benthos of Mangrove Coastal Lagoons in Southern Cuba. Int. Revue ges. Hydrobiol. 71: 779–794.

Lugo, E. L. & S. C. Snedaker, 1974. The ecology of mangroves. Annu. Rev. Ecol. Syst. 5: 39–64.

McIntyre, A. D. & R. M. Warwick, 1984. Meiofauna techniques. In N. A. Holme & McIntyre (eds), Methods for the study of marine benthos, Blackwell Scientific Publications: 217–244.

Nicholas, W. L., J. A. Elek, A. C. Stewart & T. G. Marples, 1991. The nematode fauna of a temperate Australian mangrove mudflat; its population density, diversity and distribution. Hydrobiologia 209: 13–27.

Odum, W. E., 1988. Comparative ecology on tidal fresh water and salt marshes. Annu. Rev. Ecol. Syst. 19: 147–176.

Okemwa, E. N., R. K. Ruwa & P. Polk, 1986. The autecology of the edible oyster *Crassostrea cucullata* Born, 1778: size related vertical distribution at Mkomani, Mombasa. Kenya J. Sci. B 7: 9–14.

Robertson, A. I., 1986. Leaf-burying crabs: their influence on energy flow and export from mixed mangrove forests (*Rhizophora* spp.) in northeastern Australia. J. exp. mar. Biol. Ecol. 102: 237–248.

Ruwa, R. K., 1990. The effects of habitat complexities created by mangroves on macrofaunal composition in brackishwater intertidal zones at the Kenyan coast. Discovery and Innovation 2: 49–55.

Ruwa, R. K. & P. Polk, 1986. Short communication. Additional information on mangrove distribution in Kenya: some observations and remarks. Kenya J. Sci. B 7: 41–45.

Steinke, T. D. & C. J. Ward, 1988. Litter production by mangroves. II. St. Lucia and Richards Bay. S. Afr. J. Bot. 54: 445–454.

Twilley, R. R., A. E. Snedaker & C. Pattersen-Zucca, 1986. Litter production and turnover in basin mangrove forests in southwest Florida. Ecology 67: 670–683.

Wieser, W., J. Ott, F. Schiemer & E. Gnaiger, 1974. An ecophysiological study of some meiofauna inhabiting a sandy beach at Bermuda. Mar. Biol. 26: 235–249.

Woodroffe, C. D., 1987. Pacific Island mangroves: distribution and environmental settings. Pacific Science 41: 1–4.

Hydrobiologia **247**: 109–117, 1992.
V. Jaccarini & E. Martens (eds), The Ecology of Mangrove and Related Ecosystems.

Ecomorphology of *Crassostrea cucullata* (Born, 1778) (Ostreidae) in a mangrove creek (Gazi, Kenya)

J. F. Tack,[1] E. Vanden Berghe[2] & Ph. Polk[1]
[1] *Free University of Brussels, Laboratory of Ecology and Systematics, Pleinlaan 2, 1050 Brussels 5, Belgium;* [2] *University of Nairobi, Dept. of Zoology, P.O. Box 30197, Nairobi, Kenya*

Key words: Crassostrea cucullata, Fourier analysis, shell form, shell length, regression analysis, environmental factors

Abstract

Variations of shell form and shell length were studied for oysters growing in the mangroves of Gazi Creek, Kenya, and related to different environmental factors.

For the study of the form, Fourier analysis was performed on the circumference of 85 oysters. The resulting coefficients were compared among specimens using cluster analysis. The correspondence between this classification and substrate diameter is virtually perfect (only one misclassification out of 85 oysters). For the clusters based on height above chart datum, 9 specimens, all on intermediate height levels, were misclassified. Orientation with respect to tidal current had 13 misclassifications. Mangrove species seemed to influence form only marginally, if at all.

In the study of the size of *Crassostrea cucullata*, the length of 956 oysters, growing along two transects were measured, and correlated with several environmental factors. Oyster length was not related to substrate diameter or its orientation with respect to the main current. Length was not influenced by density up to a cover of 70%. For densities higher than 70%, there was a fairly strong negative correlation ($r^2 = 0.634$, $n = 217$). Length was not correlated with height above bottom (base of the tree) for heights lower than 20 cm, while oysters growing closer to the bottom were smaller. The correlation with height above chart datum was negative but very low ($r^2 = 0.060$, $n = 957$). However, if all measurements of oysters closer than 20 cm to the bottom, and all from a density of more than 70% cover are deleted from the data set, the correlation with height increased dramatically, the slope still being negative ($r^2 = 0.859$, $n = 543$).

Introduction

Crassostrea cucullata (Born, 1778) is an edible oyster found in the upper littoral zone, following the terminology of Lewis (1964) and Hartnoll (1976). It is found on the trunks and the stilt roots of mangrove plants and rocky substrata in brackish marine environments. Day (1974) noted the zoogeographical distribution of this species in the western part of the Indian Ocean. Ecological studies of this species have been carried out over the entire region: Seychelles (Taylor, 1968), Aldabra (Taylor, 1970), Tanzania (Hartnoll, 1976), Somalia (Chelazzi & Vannini, 1980) and Kenya (Ruwa, 1984; Okemwa *et al.*, 1986).

Previous research already suggested a relationship between the form of this oyster and the substrate. Kesteven (1941) stated that the form of the oyster is partly determined by the form of the substrate. Van Someren & Whitehead (1961) distinguished different categories of form of oysters, and correlated these forms with different types of habitats, and with different types of substrate.

These observations were however, qualitative. The first quantitative work was done by the above authors. They compared oysters growing on different artificial substrates by Fourier analysis. Their conclusion was that the diameter of the substrate has an influence on the form of the shell of *Crassostrea cucullata*.

Various studies show intraspecific differences in size of molluscs in relation to environmental parameters (Comfort, 1946; Ebling *et al.*, 1962; Vermeij, 1972, 1973; Ruwa & Brakel, 1981). Okemwa *et al.* (1986) showed that such differences also exist in the mangrove oyster *Crassostrea cucullata*, growing on a rocky substratum. The present investigation was undertaken to study the effects of different environmental factors on the size of this oyster growing in a mangrove ecosystem. The results of this study may be of immediate importance to the culture of *Crassostrea cucullata*.

Material and methods

Observations were made between July and September 1989 in a small creek near Gazi, a small village 49 km south of Mombasa (Kenya). Seven species of mangrove trees occur at Gazi (Gallin *et al.*, 1989): *Avicennia marina* (Forsk.) Vierh., *Bruguiera gymnorrhiza* (L.) Lam., *Ceriops tagal* (Perrotet) C. B. Robinson, *Lumnitzera racemosa* Willd., *Rhizophora mucronata* Lam., *Sonneratia alba* J. Smith and *Xylocarpus granatum* Koenig. An elaborate description of this creek can be found in Tack (1990).

The environmental factors

Orientation with respect to the main tidal current was taken as either perpendicular or parallel. Perpendicular was further subdivided into facing the inflowing current or facing the outflowing current. The diameter of the branch serving as a substrate was measured with calipers to the nearest mm. The height above bottom was measured to the nearest cm.

To measure approximate density, we made vertical transects on each of the 10 cm vertical bands. These transects were made in four directions (facing North, East, South and West). The approximate density was expressed as the percentage of the vertical transect that was covered by oysters.

Height above chart datum was determined following the method of Okemwa *et al.* (1986). In this method, the highest level the water reaches during the tidal cycle is marked. The height of this level is taken to be the same as the one given in the tide table for the same tidal cycle, and used as an absolute reference for the rest of the measurements. From this reference, height above chart datum was determined to the nearest cm.

Shell form

Sampling

85 oysters were collected in a restricted area of 10 × 10 m. This area was chosen for the availability of all environmental conditions of interest. 'Collection' was done by making a photograph of the oyster. These photographs were taken so that the oyster filled only 1/9 of the surface of the image, to avoid distortion.

The sample taken was stratified with respect to environmental parameters. Only oysters that were growing isolated (i.e. not touching any other oyster) were included in the sample. In this way, crowding did not interfere with the analysis: all oysters included in the sample had the opportunity to grow to their full size.

The environmental parameters studied were (1) species of the mangrove tree serving as a substrate, (2) diameter of the branch or pneumatophore on which the oyster was growing, (3) height above chart datum and (4) orientation with respect to the main tidal current.

Four species of mangrove trees were present in the sampling site. The number of oysters taken from these species were more or less equal: 21 from *A. marina*, 22 from *B. gymnorrhiza*, 23 from *R. mucronata* and 19 from *S. alba*. 28 oysters were growing parallel with the current, 29 were facing the inflowing current, 28 the outflowing current.

The diameter of the substrate, a continuous variate, was dichotomised by taking oysters only growing either on small branches (less than 30 mm diameter) or large branches (more than 100 mm diameter). The sample included 58 oysters of small branches, and 27 oysters of large branches.

Height above chart datum varied between 2.20 and 3.40 m; all heights are represented in the sample (Fig. 1). It was not possible to dichotomise this variate, as was done for the diameter of the substrate. Not enough oysters were present with the right combination of environmental parameters.

The photographs of the oysters were magnified 10 times, and redrawn on paper. These drawings were digitised using a Complot series 7000 digitiser, coupled to an IBM AT. The resolution of the digitiser was 0.025 mm; points were taken at intervals of 0.8 mm around the perimeter. The average length of the digitised circumference was about 1 m.

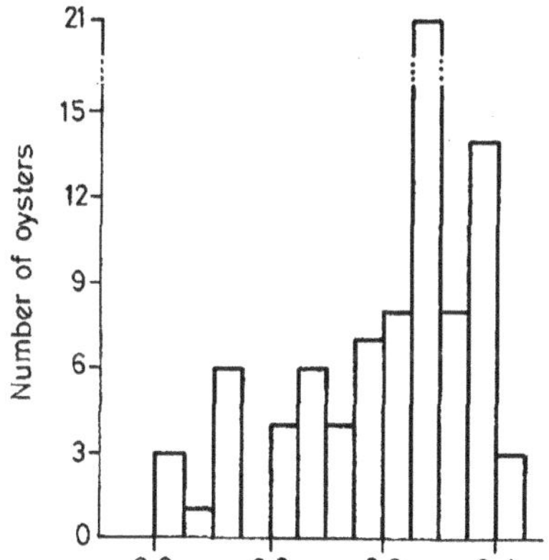

Fig. 1.

Fourier analysis

The form of closed contours can be quantified using Fourier analysis, by considering the polar coordinates of the points of the contour, with the origin of the polar coordinates chosen within the contour (e.g. Anstey & Delmet, 1972, 1973; Christophers & Waters, 1974; Ehrlich & Weinberg, 1970; Healy-Williams & Williams, 1981; Kaesler & Waters, 1972; Younker & Ehrlich, 1977). This method was also used to describe the form of oysters by the authors.

A limitation of the method involving polar coordinates is that only those forms where there is only one intersection between the boundary and a ray from the origin of the polar coordinates can be analysed. The form of most of the oysters is quite irregular, and does not satisfy this condition. Hence another method, the elliptical method, has been used in this study. For this method, both x and y coordinates of the boundary are interpreted as a function of distance along the boundary. The two resulting functions are periodical, with a period given by the length of the boundary, and can be separately analysed using the Fourier transform (Ferson *et al.*, 1985; Kuhl & Giardini, 1982).

The analysis was performed on a IBM AT. Before performing the transformation, a total of 64 equally spaced points are interpolated between the digitised points. For each of the forms to be analysed, 32 harmonics are calculated. Preliminary tests showed that, with this number of harmonics, all features of the form of the circumference could be described accurately.

Cluster analysis

The coefficients resulting from the Fourier transform were entered in a cluster analysis, to detect groups of oysters of similar form.

Ward's method (Ward, 1963; Ward & Hook, 1963) was used for clustering. In this method, the distance measure is the squared Euclidean distance. The method of aggregation is a modification of the weighted average method; the dispersion of the elements of a cluster around its centre of gravity is taken into account when determining which clusters are to be combined.

The cluster analysis was performed on a Macintosh computer, using the program SPSS/PC. The distances between different groups are given in arbitrary units. Since only relative distances are important here, this does not interfere with the interpretation of the results.

Shell length

Observations were made on *C. cucullata* growing on the trunks and stilt roots of mangrove trees covering two areas (4 × 20 m) on opposite sides of the main channel. 20 trees were growing in our study area: 12 *R. mucronata*, 2 *A. marina* and 6 *S. alba*.

Each trunk and stilt root was divided into consecutive 10 cm vertical bands starting from the bottom of the tree trunk. In each vertical band we collected, if present, an oyster growing on the northern, the eastern, the southern and the western side of the trunk or stilt root.

The shell length of all collected oysters was measured with calipers to the nearest 0.1 mm. Length was defined as the distance from the centre of the hinge to the farthest point of the opposite end. For each of the measurements, a number of environmental variables were recorded: (1) species of the mangrove tree and diameter of the branch serving as a substrate, (2) approximate density of the oysters, (3) height above the bottom (base of the tree trunk), (4) height above chart datum, and (5) orientation with respect to the main tidal current.

A regression analysis was performed to try to correlate length with each of these variables separately. These variables were taken as the main factors to study in relation to oyster length.

Results and discussion

Shell form

The dendrogram resulting from the cluster analysis is shown in Fig. 2. In this figure, each of the oysters is represented by a four-letter code, indicating the environmental conditions in which the oyster was found, as explained in the caption to this figure. Oysters less than one unit distance apart are displayed on one line.

In this dendrogram, a first division of the sample into two groups of more or less equal size occurred on the distance level 25. These two groups coincided with oysters that occurred high above datum or low above datum. Oysters growing in intermediate levels (between 3.05 and 3.13 m above datum) occurred in both groups of the cluster analysis. If the limit for discriminating 'High' and 'Low' is set at 3.07 for oysters in the field, 9 misclassifications occurred (i.e. oysters that were actually growing low in the tidal zone, but were classified with the oysters growing high in the tidal zone by the cluster analysis). A contingency table was constructed, where one of the nominal variables is the actual characteristic of the specimen, and the other the predicted characteristic based on the membership to a group in the cluster analysis. This procedure permitted the calculation of a Chi square value, indicating the significance of the association between observed and predicted characteristics. The Chi square value and its probability are given in Table 1.

Table 1 also gives the value for Lambda, Goodman and Kruskal's measure of predictive ability (Everitt, 1977). This coefficient is a measure of association between two nominal variables, and ranges from 0 to 1 (0 indicating no association, 1 a perfect association). The significance of Lambda is the same as the significance of the contingency table on which it is based.

Table 1. Statistics of the number of misclassifications. The first three tests are based on a 2 × 2 table of predicted and actual group membership. The last two lines are based on respectively a 16 × 4 and a 16 × 2 table of frequency of occurrence of the different values for the tree species in the 16 different groups of Fig. 2.

Environmental parameter	Chi square	*p*	Lambda
Height above datum	49.69	<0.001	0.84
Substrate diameter	75.99	<0.001	0.96
Orientation	31.69	<0.001	0.49
Mangrove species	64.6	<0.029	0.21
Avicennia or not	34.0	<0.003	0.11

Fig. 2. Results of cluster analysis. Distance units are arbitrary. Each four-character code represents one oyster. Group characteristics were determined on the basis of majority. a and b are considered together, and written as 'a' in the group characteristics. Upper case characters are misclassifications (not for tree species). Oysters on one line are less than 1 unit distance apart.

Each of the two above mentioned groups were separated into two subgroups, on distance levels of 9 and 14 respectively. These two divisions coincided with the oysters growing on small or on large diameters of substrate. Only one misclassification occurred for this criterion. Statistics for this division (Chi square, p and Lambda) are also given in Table 1.

The interpretation for the two other environmental parameters was not as straightforward, in the sense that there was no strict dichotomy subdividing each of the groups on levels lower than the ones discussed above. However, it was still possible to assign a group characteristic to each of the groups of the cluster analysis with respect to orientation, if only two categories are considered (parallel or perpendicular). If the group characteristic is taken to be the value of the majority in the group, 13 misclassifications occurred. As can be seen from Table 1, this is still a highly significant result, though the value for the index of predictive ability was much lower than in the two cases discussed above. The two values for perpendicular (facing outflowing or facing inflowing current) seemed to have no influence whatsoever: these values occurred together in most of the groups.

For the last environmental parameter, the species of the mangrove tree, no group characteristic could be determined, so it was not possible to construct a contingency table as before. If the randomness of the distribution of the actual value for the mangrove species over the different groups is tested using a 4×16 table, the value of Chi square turns out to be just significant. However, in view of the fact that most of the expected frequencies are very low, not too much importance should be attached to this value.

Oyster form is non-randomly distributed with regard to *Avicennia*, i.e., if a group of related forms contained an oyster growing on *Avicennia*, the other oysters of the group had a higher probability of having been collected on *Avicennia* as well. This can be tested by constructing a 2×16 contingency table, where the two categories are respectively non-*Avicennia* and *Avicennia*. The Chi square value for this table is highly significant.

It is clear from these results that the form of the oysters is influenced by both height above datum and diameter of the substrate. Height above datum had the most important influence, as reflected in the fact that the two groups were separated at the highest level of distance. Differences of form resulting from differences of diameter of the substrate were smaller. The number of misclassifications for this last environmental parameter was less in this study, but this was caused by the way in which the sampling was done (continuous sampling throughout the range for height above datum, versus sampling on the extremes for substrate diameter).

The influence of substrate diameter was already demonstrated by a preliminary study, using analogue methods on a small sample of oysters growing on artificial substrates. The influence of the height above datum on size was demonstrated earlier (Okemwa *et al.*, 1986), but no discussion of the influence of this parameter on the form had been described up to now.

Orientation with respect to main tidal current seemed to have a weaker, but still highly significant influence on the form. The species of mangrove tree had no apparent effect on the form of the oysters, if the four species were taken separately. If *Avicennia* was considered separately, the

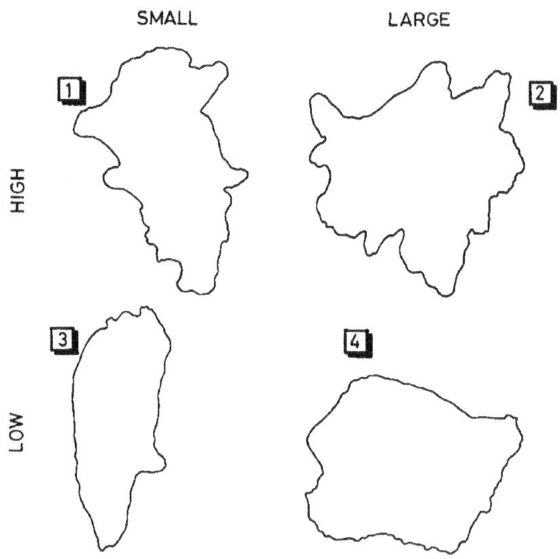

Fig. 3. Outlines of shells of oysters growing in different environmental conditions. 1: small diameter substrate, high above datum; 2: large diameter substrate, high above datum; 3: small diameter substrate, low above datum; 4: large diameter substrate, low above datum.

distribution over the different groups of the cluster analysis was clearly non-random. This would imply a different form for the oysters growing on *Avicennia*, as compared to the other species. This does not come as a complete surprise: the density of oysters was much lower on this species of mangrove tree. The mechanism of this phenomenon remains to be investigated, but it is not at all impossible that the same factor would also influence the form of the oysters. This factor could be

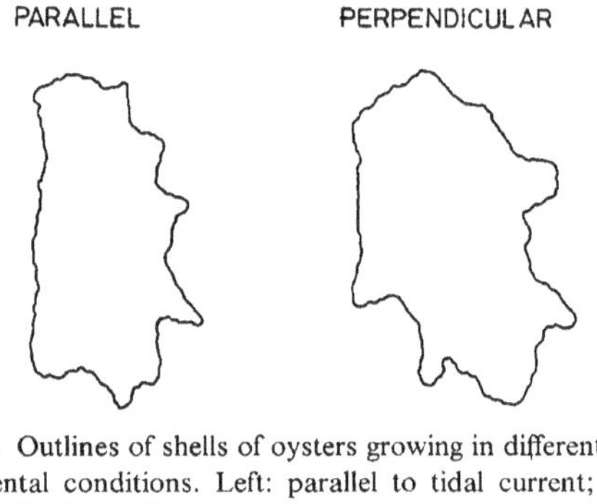

Fig. 4. Outlines of shells of oysters growing in different environmental conditions. Left: parallel to tidal current; right: perpendicular to tidal current.

a mild oyster repellent, of the texture of the bark of *Avicennia*, that is smoother than that of the other mangrove species.

Shell length

The diameter of the branch serving as substrate and the orientation with respect to the main tidal current appeared to be of no importance for shell length. Shell length was not related to density up to a cover of 70%. For densities higher than 70% there was a fairly strong negative correlation ($r^2 = 0.634$, $n = 217$), as one would expect (Fig. 5).

Length does not seem to be correlated with height above bottom for heights higher than 20 cm. At a height lower than 20 cm above the bottom, the oysters are clearly shorter than at higher levels.

The relationship between shell length of *C. cucullata* and the height above chart datum is shown in Fig. 6. From this figure it is clear that when shell length is plotted against height above datum the y values are not independently and normally distributed for any given value of x. However, according to Ricker (1973) regression analysis is very robust to deviations from its underlying assumptions. Noting the fact that we have to interpret the following results with the necessary caution we can still calculate the regression parameters. The correlation between oyster length and height above chart datum is negative and very low ($r^2 = 0.060$; $n = 957$). If all measurements of oysters closer than 20 cm to the bottom, and

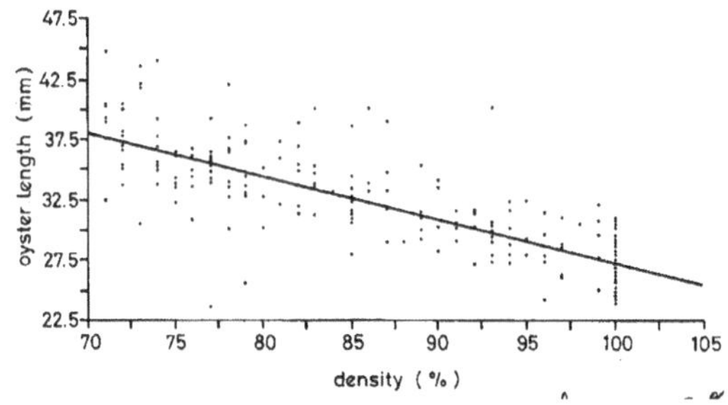

Fig. 5. Relationship between density (% cover) and oyster length (mm) for densities higher than 70% ($r^2 = 0.634$; $n = 217$).

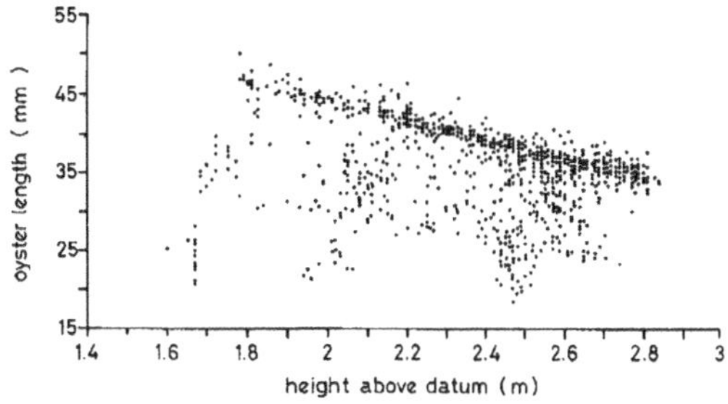

Fig. 6. Relationship between height above chart datum (m) and oyster length (mm). All measurements are included in this figure.

all from a density of more than 70% cover are deleted from the data set, the negative correlation increased dramatically ($r^2 = 0.850$; $n = 543$; Fig. 7).

The fairly strong negative correlation coefficients (r^2) demonstrate the influence of density on oyster length for densities higher than 70% cover. Other authors have demonstrated that density may affect growth rate and size in some littoral bivalves (Trevaillon *et al.*, 1970). This relationship (between size and density) is not found by Okemwa *et al.* (1986). They measured the shell length of *Crassostrea cucullata* on a rocky substratum and concluded that size is only a function of height above chart datum. However, in Okemwa *et al.*'s study, oyster density was smaller than 70% cover throughout. Combined with our results, this indicates that oyster density does not

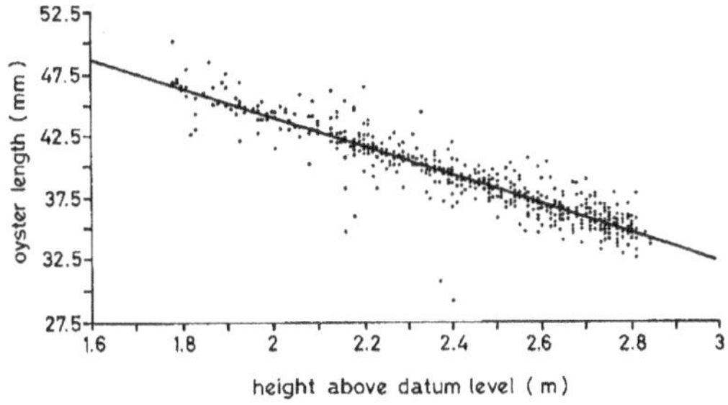

Fig. 7. Relationship between height above chart datum (m) and oyster length (mm) ($r^2 = 0.850$; $n = 543$). Measurements on oysters growing lower than 20 cm above the bottom and/or in areas with an oyster density higher than 70% cover are excluded from this data set.

affect oyster size for densities lower than 70% cover.

Shell lengths of *C. cucullata* are small at a height lower than 20 cm above the bottom as compared to oysters growing higher. This may be due to the higher sediment load closer to the muddy bottom. Because oysters are filter feeders, they may have to expend more energy in filtering the water at lower levels to get an equal amount of food as do the oysters growing at higher levels.

The correlation between oyster length and the height above datum is negative and very low. However, when we removed all oysters growing lower than 20 cm above the bottom, and all oysters growing on substrata with density greater than 70% cover, the correlation becomes very strong ($r^2 = 0.850$; $n = 543$). This high negative correlation coefficient demonstrates that *Crassostrea cucullata* exhibits size-related patterns in its vertical distribution with an up-shore reduction in shell length.

Conclusions

The form of the oysters is clearly related to both height above datum and diameter of the substrate. Changes of form associated with differences in height above chart datum were the most important, as reflected in the fact that the two groups were separated at the highest level of distance. Differences of form resulting from differences of diameter were smaller. The number of misclassifications for this last environmental factor was less in this study, but this can be attributed to the way in which the sampling was done (continuous sampling for height above datum, as opposed to dichotomised sampling for substrate diameter).

The distinction between these four groups can be related to the appearance of the oysters (Fig. 3). Oysters growing on small diameter substrates are more elongated than the ones growing on large diameter substrates. This is what one would expect: 30 mm is small compared to the size of the oyster. For the oyster to grow as wide on this type of substrate as on the larger diameter substrate, the oyster should grow with a large proportion of its shell detached from the substrate. This would interfere with the structural strength of the shell.

The outline of oysters growing higher above chart datum is 'wavier' than the lower growing oysters: their outline had more lobes, and the lobes were more pronounced. It is not clear what the reason for this difference is. Maybe it is related to the time available to the oysters for feeding: the higher above chart datum the oyster is growing, the shorter the time it will be submerged. A wavier outline will increase the size of the opening via which water can be taken in or expelled, and hence possibly increase the filtering rate. The wavier outline of the higher oysters could be a compensation for the shorter feeding time.

Orientation with respect to main tidal current seemed to have a weaker, but still highly significant correlation with form. Once again, this distinction is also apparent from a visual inspection of the outline of the oysters (Fig. 4). Oysters growing perpendicular to the tidal current often have lobes all around their outline, as opposed to oysters growing parallel to the current, that have lobes mainly on one side.

Height above chart datum, and thus percentage of time immersed, seems to be the primary factor determining the shell length of the oysters. This relationship is obscured by the influence of crowding and proximity to the bottom. Competition for space is the obvious mechanism by which crowding can influence the length of the oyster shell. The way in which proximity to the bottom is influencing oyster growth remains to be investigated, but a probable mechanism is the interference of high sediment load with the filter feeding of the oysters.

As with the correlation between form and height above chart datum, it seems logical to look for an explanation of the correlation between shell length and height in the time an oyster is submerged, and hence feeding time. Oysters growing lower have a longer proportion of the day available for feeding, hence can be expected to grow faster. If the maximum size an oyster can reach is dictated by age, rather than by size itself, an upshore reduction in size is the expected trend.

This study has only demonstrated the existence

of the relationship between environmental parameters on one hand, and form and size of the oysters on the other. A following step in this research should be the construction of a functional model, to investigate the causal relationship between morphology and the environmental parameters, and to investigate the hypotheses formulated above.

Acknowledgements

This research was conducted in the framework of the Kenyan-Belgian Project in Marine Sciences, at the Kenya Marine and Fisheries Research Institute (KMFRI) in Mombasa. We wish to thank both Mr. S. O. Allela and Dr. E. Okemwa, respectively former and current director, of this institute. We also thank Mr. R. K Ruwa for his cooperation.

References

Anstey, R. L. & D. A. Delmet, 1972. Genetic meaning of zooecial chamber shapes. Science 177: 1000–1002.

Anstey, R. L. & D. A. Delmet, 1973. Fourier analysis of zooecial shapes in fossil tubular bryozoans. Geol. Soc. Am. Bull. 84: 1753–1764.

Chelazzi, G. & M. Vannini, 1980. Zonation of intertidal molluscs on rocky shores of southern Somali. Estuar. coast. mar. Sci. 10: 569–583.

Christopher, R. A. & J. A. Waters, 1974. Fourier series as a quantitative descriptor of miospore shape. J. Paleontol. 48: 697–709.

Comfort, A., 1946. *Patella vulgata* L. relation of habitat to form. J. Conch. 22: 216–217.

Day, J. H., 1974. A guide to marine life of South African shores. Balkema. Cape Town, 300 pp.

Ebling, F. J., J. F. Sloane & H. M. Davies, 1962. The ecology of Lough Ine. XII. The distribution and characteristics of *Patella* species. J. anim. Ecol. 31: 457–470.

Ehrlich, R. & B. Weinberg, 1970. An exact method for characterisation of grain shape. J. Sed. Petrol. 40: 205–212.

Everitt, B. S., 1977. The analysis of contingency tables. Chapman and Hall. London, 128 pp.

Ferson, S., F. J. Rohlf & R. K. Koehn, 1985. Measuring shape variation of twodimensional outlines. Syst. Zool. 34: 59–68.

Gallin, E., E. Coppejans & H. Beeckman, 1989. The mangrove vegetation of Gazi Bay (Kenya). Bull. r. Soc. Belg. 122: 197–207.

Hartnoll, R. G., 1976. The ecology of some rocky shores in tropical East Africa. Estuar. coast. mar. Sci. 4: 1–21.

Healy-Williams, N. & D. F. Williams, 1981. Fourier analysis of test shape of planktonic foraminifera. Nature 289: 485–487.

Kaesler, R. L. & J. A. Waters, 1972. Fourier analysis of the ostracode margin. Geol. Soc. Am. Bull. 83: 1169–1178.

Kesteven, G. L., 1941. The biology and cultivation of oysters in Australia. I. Some economic aspects. Commonwealth of Australia. Melbourne, 32 pp.

Kuhl, F. P. & C. R. Giardini, 1982. Elliptic Fourier features of a closed contour. Computer Graphics and Image Processing 18: 236–258.

Lewis, J. R., 1964. The ecology of rocky shores. Hodder & Stoughton, London, 323 pp.

Okemwa, E. R., R. K. Ruwa & P. Polk, 1986. The autecology of the edible oyster *Crassostrea cucullata* Born, 1778: size-related vertical distribution at Mkomani, Mombasa. Kenya J. Sci. Ser. B 7: 9–14.

Ricker, W. E., 1973. Linear regressions in fishery research. J. Fish. Res. Bd Can. 30: 409–434.

Ruwa, R. K., 1984. Invertebrate faunal zonation on rocky shores around Mombasa, Kenya. Kenya J. Sci. Ser. B 5: 49–65.

Ruwa, R. K. & W. H. Brakel, 1981. Tidal periodicity and size-related variations in the zonation of the gastropod *Nerita plicata* on East African rocky shore. Kenya J. Sci. Ser. B 2: 61–67.

Tack, J. F., 1990. Ecomorfologie van de mangroveoester *Crassostrea cucullata* (Born, 1778). Unpublished M. Sc. thesis, Free university of Brussels, 99 pp.

Taylor, J. D., 1968. Coral reef associated invertebrate communities (mainly molluscan) around Mahe, Seychelles. Phil. Trans. r. Soc., Lond. Ser. B 254: 129–206.

Taylor, J. D., 1970. Intertidal zonation at Aldabra atoll. Phil. Trans. r. Soc., Lond. Ser. B 260: 173–213.

Trevaillon, A., R. R. C. Edwards & J. H. Steel, 1970. Dynamics of a benthic bivalve. In J. H. Steel (ed.), Marine Food Chains. University of California Press, Los Angeles: 285–295.

Van Someren, V. D. & J. P. Whitehead, 1961. An investigation of the biology and culture of an East African oyster, *Crassostrea cucullata*. Col. Off. Fish. Pub. 14: 1–36.

Vermeij, G. J., 1972. Intraspecific shore-level size gradients in intertidal molluscs. Ecology 53: 693–700.

Vermeij, G. J., 1973. Morphological patterns in high intertidal gastropods: adaptive strategies and their limitations. Mar. Biol. 20: 319–346.

Ward, J. H., 1963. Hierarchical grouping to optimise an objective function. J. am. Stat. Ass. 58: 236–244.

Ward, J. H. & M. E. Hook, 1963. Application of an hierarchical grouping procedure to a problem of grouping profiles. Educational and Psychological Measurement 23: 69–81.

Younker, J. L. & R. Ehrlich, 1977. Fourier biometrics: harmonic amplitudes as multivariate shape descriptors. Syst. Zool. 26: 336–342.

Hydrobiologia **247**: 119–120, 1992.
V. Jaccarini & E. Martens (eds), The Ecology of Mangrove and Related Ecosystems.

A note on the zooplankton distribution and diversity in a tropical mangrove creek system, Gazi, Kenya

Melckzedeck K. W. Osore
Kenya Marine and Fisheries Research Institute, P.O. Box 81651, Mombasa

Abstract

Gazi Creek is situated at some 40 km south of Mombasa (Kenya 4° 25′ S, 39° 50′ E) sampling stations are located in the creek mouth (stn 1), in the inner creek (stn 3) and intermediate (stn 2). Sampling is done twice a month; it starts from stn 1 through stn 2 up to stn 3. A 335 μm Mesh size net is towed in near surface water for 5 minutes and the collected sample preserved in 5% formaldehyde. Hydrographic parameters are recorded during sampling.

The work at Gazi is undertaken to survey the Zooplankton composition both qualitatively and quantitatively. 22 important taxa have been studied to determine their seasonal variation in abundance and distribution. Zooplankton population are highest in March (374 animals m^{-3}). The abundance gradually falls during the S.E. monsoon period (May–September) to the lowest value in August (30 animals m^{-3}) (see Fig. 1). Copepoda is the most abundant taxon throughout (48.5–92.4%) (see Table 1). This group is found throughout the creek. Nevertheless, it is evident that the creek mouth has a higher diversity of copepoda compared to the inner creek. Using the Margalef Index a consistently higher value has been observed in stn 1 compared to stn 3 (see Table 2). The surface water temperature decreases during the S.E. monsoon (28.0 to 25.5 °C) and it rises during the N.E. monsoon (29.0 to 35.5 °C). The trend in temperature variation corresponds broadly with zooplankton abundance (see Fig. 1), suggesting that zooplankton thrive best in warmer water. High Zooplankton counts were also observed around May (326 animals m^{-3}) during the long rains; probably due to high amounts of nutrients input. Generally, average monthly pH values vary only slightly but the pH up the creek is almost always lower than at the creek mouth. Salinity is quite constant at 35‰.

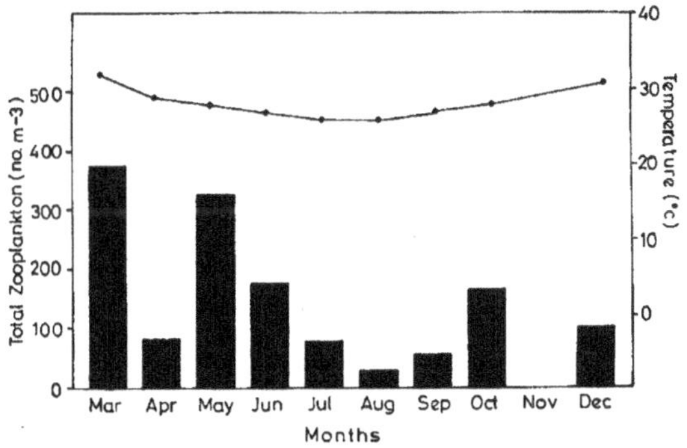

Fig. 1. Mean monthly total zooplankton variation and mean monthly surface water temperature for the three stations (average) at Gazi during March to December 1990.

Table 1. Percentage composition of important Zooplankton groups as average of the three stations at Gazi during March to December 1990.

Group	1990									
	Mar	Apr	May	Jun	Jul	Aug	Sept	Oct	Nov	Dec
Copepoda	83.36	79.10	62.56	76.49	75.05	76.85	48.49	92.42	–	84.26
Medusae	0.18	0.16	0.31	0.09	0.00	0.00	0.00	0.04	–	0.00
Brachyura zoea	0.62	11.51	18.85	1.16	5.04	2.50	26.87	0.91	–	2.02
Caridea	0.28	0.06	1.91	0.68	3.86	2.26	6.33	0.76	–	0.35
Chaetognatha	0.17	0.86	2.84	0.42	3.24	0.99	0.71	1.19	–	0.75
Ostracoda	0.13	0.58	0.16	0.09	0.28	0.82	0.63	0.19	–	0.27
Mollusca	0.29	1.55	1.68	5.08	4.86	8.41	3.10	0.70	–	1.33
Fish larvae	0.11	0.00	0.19	0.16	0.77	0.38	1.05	0.18	–	0.23
Nauplii	8.59	1.34	0.28	0.16	0.95	4.65	1.83	0.99	–	6.62
Amphipods	0.21	0.43	0.33	1.75	0.68	1.68	1.31	0.38	–	0.48
Appendicularia	0.29	2.08	0.85	0.43	0.68	0.00	0.00	0.63	–	0.11
Polychaeta	0.13	0.84	0.58	0.25	0.94	1.37	0.63	0.16	–	0.45
Isopoda	0.11	0.00	0.24	0.93	0.27	1.03	0.63	0.34	–	0.71
Fish eggs	0.10	0.41	8.06	0.43	1.92	0.82	6.14	0.18	–	1.18
Cumacea	0.16	0.06	0.00	0.34	0.50	0.72	0.26	0.25	–	0.38
Stomatopoda	0.02	0.00	0.06	0.00	0.00	0.27	0.71	0.16	–	0.00
Penaeidae	0.02	0.00	0.06	0.00	0.21	0.03	0.00	0.05	–	0.18
Cirripede nauplii	0.02	0.34	0.00	0.00	0.00	0.03	0.00	0.05	–	0.15
Brachyura megalopa	0.00	0.10	0.54	0.09	0.30	1.03	0.47	0.11	–	0.29
Siphonophora	0.00	0.06	0.64	0.09	0.18	0.00	0.00	0.00	–	0.00
Euphausiacea	0.00	0.00	1.19	0.00	0.00	0.00	0.00	0.00	–	0.00
Sergestidae	0.00	0.00	0.0	0.00	0.09	0.00	0.00	0.00	–	0.00
Others	0.05	0.32	0.00	0.00	0.00	0.03	0.54	0.18	–	1.50

Table 2. Monthly variation in diversity of Copepods in the three stations at Gazi during March to December 1990. The Margalef Index is used.

Month	Diversity		
	Stn 1	Stn 2	Stn 3
March	1.25	0.80	0.46
April	1.29	1.17	0.97
May	1.79	2.02	1.77
June	1.12	1.87	0.82
July	1.99	1.87	1.85
August	2.82	2.30	2.04
September	1.86	2.05	1.66
October	1.73	1.22	0.84
November	–	–	–
December	1.64	1.43	1.63

Hydrobiologia **247**: 121–132, 1992.
V. Jaccarini & E. Martens (eds), The Ecology of Mangrove and Related Ecosystems.

Phase shifts in coral reef communities and their ecological significance

T. J. Done
Australian Institute of Marine Science, PMB 3, Townsville MC, Queensland, 4810, Australia

Key words: coral reef, degradation, management, macroalgae, phase-shift

Abstract

Many coral reefs around the world have degraded to a degree that their present intrinsic value and utility are greatly reduced: (mass coral mortality followed by algal invasions; local depletions of reef fisheries; deficit of reef accretion compared to physical and biological erosion). Though we can sometimes identify proximal causes (outbreaks of coral predators and eroders; over-fishing; habitat destruction), we do not have a good understanding of how population, community and ecosystem structure and function differ in degraded from un-degraded reefs. The deficiencies in our understanding limit our ability to interpret the long-term significance of reef degradation, and therefore to develop scientifically based plans for conservation and management of reefs.

A particular course of action, or lack of action, based on uncritical acceptance of any of the various views of temporal variability can lead to further deterioration of specific reefs. None of these views – that reefs are either inherently robust, inherently fragile, or inherently resilient – is true over all time-space scales. This presentation reviews various models and case studies which suggest that reefs can be knocked precipitously or move slowly from one phase (coral-dominated) to another (coral-depleted and/or algal dominated). Transitions in the other direction ('recovery') involve changes (e.g. succession) in populations and communities (of all reef-associated biota, not just sessile benthos), and in reef function (e.g. community metabolism, trophodynamics) which are of great intrinsic interest but only poorly understood.

Introduction

We often hear coral reefs described in words something like these: oases of high diversity and productivity in oceanic deserts; builders of their own habitat; objects of great beauty. And yet just as frequently, it seems, we hear of yet another reef which has become 'degraded' in some way (e.g. Brown, 1987; Salvat, 1987). Reef degradation has many causes, both episodic disturbances of short duration (e.g. storms, outbreaks of coral predators; brief exposure to toxic pollutants; siltation caused by dredging), and chronic disturbances of long duration, frequently anthropogenic. Hatcher *et al.* (1989) reviewed anthropogenic effects on coral reefs under 9 categories: sedimentation; sewage pollution, thermal pollution, radioactive pollution; hydrodynamic influences; physical disturbance; extractive activities; introductions; tourism. The scale of reef degradation ranges from the immediate vicinity of a marina development or sewage outfall, to a significant proportion of the reefs within large systems such as the Great Barrier Reef (GBR) or the Florida Keys. For example, populations of the coral eating crown-of-thorns starfish *Acanthaster planci* seeded successive reefs in a north/south sequence over the decade 1980–90, causing serious damage to 65% of reefs surveyed in a 500 km section of the 2000 km long GBR (Moran *et al.*, 1988). By the time those in the south were attacked, there had been substantial coral recovery in the north (Done *et al.*, in prep.).

In this presentation, I discuss what degradation means. What are its symptoms and what is its significance in relation to 'normal' temporal variability in coral reef ecosystems?

The one universal symptom of coral reef degradation is mass coral mortality followed by an invasion of coral skeletons by algae. This is the 'phase shift' referred to in the title of this presentation. Associated with such a change, there may be significant depletions of reef fisheries, and a deficit of reef accretion compared to physical and biological erosion. Thus the reef is degraded in terms of its physical attractiveness, its fisheries yield and its ability to retain its integrity as a breakwater protecting adjacent shorelines. Clearly, the potential social and economic implications of reef degradation will depend on how long the reef remains degraded.

The following two quotations define some of the important ecological issues and points of disagreement about the significance of phase shifts in coral reefs. Referring to the entire Great Barrier Reef system (some 2900 reefs stretched over 2000 km), Davies (1988) says:

> 'Its uniqueness is ..... related to its robustness and its ability *always* to regenerate after catastrophes' (italics added).

Referring to algal invasions of previously coral dominated areas of individual reefs, Hatcher *et al.* (1989) state:

> 'Evidence is accumulating that coral reef benthic community structure can assume at least two stable forms: one coral-dominated and one macroalgae-dominated'.

A switch to macroalgal dominance which lasts years to decades constitutes a fundamental change in the benthic community structure of the reef with implications for the trophic structure of the reef and its resource value to human populations (Dahl, 1974). A more protracted switch (decades to centuries) has major implications for reef maintenance and growth, whenever calcium carbonate losses to physical and biological erosion are not replaced by calcifying organisms such as corals and coralline algae.

Ecological and human consequences of phase change

The implications that extensive changes from coral to algal dominance have for the reef's trophic structure and resource value to human populations have received little systematic study. Most or all the increased algal production may be dislodged by storms, thence being exported from the reef (Carpenter, 1990a) or entering benthic microbial communities on the reef (Moriarty *et al.*, 1985; Hansen *et al.*, 1987). In poorly flushed areas, accumulation of algal detritus could lead to excess biological oxygen demand and anoxic conditions inimical to survival of some higher reef biota (Johannes & Betzer, 1975). Not surprisingly, the worst reported community calcification rates on reef flats were recorded on coral/algal reef flats dominated by macroalgae – viz. at Tulear, Madagascar, and Moorea (Pichon & Morrisey, 1985; Sournia, 1976; both cited in Kinsey, 1985).

Increased macroalgal production may affect harvestable secondary production of herbivorous fish and invertebrates, and their fish predators. Fish production can be increased, either as an increase in population size (Carpenter, 1990b) or fish growth rates, in situations where food was previously limiting. However it may be unaffected in at least two circumstances: first, where food chains are based much more on turfs and/or pelagic zooplankton washed onto the reef's windward margin (Hamner *et al.*, 1988), than on macroalgae: and second, where fish community structure is strongly determined by recruitment and comparatively weakly determined by food availability (Doherty & Williams, 1989). In either case, the significance of the loss of live coral may relate much more to changes in habitat structure (Bell & Galzin, 1984) than to trophodynamic considerations. Effects on motile reef invertebrates may be equally unpredictable, and for the same sorts of reasons.

Time scales of decades to centuries are crucial in the context of a reef's ability to maintain its structural and biological integrity over geological time (Buddemeier & Hopley, 1988). It is true that coral reefs around the world have developed as and where they are, not despite, but because of, what Davies (*op. cit.*) refers to as catastrophes: i.e. continental drift, subsidence and sea-level change. It is impossible not to be impressed with these facts, and it is tempting to believe that coral reefs will persist despite anything human-kind may do to tropical seas. However our opinions – whether optimistic or pessimistic – should not be based only on the past 'track-record' of coral reefs. They should be based on an understanding of the ecological processes which determine whether in fact coral reefs, rather than algal-covered carbonate deposits derived from 'ex-reefs', can exist in environments created by an exponentially increasing human population.

In the remainder of this presentation, I refer to commonly used ecological 'models' relevant to the determination of benthic community structure on coral reefs. I then present some case studies which show how local ecological idiosyncracies of particular reefs – physical; nutritional; biological – can influence the course of benthic community development.

Models for community structure in reef corals and algae

Some characteristics of minimally disturbed coral communities on coral reefs are illustrated in Fig. 1. As time passes, the largest corals get larger; coral cover increases; β species diversity (scale of hectares) rises and asymptotes. These net increases occur despite localised colony mortality and recolonisation, which cause fluctuations in percent coral cover and α diversity (scale of square metres to decametres). The reviews of Connell

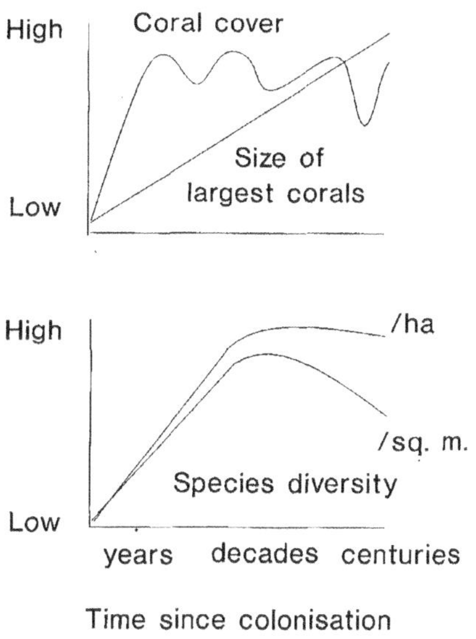

Fig. 1. Changes in some coral community attributes over whole reef scale versus increasing time since colonisation and without major disturbance. Partly after Connell (1978).

(1978) and Connell & Keogh (1985) discuss the implications of differences in scale, intensity and frequency of disturbance in relation to different life history strategies of corals.

Useful models for algal production and community structure on coral reefs have been provided by Littler & Littler (1985) and Steneck (1988). Spatial patterns of algal composition and production are strongly influenced by patterns of grazing, disturbance, and competition with corals, the potential upper levels of production being determined by ambient nutrient levels (Littler and Littler, *op. cit.*). There appears to be a systematic causative relationship between the intensity of grazing from scraping and denuding herbivores, and the structure of algal assemblages (Hixon, 1983; see also Fig. 2). It follows from Fig. 2 that if pressure of fishing substantially changes grazing intensity, a consequence of that fishing may be changes in biomass and community composition of the algal assemblage.

The types of changes which have been observed in coral reef benthic communities are shown in Fig. 3 (after Bazzaz, 1983). Reefs can be knocked

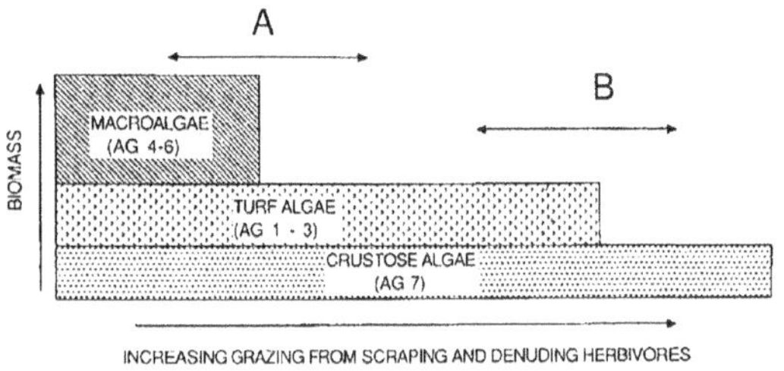

Fig. 2. Herbivore-induced changes in algal community structure, deduced from exclusion studies and natural herbivore mortalities. 'AG' refers to algal functional groups defined by Steneck (1988) as follows: 1 microalgae; 2 filamentous (simple); 3 foliose or sheet; 4 corticated or coarsely branched; 5 leathery; 6 articulated or jointed calcareous; 7 crustose. The most common herbivore-mediated shifts are between macroalgae and turfs (marked 'A') and between turfs and crusts (marked 'B'). Reproduced from Steneck (1988) with permission of the organizing committee of the Sixth International Coral Reef Symposium.

precipitously or move slowly from one phase (coral-dominated) to another (coral-depleted and/or algal dominated). For a period of time, algae become the dominant form of benthic cover. Disturbed areas can recolonise without passing through a macroalgal phase, or they may enter a macroalgal phase which they retain indefinitely, either because ongoing conditions favour algal growth over coral growth (Lighty, 1982; Lapointe, 1989) or because of a relatively brief period in which algal growth is enhanced, and is then sustained long after any measurable change in environment exists (Dollar & Grigg, 1981; Hatcher, 1984). Hatcher (*op. cit.*) lists the following examples of algal enhancing circumstances: (1) exclusion of grazers; (2) increases in availability of potentially limiting nutrients; (3) reduction of

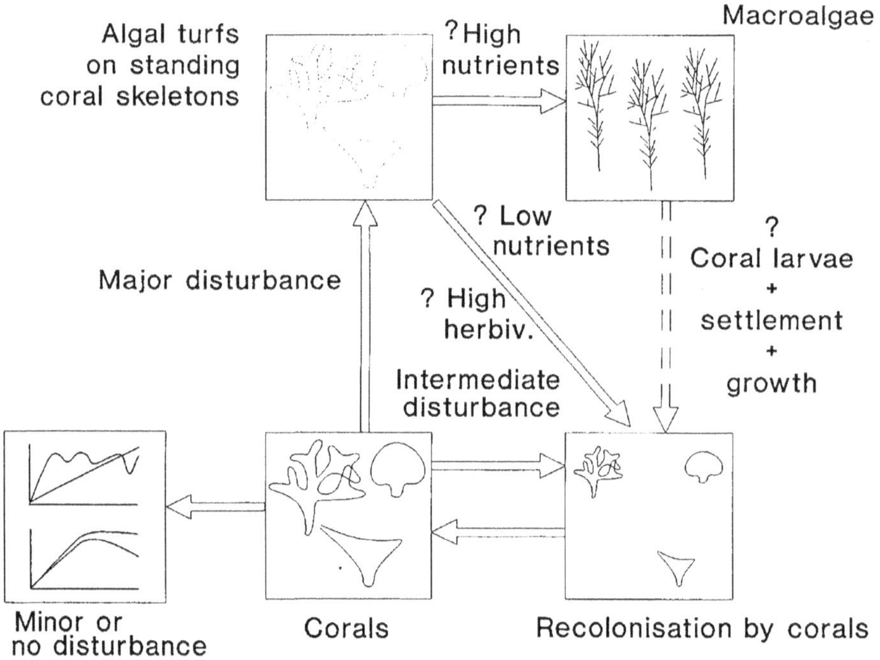

Fig. 3. Effects of different intensities of disturbance on coral reef benthic community structure (after Bazzaz, 1983). Coral communities which are undisturbed or subjected to minor disturbance (bottom left) undergo types of changes indicated in Fig. 1. After intermediate disturbance, the coral community nevertheless retains coral dominance, whereas major disturbance, by definition, causes a 'phase-shift' from coral dominance to algal dominance. The broken arrow indicates that reversion to coral dominance will not necessarily take place, and the question marks indicate location-dependant variables which will influence the trajectory of the benthic community.

competition with other benthic organisms (e.g. corals) by inhibiting their growth or killing them; (4) clearing or modification of large areas of substratum in a manner, or at a time which favoured colonisation by a formerly rare alga. Humans may be directly involved, either by affecting ambient nutrient or sediment loads, or by changing fish and invertebrate community structure by fishing.

Coral to algal phase transitions and the extent of coral recovery

The following six examples describe coral-algal(-coral) phase changes in Hawaii, Jamaica, Reunion, Moorea, and Australia. In some, the causes of the coral to algal shift are clear while in others, they are more conjectural. The examples show that there are no *a priori* reasons to expect that corals will necessarily re-establish populations in areas they dominated previously. Recovery of coral dominance requires a substantial supply of propagules (larvae, fragments), followed by prolific settlement and growth. Any given reef or part of a reef may be 'starved' of water-borne propagules as a consequence of a general scarcity of larvae in the water column (say as a result of isolation from source reefs – e.g. Glynn, 1985) or of topographic reef features which make it unlikely that larvae will be carried to the site in any given year (e.g. Black, 1988). Moreover, even when physical and nutritional conditions (including irradiance) are optimal for coral growth, the establishment of populations of corals may be essentially dictated by the presence, absence or abundance of 'other organisms', for example sea urchins and/or fish (see below).

Kaneohe Bay, Hawaii

In Kaneohe Bay, Hawaii, particularly in its poorly flushed southern basin, cause and effect have been clearly established. Several decades of siltation (due to dredging and land clearing) and eutrophication (due to sewage pollution) killed off most of the coral by the early 1970s (Smith *et al.*, 1981). The reefs became colonised by a non-reef-building assemblage dominated by macroalgae (*Dictyosphaeria cavernosa*) and a variety of suspension and filter feeding organisms (Smith *et al.*, *op. cit.*). In 1977–78, the sewage was diverted away from the bay and by 1982 there was some decline in *Dictyosphaeria* and some increase both in numbers of colonies and percent cover of corals (Maragos *et al.*, 1985). This example demonstrates the capacity of corals to re-establish populations once a major perturbation is removed.

Discovery Bay, Jamaica

At Discovery Bay, Jamaica, a shift from coral to algal dominance has been well documented. The transitions which have taken place reflect both the nature and the chronology of recent disturbances, and an effect of over-fishing. In 1981, Hurricane Allen stripped off and smashed down most branching and much encrusting coral (Woodley *et al.*, 1981). Turf algae biomass was kept low by a dense population of the omnivorous sea urchin *Diadema antillarum* on the reef, and some recruitment of corals took place in 1980–83 (Hughes *et al.*, 1987). However, in 1982 there was mass mortality among the *D. antillarum* (Hughes *et al.*, *op. cit.*), and a dense carpet of macroalgae dominated by *Dictyota*, *Padina* and *Halimeda* species established itself, smothering the recruits and other small corals. Up to 1990, this algal carpet effectively excluded further coral settlement (Hughes, 1989 and pers. comm.).

The urchin density may have been high, and thus prone to the disease which finally decimated their populations (Carpenter, 1990a), because of overfishing of the fish predators and/or competitors of the urchins (Woodley, 1979). This hypothesis has been supported by work on Kenyan reefs, where urchin (*Echinometra mathaei*) densities are lowest in marine reserves protected from fishing for the last 2 decades (Muthiga & McClanahan, 1987). Fish predation rates upon the urchin are highest (McClanahan & Muthiga, 1989) in marine parks. However Samoilys (1988) considered that siltation and dynamite fishing also directly affected the reefs and their fish populations.

There is still much to be learnt about the degree to which overfishing can change the abundance and composition of herbivorous fish, predators of herbivorous fish or invertebrates, let alone downstream effects on benthic community structure (Hay, 1984; McClanahan & Muthiga, 1989). Munro & Smith (1984) attributed major changes in total fish catch and composition in the Caribbean, including changes in the proportion of herbivorous fish, to sustained, intense and relatively indiscriminate fishing. In principle, fishing-induced changes in fish community structure may be expected to have significant effects on benthic communities. However fishing pressure is thought unlikely to affect reef fish community structure in systems with strong recruitment limitation of fish populations (Munro & Williams, 1985; Doherty & Williams, 1989).

La Saline Reef, Reunion

Overfishing is also thought to have played a part in the degradation of the reef at La Saline in Reunion. At that reef, chronic pollution and overfishing have been blamed for a marked coral-algal shift. There has been a proliferation of filamentous and fleshy algae *Vaughaniella* and *Gracilaria crassa* and the bioeroding sponge *Cliona inconstans* in previously coral-dominated areas (Cuet *et al.*, 1988). The shift to algal dominance is believed to result partly from chronic nutrient pollution of groundwater, but also from undergrazing by urchins, which were rare, and by fish, also rare as a result of daily netting of the reef. The algae killed corals by overgrowth, and it was predicted that it would also inhibit settlement of coral larvae.

Moorea, French Polynesia

At Moorea (French Polynesia), nutrients and overfishing have again been implicated in reef degradation. In addition, the importance to coral recovery of hydrodynamics and of other populations (viz. urchins and damselfish) have also been recognised. In the period 1971–1981, the macroalgae *Boodlea siamensis* (Chlorophyta), *Sargassum* sp. and *Turbinaria ornata* (Phaeophyta) expanded their ranges from the fringing reef to the coral dominated inner and outer barrier reef (Payri & Naim, 1982). This took place during a period of dredging of the coral reef for building material (Gabrie *et al.*, 1985) and rapid development of agriculture and tourist accommodation adjacent to the shore (Salvat, 1987). It had persisted until 1989 (Done *et al.*, 1991) at which time much of the substratum occupied by macroalgae comprised standing skeletons of corals killed by *Acanthaster planci* around 1982 (Bouchon, 1985; Faure, 1989).

It is very easy to point the finger of blame at the coastal development and dredging, but difficult to prove the connections. There was no detectable elevation of nutrient levels in the strongly flushed lagoonal and inner barrier reef waters in a set of samples collected in 1976 (Ricard, 1982). Levels of nitrate + nitrite and reactive phosphate were much less than occur in moderately flushed lagoons of some GBR reefs 50 km and more from shore (Furnas *et al.*, 1990). This suggests that if nutrient enhancement is necessary to sustain the algal biomass at Moorea, it may occur in pulses, such as during dredging of storms (resuspension of reef sediments, perhaps enriched by septic tank runoff and groundwater), and wet season runoff of silt-laden water, (which retains its integrity to the outer reef passes (R. Steger, pers. comm.). Alternatively, or in addition, there may be high nutrient fluxes, despite low ambient concentrations, as a result of very rapid uptake by benthic algae and phytoplankton (Atkinson, 1988).

Coral recolonisation at Moorea was very patchy by 1987–1989 (Done *et al.*, 1991). It was high on the outside of the northern barrier reef and low on the inner barrier reef. In order to understand coral recovery on the inner barrier reef, one needs to understand something of the hydrodynamics of the area and the natural history and population dynamics of some key species, notably sea urchins and damselfish. On some areas of the reef platform, sand abrasion is a major problem for small corals. Only massive *Porites* corals were abundant in 1987–9, both because they were not eaten by the starfish, and because they can colonise the sand-blasted areas by fragmentation (Highsmith, 1980).

Poor coral settlement and survival were also noted in patches with dense urchin populations (especially *Diadema setosum* and *Echinometra mathaei*). However there are refuges from urchin grazing. These are damselfish (*Stegastes nigricans*) territories which cover about 40% of the habitat. Corals (notably *Acropora* and *Pocillopora* spp.) were much more abundant and diverse inside than outside the territories, probably in large part due to *S. nigricans*' active exclusion of urchins and herbivorous fish likely to graze off small corals. This finding is precisely the opposite to the observations on other damselfish species (e.g. Potts, 1977; Lobel, 1980; Kaufman, 1977), where there is a net detrimental effect on coral survival inside territories.

Nearshore and fringing reefs, Great Barrier Reef

Nearshore and island fringing reefs on the Great Barrier Reef (GBR) are dominated over much of their shallow, wave-swept slopes and flats by brown macroalgae, mainly *Sargassum* spp. The significance of this prolific growth is unclear. In the Caribbean, where *Sargassum*-dominated assemblages occupy a similar habitat, heavy wave-action is thought to favour *Sargassum* through exclusion of grazing sea urchins and fish (Adey *et al.*, 1977). Within recent years to decades, *Sargassum* dominated algal assemblages have both invaded large areas (hectares) of Great Barrier Reef fringing reefs previously occupied by living coral, and have themselves been displaced by corals in other areas (T. J. Done & K. F. Navin, unpubl.). The annual life cycle of *Sargassum* gives it an advantage in the wave-swept rubble banks, but puts it at a disadvantage when the alga is in direct competition for space against corals. Fast lateral growth of encrusting *Montipora* and tabulate *Acropora* corals enable them to overgrow *Sargassum* holdfasts during their seasonal (summer) decline (Done & Navin, *op. cit.*).

High macroalgal biomass on GBR nearshore reefs may be indicative of high ambient nutrient concentrations and/or low levels of herbivory (Russ, 1984; Williams & Hatcher, 1983). A case has been made for eutrophication of GBR lagoon waters (Bell *et al.*, 1989), but it is also possible that current standing crops of macroalgae may not require an anthropogenic subsidy of nutrients. In a shallow (< 10 m) nearshore GBR station, soluble nutrients (nitrate + nitrite, reactive phosphate) were strongly correlated with previous day's wind speed, suggesting the nutrients were released from bottom sediments disturbed by waves (Walker & O'Donnell, 1981). However the degree to which runoff and sewage outfalls have enriched the sediment pore-water has not been investigated, even though Walker and O'Donnell's station is within a few kilometres of the sewage outfall of the largest city adjacent to the Great Barrier Reef.

Offshore reefs, Great Barrier Reef

On the Great Barrier Reef, coral to macroalgal phase shifts have been caused by major episodic disturbances such as *Acanthaster planci* and cyclones. *A. planci* infestations killed close to 100% of a high cover of corals over large areas early in the 1980s (Endean & Cameron, 1985; Done, 1985). Algal biomass per hectare increased greatly as a result of increased area for settlement (viz. skeletons of dead corals). Unfortunately, there has been no large scale systematic study of algal community dynamics following *A. planci* damage. However it appears that fish grazing, which usually keeps algal biomass low on these reefs (Russ, 1984; Williams & Hatcher, 1983) was no longer able to do so, and that there was an increase in abundance of fleshy macrophytes following the outbreak (Done, pers. obs.).

Nevertheless, there has been significant reversion towards coral dominance in the 6–10 years since the outbreak. Percent coral cover and colony size frequency distributions determined in

1990 (Done *et al.*, in prep.) are consistent with good settlement and growth of corals each year since the *A. planci* outbreak, and a high coral cover is predicted by the year 2000 (Done *et al.*, 1988). However on the inner lagoon of one reef (only 200–300 m from the well colonised outer slope), there had been a very poor recovery (Done *et al.*, in prep.), possibly due to low densities of planulae arriving in the lagoon, poor settlement, and/or post-settlement survival (see Babcock, 1988).

Summary of examples

Of all the examples, the outer slopes of offshore GBR reefs may provide the best ecological circumstances for coral recovery. There is an adequate supply of planktonic larvae from upstream reefs and communities (Oliver & Willis, 1987; Williams *et al.*, 1984), and sufficiently good early coral growth and survival throughout the first 6–10 years following the disturbance. By contrast, some reefs or sections of reefs may rarely be reached by dispersive coral larvae due to their remoteness from source reefs (Glynn, 1985), or unfavourable reef-scale circulation patterns (Black, 1988; cf. Sammarco & Andrews, 1988). In several other examples (e.g. Discovery Bay; Moorea) it was shown that it may be necessary to understand something of the natural history and population dynamics of other reef biota (e.g. urchins and damselfish), the direct effect of fishing on fish community structure, and indirect effects on benthic community structure.

Temporal perspectives on phase change

While degradation may seriously affect the immediate amenity of a coral reef, it may not be significant when viewed from a geological perspective.

> 'Events that loom large in the eyes of contemporary observes....may be undetectable in the reef's sedimentary record and trivial in the reef's geological history if the killed areas are promptly recolonised' (Buddemeier & Hopley, 1988).

This is a perceptive statement, both acknowledging that 'prompt recolonisation' will not necessarily follow disturbance, and at the same time counselling conservationists, biologists and managers to seek a longer-term view of the implications of reef change. In some circumstances, corals of any type are very slow to recolonise damaged areas, there being little recovery after a decade or more, e.g. following coral mining at Maldives (Brown & Dunne, 1988), and following coral death at Moorea (Done *et al.*, 1991). However there are other examples where damaged areas have been promptly recolonised by reef building corals (Pearson, 1981; Colgan, 1987; Done *et al.*, 1988), and it is apparently on the basis of the latter type of example that some authors have been lured to suggest that corals are most likely to quickly (decades) establish dominance in any suitable surface which becomes available (e.g. Hopley & Kinsey, 1988 – writing in reference to inundation of coral reef tops with rising sea level).

However, the definition of that point in time when recovery may be regarded as 'complete' varies. Restitution of pre-disturbance percent coral cover is the most cited (see Pearson, 1981) but least stringent criterion of recovery. The most stringent – 'total recovery' by both ecological and geological standards – would be the restitution of prior diversity, mix of colony growth forms (if not species), colony size frequency distributions and structural extent and complexity of reef framework (after Johannes, 1975). These can take considerably longer to recover than percent coral coverage (see Fig. 1). For example, Done (1988a) and Endean *et al.* (1988) suggest that recovery of size-frequency distributions is partially killed and injured populations of slow growing corals (genus *Porites* and other massive corals) can take a century or more in the same habitats in which percent coral cover has recovered in the first 1–2 decades (Done *et al.*, 1988).

Problems and scale of management

One management option which is widely acceptable is the creation of marine preservation areas including coral reefs. Areas which are protected from human exploitation can act as reservoirs for unprotected reefs downstream, where species are depleted by fishing, collecting etc. However the same current systems which carry the larvae of corals and fish also carry pollutants, sediments, and the larvae of destructive organisms such as *Acanthaster planci* and *Drupella* spp. (coral-eating gastropods). It is therefore prudent from a reef conservation perspective to create as many reserves as possible, and to make them as large as possible. Experience on the Great Barrier Reef is that major coral kills can take place on reefs scattered over hundreds of kilometres of the reef tract, equivalent to the length of the entire Kenyan coast.

Conclusions

While the symptoms of widespread coral death and algal invasion are usually obvious, the causes are frequently more a matter for conjecture than direct observation. Where a cause is definitely identified – say, overfishing, siltation caused by dredging or land clearing, sewage pollution, toxic wastes – it may be in a country's best immediate economic interests to prevent that impact. As one example, Hodgson & Dixon (1988) found the revenue loss to dive tourism caused by siltation of a Philippine reef area would be greater than revenue gained by the logging watersheds which would cause the siltation in the first place. As a second example, Done (1988b) suggested in relation to Mauritian fringing reefs, that one must weigh the cost of protective management against the potential for reef erosion and damage to the shore and structures.

A scientist offering advice to government on the best course of action in relation to coral reef management walks a precarious path. Knowledge of general reef character, history, structure and processes – as can be gleaned from personal experience and from the literature – is just the beginning. The uncertainties associated with interpreting what has happened to a specific reef in the past, and predicting how it will behave in the future, must also be confronted. On the time and space scales relevant to coral reef management, it is not appropriate to make decisions based on a presumption that a specific reef or system of reefs has 'ability always to regenerate after catastrophes' (Davies, 1988). Neither is it correct to assume that a degraded reef will remain so forever, or that those reefs which do regenerate will do so in time scales which are convenient to human users of the reef.

This presentation has touched lightly on the many issues – physical, chemical and biological – which must be addressed in trying to understand the causes and consequences of coral to algal phase shifts in coral reefs. Clearly, a multidisciplinary approach is desirable for in-depth understanding of all complex natural systems. The initiatives of UNESCO – COMARAF in creating 'The Regional Project for Training and Research on African Coastal Systems', and the University of Nairobi in organizing this Symposium are both welcome and timely.

Acknowledgements

I am grateful to the sponsors, executive committee and co-ordinators of the International Symposium on the Ecology of Mangrove and Related Ecosystems, 24–30 September 1990, Mombasa, Kenya, for providing the forum for this presentation. Els Martens' tireless efforts throughout all stages of organisation and publication are especially appreciated. Comments on the manuscript by Jeremy Woodley and an anonymous reviewer were much appreciated. This is Australian Institute of Marine Science Contribution No. 536.

References

Adey, W. H., P. J. Adey, R. Burke & L. Kaufman, 1977. The Holocene reef systems of Eastern Martinique, French West Indies. Atoll Res. Bull. 218: 1–40.

Atkinson, M. J., 1988. Are coral reefs nutrient limited? Proc. 6th int. Coral Reef Symp. 1: 157–166.

Babcock, R. C., 1988. Fine-scale spatial and temporal patterns in coral settlement. Proc. 6th int. Coral Reef Symp. 2: 635–639.

Bazzaz, F. A., 1983. Characteristics of populations in relation to disturbance in natural and man-modified ecosystems. In H. A. Mooney & M. Godron (eds), Disturbance and Ecosystems. Components of Response. Springer-Verlag, Berlin: 258–275.

Bell, J. & R. Galzin, 1984. Influence of live coral cover on coral reef fish communities. Mar. Ecol. Prog. Ser. 15: 265–274.

Bell, P. R. F., P. F. Greenfield, D. Hawker & D. Connell, 1989. The impact of waste discharge on coral reef systems. Wat. Sci. Tech. 21: 121–130.

Black, K. P., 1988. The relationship of reef hydrodynamics to variations in numbers of planktonic larvae on and around coral reefs. Proc. 6th int. Coral Reef Symp. 2: 125–130.

Bouchon, C., 1985. Quantitative study of scleractinian coral communities of Tiahura Reef, (Moorea, French Polynesia). Proc. 5th int. Coral Reef Symp. 6: 279–290.

Brown, B. E., 1987. Worldwide death of corals: natural cyclic events or man-made pollution? Mar. Pollut. Bull. 18: 9–13.

Brown, B. E., 1988. The environmental impact of coral mining on coral reefs in the Maldives. Env. Cons. 15: 159–165.

Buddemeier, R. W. & D. Hopley, 1988. Turn-on and turn-offs: causes and mechanisms of the initiation and termination of coral reef growth. Proc. 6th int. Coral Reef Symp. 1: 253–261.

Carpenter, R. C., 1990a. Mass mortality of *Diadema antillarum*. 1. Long-term effects on sea urchin population-dynamics and coral reef algal communities. Mar. Biol. 104: 67–77.

Carpenter, R. C., 1990b. Mass mortality of *Diadema antillarum*. 2. Effects on population densities and grazing intensity of parrotfishes and surgeonfishes. Mar. Biol. 104: 79–86.

Colgan, M., 1987. Coral reef recovery of Guam (Micronesia) after catastrophic predation by *Acanthaster planci*. Ecology 68: 1592–1605.

Connell, J. H., 1978. Diversity in tropical rainforests and coral reefs. Science 199: 1302–1310.

Connell, J. H. & M. J. Keogh, 1985. Disturbance and patch dynamics of subtidal marine animals on hard substrata. In S. T. A. Pickett & P. S. White (eds), The Ecology of Natural Disturbance and Patch Dynamics. Academic Press Inc., Orlando: 125–151.

Cuet, P., O. Naim, G. Faure & J. Y. Conan, 1988. Nutrient-rich groundwater impact on benthic communities on la Saline reef (Reunion Island, Indian Ocean): Preliminary results. Proc 6th int. Coral Reef Symp. 2: 207–212.

Dahl, A. L., 1974. The structure and dynamics of algae in the coral reef ecosystem. Proc. 2nd int. Coral Reef Symp. 1: 21–25.

Davies, P. J., 1988. Evolution of the Great Barrier Reef – reductionist dream or expansionist vision. Proc. 6th int. Coral Reef Symp. 1: 9–17.

Doherty, P. J. & D. McB. Williams, 1989. The replenishment of coral reef fish populations. Oceanogr. mar. biol. ann. Rev. 26: 487–551.

Dollar, S. J. & R. W. Grigg, 1981. Impact of a kaolin clay spill on a coral reef in Hawaii. Mar. Biol. 65: 269–276.

Done, T. J., 1985. Effects of two *Acanthaster* outbreaks on coral community structure: the meaning of devastation. Proc. 5th int. Coral Reef Symp. 5: 315–320.

Done, T. J., 1988a. Simulation of recovery of pre-disturbance size structure in populations of *Porites*. Mar. Biol. 100: 51–61.

Done, T. J., 1988b. An assessment of coral reefs in Mauritius. COMARAF Ser. Doc. 2.

Done, T. J., K. Osborne & K. F. Navin, 1988. Recovery of coral communities post-*Acanthaster*: Progress and prospects. Proc. 6th int. Coral Reef Symp. 2: 137–142.

Done, T. J., P. K; Dayton, A. E. Dayton & R. Steger, 1991. Regional and local variability in recovery of shallow coral communities: Moorea, French Polynesia and central Great Barrier Reef. Coral Reefs 9: 183–192.

Endean, R. & A. M. Cameron, 1985. Ecocatastrophe on the Great Barrier Reef. Proc. 5th int. Coral Reef Symp. 5: 309–314.

Faure, G., 1989. Degradation of coral reefs at Moorea Island (French Polynesia) by *Acanthaster planci*. J. Coast. Res. 5: 295–306.

Furnas, M. J., A. W. Mitchell, M. Gilmartin & N. Relevante, 1990. Phytoplankton biomass and primary production in semi-enclosed reef lagoons of the Central Great Barrier Reef, Australia. Coral Reefs 9: 1–10.

Gabrie, C., M. Porcher & M. Masson, 1985. Dredging in French Polynesian reefs: towards a general policy of resource exploitation and site development. Proc. 5th int. Coral Reef Symp. 4: 271–277.

Glynn, P. W., 1985. Corallivore population sizes and feeding effects following el Niño (1982–1983) associated coral mortality in Panama. Proc. 5th int. Coral Reef Symp. 1: 183–188.

Hamner, W. M., M. S. Jones, J. H. Carleton, I. R. Hauri & D. McB. Williams, 1988. Zooplankton, planktivorous fish, and water currents on a windward reef face: Great Barrier Reef, Australia. Bull. mar. Sci. 42: 459–479.

Hansen, J. A., D. M. Alongi, D. J. W. Moriarty & P. C. Pollard, 1987. The dynamics of benthic microbial communities at Davies Reef, central Great Barrier Reef. Coral Reefs 6: 63–70.

Hatcher, B. G., 1984. A maritime accident provides evidence for alternate stable states in benthic communities on coral reefs. Coral Reefs 4: 199–204.

Hatcher, B. G., R. E. Johannes & A. I. Robertson, 1989. Review of research relevant to the conservation of shallow water tropical marine ecosystems. Oceanogr. mar. biol. ann. Rev. 27: 337–414.

Hodgson, G. & J. A. Dixon, 1988. Logging versus fisheries

and tourism in Palawan. Occasional Papers of the East-West Environment and Policy Institute, 7.

Hopley, D. & D. W. Kinsey, 1988. The effects of a rapid short-term sea-level rise on the Great Barrier Reef. In G. I. Pearman (ed.), Greenhouse. Planning for climatic change. E. J. Brill, Leiden: 189-201.

Hay, M., 1984. Patterns of fish and urchin grazing on Caribbean coral reefs: are previous results typical? Ecology 65: 446–454.

Highsmith, R. C., 1980. Passive colonization and asexual colony multiplication in the massive coral *Porites lutea* Edwards. J. exp. mar. Biol. Ecol. 47: 55–67.

Hixon, M. A., 1983. Fish grazing and community structure of reef corals and algae: a synthesis of recent studies. In M. L. Reaka (ed.), The ecology of deep and shallow coral reefs. NOAA, Rockville, Maryland. 1: 79–87.

Hughes, T. P., 1989. Community structure and diversity of coral reefs: the role of history. Ecology 70: 275–279.

Hughes, T. P., D. C. Reed & M. J. Boyle, 1987. Herbivory on coral reefs: community structure following mass mortalities of sea urchins. J. exp. mar. Biol. Ecol. 113: 39–59.

Johannes, R. E., 1975. Pollution and degradation of coral reef communities. In E. J. Ferguson-Wood & R. E. Johannes (eds), Tropical marine pollution, Elsevier Scientific Publ., New York. Elsevier Oceanography Series 12: 15–51.

Johannes, R. E. & S. B. Betzer, 1975. Introduction: Marine communities respond differently to pollution in the tropics than at higher latitudes. In E. J. Ferguson-Wood & R. E. Johannes (eds), Tropical marine pollution, Elsevier Scientific Publ., New York. Elsevier Oceanography Series 12: 1–12.

Kaufman, L., 1977. The threespot damselfish: effects on benthic biota of Caribbean coral reefs. Proc. 3rd int. Coral Reef Symp. 1: 559–564.

Kinsey, D. W., 1985. Metabolism, calcification and carbon production. I System level studies. Proc. 2nd int. Coral Reef Symp. 4: 505–626.

Lapointe, B. E., 1989. Caribbean coral reefs: Are they becoming algal reefs? Sea Frontiers 35: 82–91.

Larkum, A. W. D., 1983. The primary productivity of plant communities on coral reefs. In D. J. Barnes (ed.), Perspectives on Coral Reefs. The Australian Institute of Marine Science, Brian Clouston Publisher, Manuka: 221–239.

Lighty, R. G., 1982. Fleshy-algal domination of a modern Bahamian barrier reef: example of an alternate climax reef community. Abstract – Proc. 4th int. Coral Reef Symp. 1: 722.

Littler, M. M. & D. S. Littler, 1985. Factors controlling relative dominance of primary producers on biotic reefs. Proc. 5th int. Coral Reef Symp. 4: 35–39.

Lobel, P. S., 1980. Herbivory by damselfish and their role in coral reef community ecology. Bull. mar. Sci. 30: 273–289.

Maragos, J. E., C. Evans & P. Holthus, 1985. Reef corals in Kaneohe Bay six years before and after termination of sewerage discharges. Proc. 5th int. Coral Reef Symp. 4: 189–194.

McClanahan, T. R. & N. A. Muthiga, 1988. Changes in Kenyan coral reef community structure and function due to exploitation. Hydrobiologia 166: 269–276.

McClanahan, T. R. & N. A. Muthiga, 1989. Patterns of predation on a sea urchin, *Echinometra mathaei* (de Blainville), on Kenyan coral reefs. J. exp. mar. Biol. Ecol. 126: 77–94.

Moran, P. J., R. H. Bradbury & R. E. Reichelt, 1988. Distribution of recent outbreaks of the crown of thorns starfish (*Acanthaster planci*) along the Great Barrier Reef: 1985–1986. Coral Reefs 7: 125–137.

Moriarty, D. J. W., P. C. Pollard, W. G. Hunt, C. M. Moriarty, T. Wassenberg, 1985. Productivity of bacteria and microalgae and the effect of grazing by holothurians on a coral reef flat. Mar. Biol. 85: 293–300.

Munro, J. L. & D. McB. Williams, 1985. Assessment and management of coral reef fisheries: biological, environmental and socioeconomic aspects. Proc. 5th int. Coral Reef Symp. 4: 543–578.

Munro, J. L. & I. R. Smith, 1984. Management strategies in multispecies complexes in artisanal fisheries. Proc. Gulf Caribb. Fish. Inst. 36: 127–141.

Muthiga, N. A. & T. R. McClanahan, 1987. Population changes of a sea urchin (*Echinometra mathaei*) on an exploited fringing reef. Afr. J. Ecol. 25: 1–8.

Oliver, J. K. & B. L. Willis, 1987. Coral-spawn slicks in the Great Barrier Reef: preliminary observations. Mar. Biol. 94: 521–529.

Payri, P. E. & O. Naim, 1982. Variations entre 1971 et 1980 de la biomass et de la composition des populations de macroalgues sur le recif corallien de Tiahura (Ile de Moorea, Polynesie Francaise) Cryptogamie: Algologie 3: 229–240.

Pearson, R. G., 1981. Recovery and recolonisation of coral reefs. Mar. Ecol. Prog. Ser. 4: 105–122.

Pichon, M. & J. Morrissey, 1985. Premieres mesures de bilan metabolique dans l'ecosysteme recifal de Tulear (Madagascar). C.r. Acad. Sci. Paris 300: 99–101.

Potts, D. C., 1977. Suppression of coral populations by filamentous algae within damselfish territories. J. exp. mar. Biol. Ecol. 28: 207–216.

Ricard, M., 1982. Mean features of phytoplankton and primary productivity of Moorea Island (French Polynesia). Abstract. Proc. 4th int. Coral Reef Symp. 1: 453.

Russ, G., 1984. The distribution and abundance of herbivorous grazing fishes in the central Great Barrier Reef. I. Levels of variability across the entire continental shelf. Mar. Ecol. Prog. Ser. 20: 23–34.

Russ, R. G. & A. C. Alcala, 1989. Effects of intense fishing pressure on an assemblage of coral reef fishes. Mar. Ecol. Prog. Ser. 65: 13–27.

Sale, P. F., 1988. What coral reefs can tell us about ecology. Proc. 6th int. Coral Reef Symp. 1: 19–27.

Salvat, B. (ed.), 1987. Human Impacts on Coral reefs: Facts and Recommendations. Antenne Museum Ecole Pratique des Hautes Etudes, French Polynesia, 253 pp.

Sammarco, P. W. & J. C. Andrews, 1988. Localised dispersal

and recruitment in Great Barrier Reef corals: the Helix Experiment. Science 239: 1422–1424.

Samoilys, M. A., 1988. Abundance and species richness of coral reef fish on the Kenyan coast: the effects of protective management and fishing. Proc. 2nd int. Coral Reef Symp. 2: 261–266.

Smith, S. V., W. J. Kimmener, E. A. Laws, R. E. Brock & T. W. Walsh, 1981. Kaneohe Bay sewage diversion experiment: perspectives on ecosystem responses to nutritional perturbation. Pac. Sci. 35: 279–395.

Sournia, A., 1976. Oxygen metabolism of a fringing reef in French Polynesia. Helgol. wiss. Meeres. 28: 401–410.

Steneck, R. S., 1988. Herbivory on coral reefs: a synthesis. Proc. 6th int. Coral Reef Symp. 1: 37–49.

Walker, T. & G. O'Donnell, 1981. Observations on nitrate, phosphate and silicate in Cleveland Bay, northern Queensland. Aust. J. mar. Freshwat. Res. 32: 877–887.

Williams, D. McB. & A. I. Hatcher, 1983. Structure of fish communities on outer slopes of inshore, mid-shelf and outer shelf reefs of the Great Barrier Reef. Mar. Ecol. Prog. Ser. 10: 239–250.

Williams, D. McB., E. Wolanski & J. C. Andrews, 1984. Transport mechanisms and the potential movement of planktonic larvae in the central region of the Great Barrier Reef. Coral Reefs 3: 229–236.

Woodley, J. D., 1979. The effects of trap-fishing on reef communities in Jamaica. Proc. Is. Mar. Labs. Carib. 13: 27.

Woodley, J. D. & 19 others, 1981. Hurricane Allen's impact on Jamaican coral reefs. Science 214: 749–755.

Hydrobiologia **247**: 133–138, 1992.
V. Jaccarini & E. Martens (eds), The Ecology of Mangrove and Related Ecosystems.

The incidence of hurricanes on the north coast of Jamaica since 1870: are the classic reef descriptions atypical?

J. D. Woodley
Discovery Bay Marine Laboratory, P.O. Box 35, Discovery Bay, Jamaica, West Indies

Key words: Acropora, disturbance, hurricanes, Jamaica, coral reefs

Abstract

When the coral reefs of north Jamaica were described and studied in the 1950's, 60's and 70's, their shallow zones were dominated by dense thickets of *Acropora palmata* and *A. cervicornis*. Those stands were demolished by hurricanes in 1980 and 1988, but these seemed to be unusual events.

Published records of storm tracks indicate that, in the last 120 years, about 35 hurricanes have passed within 364 km (200 nautical miles) of Discovery Bay, Jamaica. In the absence of information on their intensity, an approximate measure of their impact is the reciprocal of their closest distance. Recent experience suggests that passage within about 65 km, whether to the north or the south, generates waves capable of destroying *Acropora* stands. There have been 13 years, out of the last 120, in which hurricanes have passed that close to Jamaica, with a median interval of 6.5 years. Large colonies of *A. palmata* (over 1 m high) might take about 12 years to develop. The total number of years in which the reefs have been free of major disturbance for longer than 12 years is only 33, and 24 of them (1956–80) were in the unusually long interval of 36 years, from 1944 to 1980. The luxuriant *Acropora* stands of the classic descriptions may therefore be atypical; one extreme of a variable condition.

Introduction

Among tropical coastal ecosystems, coral reefs are the most exposed to wave action. The intensity and direction of prevailing wave regimes partly determine their morphology, community composition and zonation. Many reefs, however, occasionally experience exceptionally large waves generated by intense tropical revolving storms (cyclones, hurricanes, typhoons). These are brief events, but the changes wrought by their violent impact may persist for many years. What is their pattern of occurrence at a given locality?

Contribution Number 500 of the Discovery Bay Marine Laboratory, University of the West Indies.

Hurricanes (and I use the Caribbean term) sweep narrow trails of extreme impact. There has been an average of only five per year in the Western Atlantic over the past century (Neumann *et al.*, 1981). It is like a lottery where, in any year, a given site will have a good chance of escaping hurricane influence. To a local observer, on that short time-scale, it is a rare, unpredictable event. On larger scales of time and space, however, hurricanes are common and almost ubiquitous; a map showing a hundred years of Caribbean tropical storms and hurricanes is black with their tracks (Neumann *et al.*, 1981).

No reef in the area escapes their influence, but the temporal structure of that influence depends on the time-scale of the processes affected by storms. Thus, in relation to geological processes of reef growth or sedimentation, on a time-scale of hundreds or thousands of years, hurricanes can be regarded as a continuous force. One could recognize different intensities of that force due to differences in hurricane frequency in space or over time (e.g., Hubbard, 1988). On shorter time scales, the occurrence of hurricanes is irregular. Their influence on processes measured on a scale of the same order as the interval between hurricanes, such as the growth of corals, may be better understood in terms of the time elapsed since the previous storm, rather than their average frequency (Kjerfve *et al.*, 1986).

Corals differ in their growth strategies; that is to say, in their growth rates, recruitment rates, competitive ability, fragility and regenerative powers. As a result, they differ in their population responses to storm impacts. *Montastraea annularis* (Ellis & Solander) is a massive, slow-growing frame-builder on Caribbean reefs, resistant to storms, but susceptible to overgrowth by faster-growing branching corals such as *Acropora palmata* (Lamarck) and *A. cervicornis* (Lamarck). These are strong enough to resist moderate wave energies and often dominate shallow reef zones that are named after them (Goreau, 1959). On the north coast of Jamaica, Goreau described the breaker zone as 'populated almost exclusively by huge tree-like colonies of *Acropora palmata*', and the upper part of the seaward slope as dominated by 'immense beds of staghorn coral., *Acropora cervicornis*'. South coast reefs had been hit by Hurricane Charlie in 1951; the *Acropora* spp. were less dominant and there was more rubble. Goreau suggested, correctly but cautiously, that 'storms can play an important part in altering the basic pattern of species dominance and zonal succession in the upper parts of a climax reef'.

For another 20 years, the two *Acropora* species remained conspicuous dominants on the north coast of Jamaica. *A. palmata* formed a continuous thicket, up to 2 m high, from the surface to −6 m depth, while *A. cervicornis* covered the shallow terrace to a depth of about −22 m, punctuated by *M. annularis* and other massive corals. At normal wave base, in the Mixed or Buttress Zone around −5 to −10 m, massive corals were much commoner. In 1980, the huge waves generated by Hurricane Allen smashed the *Acropora* thickets and most of their fragments died (Knowlton *et al.*, 1981). The *Acropora* zones were transformed into rubble like that which Goreau had seen on the south coast. Allen was the strongest Caribbean storm of the century (at the time), and the first to affect the north coast for many years (Woodley *et al.*, 1981). It was generally believed that it was only a matter of time before the north coast reefs returned to their former luxuriance. However, not only was the recovery of *Acropora* populations unexpectedly slow (Knowlton *et al.*, 1990), but only eight years later they suffered another violent disturbance, from Hurricane Gilbert. The physical impact of this storm was at least as great as that of Allen, but the damage to branching corals was much less spectacular because they had not had time to regain their previous abundance (Woodley, 1989).

There are good records of hurricanes in the Caribbean, and it is possible to determine their pattern of occurrence with reasonable accuracy for the last century or so. Important sources are a set of maps showing tropical cyclones of the North Atlantic, 1871–1980 (Neumann *et al.*, 1981), and the very detailed Jamaica Weather Reports which began in 1881. In this paper, I use those sources to analyse the tracks of hurricanes that might have damaged reefs at Discovery Bay, in the centre of the north coast of Jamaica, where many reef studies have been made.

Methods

Table 1 presents data on 39 storms that passed closer than 364 km (200 nautical miles) to Discovery Bay since 1870. Their initial selection was made from the maps of Neumann *et al.* (1981). On the earliest maps, tropical storms (central winds less than 33 ms^{-1} [74 mph]) and hurricanes are not distinguished. From 1886 to 1898, storms known to have reached hurricane intensity at some time are distinguished from the others, while from 1899 onwards the various stages of storm development are indicated along each track. For the present work, the intensity category of storms passing close to Jamaica from 1880 onwards was determined from the Jamaica Weather Reports. These Reports were also used for more accurate definition of hurricane tracks where they differed from the small-scale maps of Neumann *et al.* The Reports on hurricanes are often remarkably detailed; especially those written by Maxwell Hall, amateur founder of the weather service in 1881 and Government Meteorologist until at least 1919. He collated the reports of other amateur observers across the island and interpreted them with great skill and enthusiasm.

The adjusted tracks of the 35 remaining storms that were, or might have been, hurricanes when they passed Jamaica, are illustrated for 12 successive decades in Fig. 1. In the first decade, no records were available for 1870, and three of the four storms might not have reached hurricane intensity in this area. The same can be said for the storms of 1889, 1894, 1895, 1909 and 1910 (Table 1). Further evidence on the category of these storms might be obtained from Jamaican (or Cuban) newspapers, but none is mentioned in Hall's historical review (1916) of hurricanes and other phenomena observed in Jamaica.

The height of the waves radiating from the eye of a hurricane decreases with distance (Bretschneider & Tamaye, 1976). A crude comparative indicator of their destructive power is the reciprocal of that distance. If no other information is available on the intensity of different hurricanes, their impacts can be compared in this way, as a first approximation. Table 1 lists the track of each storm, its direction (aspect) from Discovery Bay when closest, its distance at that point (r), and the reciprocal of that distance (multiplied by 1000). Those reciprocals are plotted in Fig. 2, to an arbitrary maximum of 33 (that corresponds to a hurricane centre passing 30 km away, which is close enough to cause catastrophic damage, so there is little point in discriminating it from even closer encounters).

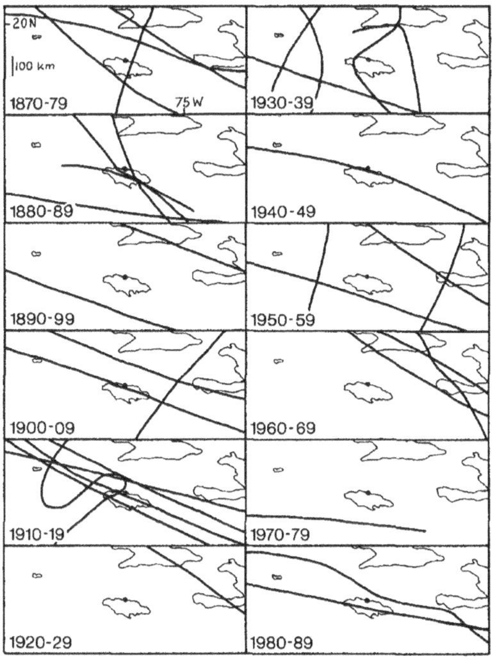

Fig. 1. Maps showing probable hurricane tracks near Jamaica in 10 year periods, 1870–1989. Location of Discovery Bay shown by spot on the north coast. Data from sources in Table 1.

Results and discussion

Figure 1 gives a good impression of the irregularity of hurricane influence on a time scale of decades. But many of these storms were sufficiently distant that their impact at Discovery Bay must have been relatively slight. Figure 2 compares the reciprocals of their distance, as described above. To calibrate them, we can consider those storms for which the impact on reefs at Discovery Bay is known, and ask whether or not they destroyed *Acropora* stands. The best known is Hurricane Allen (1980), for which the catastrophic impact is well-documented (Woodley *et al.*, 1981). That destruction was effected from a minimum distance of 45 km to the north. Gilbert (1988) was equally violent, passing 40 km to the south. Carmen (1974) was 145 km south, and generated waves that destroyed experimental cages in Discovery Bay, but did not destroy *Acropora* formations.

Table 1. Data on storms that might have reached hurricane intensity within 364 km (200 nautical miles) of Discovery Bay, Jamaica since 1870. Direction (aspect) and distance (*r*) from Discovery Bay, and the reciprocal of that distance (multiplied by 1,000) are shown. Storm data were taken from Neumann *et al.* (1981) and modified from other sources as indicated. The Jamaica Weather Reports (JWR) were published twice; first as monthly and occasional bulletins, then in collected volumes entitled Jamaica Meteorological Observations every 6–8 years. Tropical storm (TS).

Year	#/Name	Date	Track	Aspect	*r* (km)	1000/*r*	Notes
1873	5	28 Sept	WNW	N	136	7	TS or Hurricane?
1874	7	2 Nov	N	E	22	45	Hurricane (Hall, 1916)
1875	2	12 Sept	NW	NE	200	5	TS or Hurricane?
1879	6	12 Oct	WNW	S	90	11	TS or Hurricane?
1880	3	19 Aug	NW	E	20	50	Track adjusted (Hall, 1889)
1884	3	8 Oct	N by E	E			Not a hurricane (JWR 46)
1886	3	27 June	WNW	S	20	50	Track adjusted (JWR 67)
1886	6	20 Aug	NW	E	30	33	Track adjusted (JWR 69)
1886	7	18 Sept	NW	N			Not a hurricane (JWR 70)
1889	6	15 Sept	W	S	227	4	TS or Hurricane?
1894	3	23 Sept	WNW	NE	320	3	TS or Hurricane?
1895	2	25 Aug	WNW	S	180	6	TS or Hurricane?
1895	5	19 Oct	NW	SW			Not a hurricane (JWR 175)
1896	4	26 Sept	W by N	S			Not a hurricane (JWR 204)
1903	2	11 Aug	W by N	S	20	50	Track adjusted (Hall, 1911)
1905	4	5 Oct	NE	E	220	5	
1909	5	24 Aug	WNW	N	90	11	Not mentioned in JWR
1910	2	9 Sept	WNW	N	36	28	Further off? (Hall, 1911)
1912	6	18 Nov	N	W	13	77	Track adjusted (JWR 411)
1915	2	13 Aug	WNW	N	29	34	Track adjusted (JWR 446)
1916	4	15 Aug	WNW	S	58	17	Track confirmed (JWR 460)
1917	3	23 Sept	WNW	S	15	67	Track adjusted (JWR 474)
1928	2	11 Aug	NW	NE	320	3	
1932	10	9 Nov	NE	NW	330	3	
1933	19	30 Oct	NE	NW	50	20	Track confirmed (JWR 685)
1935	4	27 Sept	NNW	W	160	6	
1935	5	22 Oct	NNW	NE	160	6	
1938	2	11 Aug	W by N	S	180	6	
1944	4	20 Aug	W by N	S	4	250	Track adjusted (JWR 828)
1950	King	16 Oct	N by E	W	190	5	
1951	Charlie	17 Aug	W by N	S	90	11	Track adjusted (JWR (1951))
1954	Hazel	12 Oct	NNE	E	360	3	
1958	Ella	1 Sept	NW	NE	255	4	
1963	Flora	4 Oct	NW	NE	345	3	
1964	Cleo	25 Aug	WNW	NE	135	7	
1966	Inez	30 Sept	WNW	NE	320	3	
1974	Carmen	31 Aug	W	S	145	7	
1980	Allen	6 Aug	WNW	N	45	22	Taylor and Staff (1981)
1988	Gilbert	11 Sept	W by N	S	40	25	Jamaica Met Office, 1989

Nor did that happen during any of the distant hurricanes from 1954–1966, when T. F. Goreau was working in Jamaica, or he would surely have documented it. Hurricane Charlie (1951) passed only 90 km south of Discovery Bay. It caused much damage in Kingston and to the reefs of the Port Royal Cays, but little on the north coast (as indicated by Goreau, 1959), where maximum winds were only 22 ms^{-1} (50 mph).

Assuming that all these hurricanes were of equal size and strength (which they were not), the threshold distance for the destruction of *Acropora*

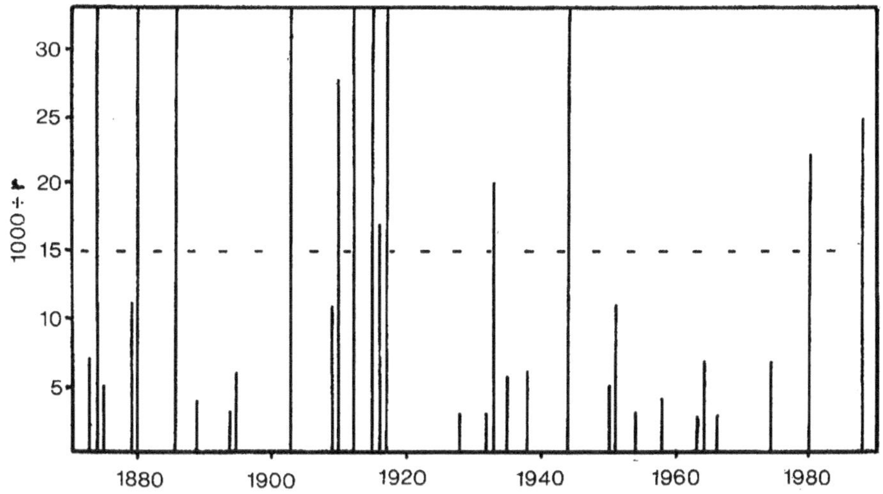

Fig. 2. Reciprocals (× 1000) of the closest approach of hurricanes to Discovery Bay, 1870–1989, to a maximum of 33 (equivalent to 30 km). Dotted line at 15 represents 65 km.

stands would be somewhere between 45 and 90 km; say about 65 km ($1000/r = 15$, indicated by the dotted line across Fig. 2). There are 13 years in which hurricanes passed closer than that to Discovery Bay, and it is instructive to look at their distribution in time (Fig. 2, above the dotted line). It is highly irregular; the intervals between years of potentially serious hurricane impact range from 1 to 36, with a mean near 11, but a median of 6.5 (Fig. 3). That means wide variation, over the last 120 years, in the time available for *Acropora* thickets to regenerate, before they were knocked down again. For much of the time, *Acropora* rubble would have been as prevalent on the north coast as it is today, or as it was at Port Royal in the 1950's.

If we assume that *Acropora palmata* colonies grow at 12 cm per year, it would take them 12 years to reach a height of one metre, growing up at 45°. Of the 12 intervals between the 13 storms, only 3 exceed 12 years, and there have been only 33 years (since 1874) that reefs at Discovery Bay have been free of major disturbance for 12 years or longer. 24 of these (1956–1980) lie within the single longest interval of 36 years between 1944 and 1980. It was during this unusual period, when *Acropora* thickets flourished, that most descriptions of Jamaican north coast reefs were written (Goreau, 1959; Kinzie, 1973; Woodley & Robinson, 1977). Perhaps they described an atypical situation; one extreme of a variable condition.

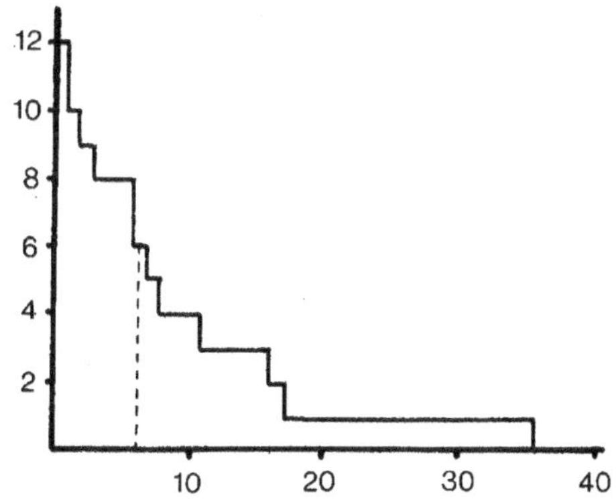

Fig. 3. Histogram showing the number (vertical axis) of continuous periods of years (horizontal axis) during which no hurricane approached closer than 65 km to Discovery Bay, in the period 1874–1988. Dotted line indicates median interval of 6.5 years.

Another perspective is provided by a comment in Jamaica Weather Report 685 (1934) on occasions when Jamaica has been free of hurricane impact for an unusually long time. In chronological order these are 1655–1689 (34 years), 1786–1812 (26 years), 1844–1874 (30 years) and 1917–1933 (16 years). The next period would be 1951–1980 (29 years) or, using the criterion of 'traversing Jamaica', 1944–1988 (44 years).

This analysis has left out any consideration of differences in magnitude between the hurricanes, largely because such information is not readily available. It would be possible to gather data about the important storms, from the Jamaica Weather Reports and from contemporary newspapers, to hindcast their impacts on corals. Kjerfve *et al.* (1986) calculated that turbulent flows of 2–4 ms^{-1} would bring down *A. palmata* colonies, depending whether or not they were bio-eroded. Equivalent values for *A. cervicornis* were 0.35–1.6 ms^{-1}. Only when that analysis has been made shall we know if the classic descriptions of Jamaican reefs are indeed atypical.

References

Bretschneider, C. L. & E. Tamaye, 1976. Hurricane wind and wave forecasting techniques. Look Lab/Hawaii 6: 1–29.

Goreau, T. F., 1959. The ecology of Jamaican coral reefs I. Species composition and zonation. Ecology 40: 67–90.

Hall, M., 1889. Introduction. Jamaica Meteorological observations 1. Jamaica Government Printing Establishment, Kingston.

Hall, M., 1911. Notes on the storms which passed over or near Jamaica, between 1902 and 1910 inclusive. Jamaica Meteorological Observations 4. Jamaica Government Printing Office, Kingston.

Hall, M., 1916. Notes of hurricanes, earthquakes, and other physical occurrences in Jamaica. Jamaica Weather Report No. 465. In Jamaica Meteorological Observations 5. Jamaica Government Printing Office, Kingston, 1918.

Hubbard, D. K., 1988. Controls of modern and fossil reef development: common ground for biological and geological research. Proc. Sixth Int. Coral Reef Symp. Townsville 1: 243–252.

Kinzie, R. A., 1973. The zonation of West Indian gorgonians. Bull mar. Sci. 23: 93–155.

Kjerfve, B., K. E. Magill, J. W. Porter & J. D. Woodley, 1986. Hindcasting of hurricane characteristics and observed storm damage on a fringing reef, Jamaica, West Indies. J. mar. Res. 44: 119–148.

Knowlton, N., J. C. Lang & B. D. Keller, 1990. Case-study of natural population collapse: post-hurricane predation on Jamaican staghorn corals. Smithsonian Contributions to the Marine Sciences 31, 25 pp. Smithsonian Institution Press, Washington D.C.

Knowlton, N., J. C. Lang, M. C. Rooney & P. Clifford, 1981. Evidence for delayed mortality in hurricane-damaged Jamaican staghorn corals. Nature 294: 251–252.

Neumann, C. J., G. W. Cry, E. L. Caso & B. R. Jarvinen, 1981. Tropical Cyclones of the North Atlantic Ocean, 1871–1980. U.S. Department of Commerce, NOAA.

Taylor, G. & Staff, 1981. Annual data and verification tabulation Atlantic tropical cyclones 1980. National Hurricane Center, NOAA tech. Mem. NWS NHC 15.

Woodley, J. D., 1989. The effects of Hurricane Gilbert on coral reefs at Discovery Bay. Appendix 9 in P. R. Bacon (ed.), Assessment of the economic impacts of Hurricane Gilbert on coastal and marine resources in Jamaica. UNEP Regional Seas Reports and Studies 110.

Woodley, J. D., E. A. Chornesky, P. A. Clifford, J. B. C. Jackson, L. S. Kaufman, N. Knowlton, J. C. Lang, M. P. Pearson, J. W. Porter, M. C. Rooney, K. W. Rylaarsdam, V. J. Tunnicliffe, C. M. Wahle, J. L. Wulff, A. S. G. Curtis, M. D. Dallmeyer, B. P. Jupp, M. A. R. Koehl, J. E. Neigel & E. M. Sides, 1981. Hurricane Allen's impact on Jamaican coral reefs. Science 214: 749–755.

Woodley, J. D. & E. Robinson, 1977. Field Guide to the Modern and Ancient Reefs of Jamaica. Atlantic Reef Committee, Miami.

Hydrobiologia **247**: 139–140, 1992.
V. Jaccarini & E. Martens (eds), The Ecology of Mangrove and Related Ecosystems.

Short note

Importance of the Lac-lagoon (Bonaire, Netherlands Antilles) for a selected number of reef fish species

G. van der Velde,[1] M. W. Gorissen,[1] C. den Hartog,[1] T. van 't Hoff[2] & G. J. Meijer[1]
[1] *Laboratory of Aquatic Ecology, Catholic University of Nijmegen, Toernooiveld, 6525 ED Nijmegen, The Netherlands;* [2] *Stinapa, Bonaire, Netherlands Antilles*

Mangroves and seagrass beds are known to function as nursery for juveniles of reef fishes. However, the fish fauna of these often linked ecosystems is mostly studied separately due to methodological problems. Therefore a selected number of fish species was studied both in Lac-lagoon (Bonaire, Netherlands Antilles) and the coral reef around Bonaire.

Lac-lagoon is the largest lagoon of Bonaire with an area of about 8 km^2. It is a protected wetland according to the Ramsar convention. It consists of a shallow basin with clear nutrient-poor water bordered by mangroves and mudflats. The bottom of the lagoon is mainly covered by seagrasses, especially *Thalassia testudinum* Banks ex König. The investigations were focussed on the importance of the Lac-lagoon for the fishes on the reefs. The reefs form the main attraction for the tourists visiting Bonaire.

Table 1. Importance of various biotopes for juveniles of selected fish species and the importance of Lac-lagoon nursery for the fish fauna of Lac-lagoon itself and the reef.

No.	Species	Juveniles				Nursery	
		Mangroves	Thalassia	Shallow Reef	Deep Reef	Nursery for Lac	Nursery for reef
1	French grunt	xx	xxx	x	–	–	xx(x)
2	Bluestriped grunt	–	xxx	–	–	–	xxx
3	Caesar grunt	–	–	–	–	–	–
4	Smallmouth grunt	–	–	xx	–	–	–
5	Black margate	–	–	–	–	–	–
6	Gray snapper	xxx	xx	–	–	xxx	(x)
7	Schoolmaster	xxx	xx	–	–	xx	x
8	Mahogany snapper	–	x	xx	–	–	x
9	Yellowtail snapper	(x)	xxx	–	–	–	xxx
10	Great barracuda	xxx	xx	–	–	x	xx
11	Sergeant major	(x)	–	xx	–	–	x
12	Foureye butterfly	xxx	xx	x	x	–	xx(x)
13	Ocean surgeon	–	x	xx	–	–	x
14	Doctorfish	(x)	xxx	x	–	–	x
15	Blue tang	(x)	x	x	x	–	x
16	Stoplight parrotfish	–	xxx	x	x	–	xx

– not important
( ) doubtful
x minor importance
xx medium importance
xxx major importance

Sixteen species of reef fishes were selected based on easy identification and economic value (attractive for diving tourists, fishery) for a study of abundance, distribution and size classes. During May–November 1981 abundance and size classes were estimated by means of a visual census technique; estimations in the seagrass beds were checked by catches with drop nets.

Six 'biotopes' have been studied, viz. seagrass beds (*T. testudinum*) and mangroves in the lagoon and four depth zones on the reef. The major results are summarized in Table 1.

Mutual comparison indicates that the seagrass beds form the most important biotope for the juvenile stages of five fish species, viz. French grunt (*Haemulon flavolineatum* (Desmarest)), Bluestriped grunt (*H. sciurus* (Shaw)), Yellowtail snapper (*Ocyurus chrysurus* (Bloch)), Stoplight parrot fish (*Sparisoma viride* (Bonnaterre)) and Doctor fish (*Acanthurus chirurgus* (Bloch)), while mangroves are important for four other species, viz. Schoolmaster (*Lutjanus apodus* (Walbaum)), Gray snapper (*L. griseus* (L.)), Great barracuda (*Sphyraena barracuda* (Bonnaterre)) and Foureye butterfly (*Chaetodon capistratus* (L.)).

For five species the lagoon seemed to be less important, viz. Smallmouth grunt (*Haemulon chrysargyreum* Günther), Mahogany snapper (*Lutjanus mahogoni* (Cuvier)), Sergeant major (*Abudefduf saxatilis* (L.)), Ocean surgeon (*Acanthurus bahianus* Castelnau) and Blue tang (*A. coeruleus* Bloch & Schneider): their juveniles mainly use the shallow reef. Of two species, Caesar grunt (*Haemulon carbonarium* Poey) and Black margrate (*Anisotremus surinamensis* (Bloch)), of which adults were only found on the reef, too few juveniles were found to draw conclusions.

Hydrobiologia **247**: 141–161, 1992.
V. Jaccarini & E. Martens (eds), The Ecology of Mangrove and Related Ecosystems.

Hydrodynamics of mangrove swamps and their coastal waters

Eric Wolanski
Australian Institute of Marine Science, P.M.B. No. 3, Townsville M.C., Qld. 4810, Australia

Abstract

Mangrove swamps help control the tidal hydrodynamics of many tropical estuaries. They generate an asymmetry of the tidal currents in both the tidal creeks and the mangrove swamps. This results in self-scouring of the tidal channels. Mangrove land reclamation results in siltation of the channel. Mangrove swamps control the flushing rates of the estuaries through the lateral trapping effect. Lateral trapping leads to the aggregation of mangrove litter along slick lines. Evapotranspiration plays a role in the hot dry season by forming a salinity maximum zone which isolates the estuary from the coastal waters for several months of the year. In the absence of runoff, evapotranspiration in the hot dry season generates an inverse estuarine circulation which can trap high salinity mangrove water, and mangrove detritus, along the bottom of a mangrove creek. This bottom layer can become anaerobic. Groundwater flow appears to play a key role in the nutrient budget of mangrove creeks, exporting salt left behind by evapotranspiration, and inhibiting runoff after rainfall. Particulates and dissolved nutrients outwelled from mangrove swamps to coastal waters are retained in a coastal boundary layer. This coastal boundary layer water can be trapped along the shore for long periods if the coast is straight and mangrove-fringed and the coastal waters are shallow. Headlands inhibit coastal trapping because they enhance mixing. Nutrient-rich coastal boundary layer waters may be ejected offshore as tidal jets peeling off headlands and locally enriching offshore waters.

Introduction

A typical mangrove swamp is Coral Creek (Fig. 1). It is 5 km long, with fairly steep banks, and maximum depth decreasing with distance from the coast. The cross-sections can be represented by polygons (Fig. 1b). The fringing mangroves are typically 100 to 300 m wide on either side of the creek. The surface of the substrate slopes gently upwards from the creek, with elevation near the creek of about 0.3 m above mean sea level, and maximum elevation rarely exceeding 1.3 m.

When many mangrove creeks exist one next to the other, they form a large mangrove swamp. Hinchinbrook Channel (Fig. 2) is an example of such a vast mangrove swamp. The channel is 44 km long channel with 109 km^2 of open water surrounded by 164 km^2 of fringing mangrove swamps. Since the mangrove creeks draining the swamp are oriented perpendicular to the Channel, the swamp is up to 5 km wide from the dry land to the edge of the Channel.

Tidal circulation in a creek-mangrove swamp system

The barotropic tidal circulation in the Coral Creek tidal creek-mangrove system was first modeled by Wolanski *et al.* (1980). They noticed that peak tidal velocities often exceed 1 $m\ s^{-1}$ in the tidal creek but never exceeded 0.07 $m\ s^{-1}$ in the heavily vegetated mangrove swamps 50 m from

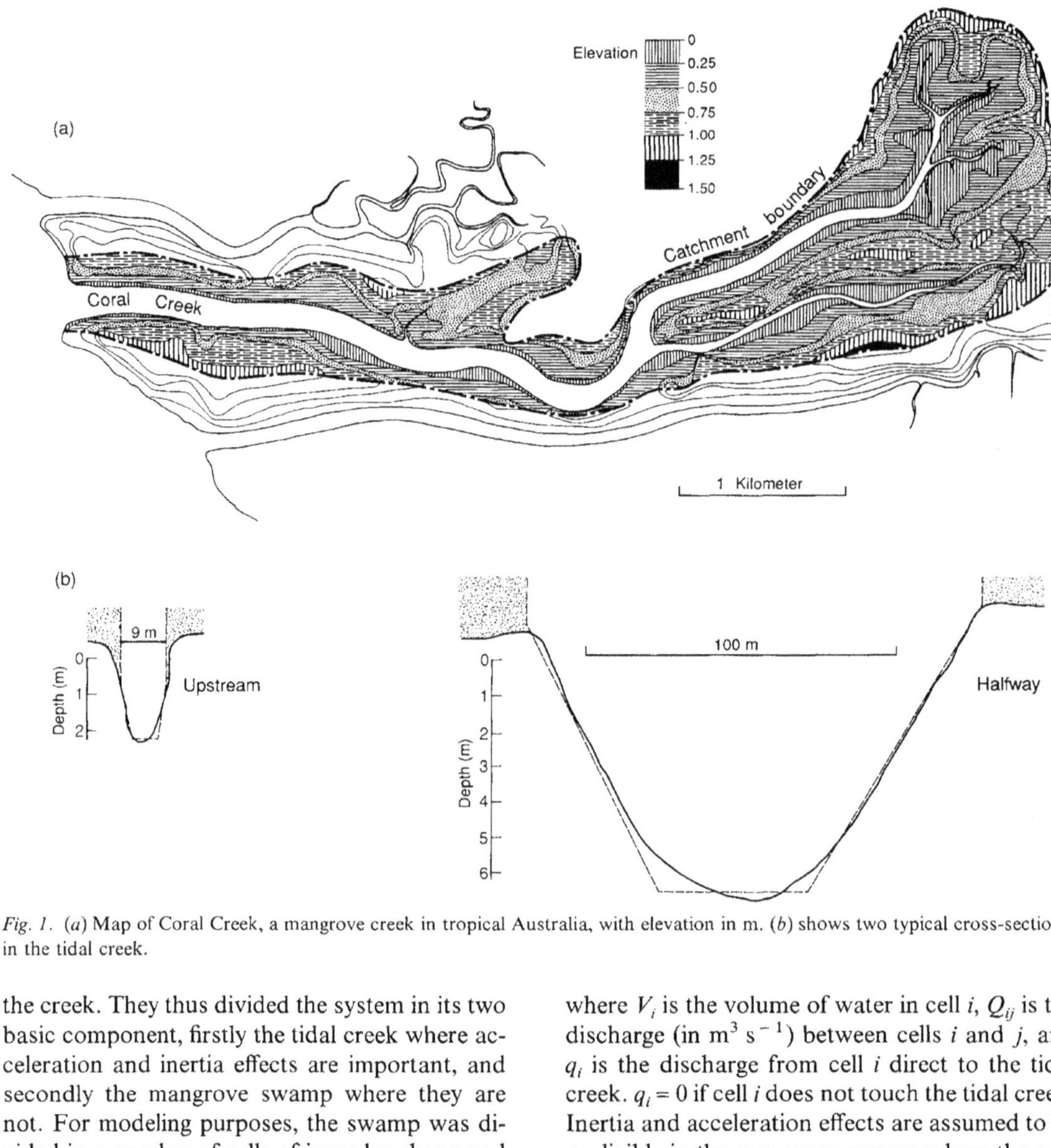

Fig. 1. (*a*) Map of Coral Creek, a mangrove creek in tropical Australia, with elevation in m. (*b*) shows two typical cross-sections in the tidal creek.

the creek. They thus divided the system in its two basic component, firstly the tidal creek where acceleration and inertia effects are important, and secondly the mangrove swamp where they are not. For modeling purposes, the swamp was divided in a number of cells of irregular shape and size fitted to the topography of the substrate (Fig. 3a). The water circulation in the swamp was calculated from the continuity equation, one such equation for each cell i,

$$\frac{\partial V_i}{\partial t} = \Sigma Q_{ij} - q_i \qquad (1)$$

where V_i is the volume of water in cell i, Q_{ij} is the discharge (in $m^3\ s^{-1}$) between cells i and j, and q_i is the discharge from cell i direct to the tidal creek. $q_i = 0$ if cell i does not touch the tidal creek. Inertia and acceleration effects are assumed to be negligible in the mangrove swamp where the momentum equation reduces to a simple balance between the frictional slope and the free surface slope,

$$S_f = \frac{\partial \eta}{\partial x} \qquad (2)$$

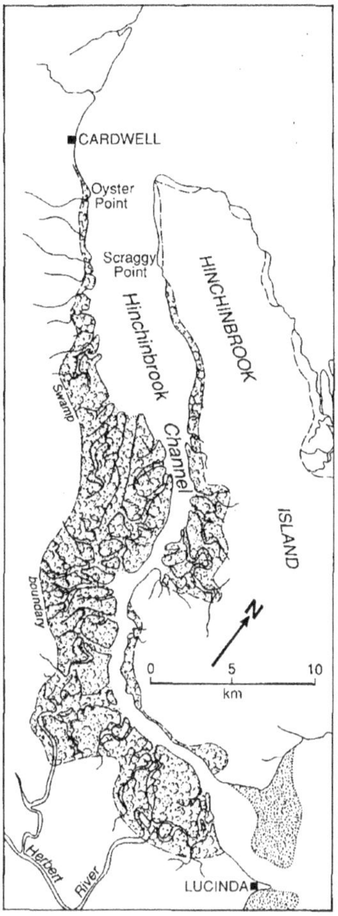

Fig. 2. Map of Hinchinbrook Channel and its fringing giant mangrove swamp, also in tropical Australia.

where η is the elevation of the water surface, x the distance and S_f the frictional slope. S_f was computed by Manning's formula and the Manning coefficient was set equal to 0.2 as characteristic of flow through dense vegetation (Petryk & Bosmajan, 1975; Wolanski *et al.*, 1980; Burke & Stolzenbach, 1983).

The tidal creek was divided in a number of cross-sections shown in Fig. 3b. In the tidal creek, acceleration and inertia effects are important and the one-dimensional barotropic equations of motion for unsteady open channel flow are

continuity:

$$\frac{\partial Q}{\partial x} + \frac{\partial A}{\partial t} = q_i \tag{3}$$

momentum:

$$\frac{\partial Q}{\partial t} + \frac{\partial}{\partial x}(Q^2/A) + gA\frac{\partial \eta}{\partial x} + gAS_f + \Delta = 0 \tag{4}$$

where Q is the discharge, t is the time, A is the cross-sectional area of the creek, and S_f the frictional slope calculated using a Manning's roughness coefficient $n = 0.025$ as characteristic of open channel flows. Note that in equation (4) is included a term Δ. This term is the momentum loss due to water leaving the creek with its momentum at flood tide to enter the swamp. This momentum is rapidly dissipated in the first few meters of the swamp by friction in the heavily vegetated swamp. At falling tide, the momentum of the water leaving the swamp to flow in the tidal creek is so small that one can safely assume $\Delta = 0$.

The two models, one for the swamp and one for the tidal creek, are linked by the condition that the water levels are the same in the creek and in a swamp cell where they touch each other. The other condition that the outflow from a mangrove cell is the inflow into the tidal creek, and vice-versa, is taken into account through the term q in equations (1) and (3).

Since freshwater runoff is negligible in Coral Creek, the model needs only one open boundary specification, namely a time series of sea level at the mouth.

The model was verified against field observations of currents at two points in the creek: the mouth and a mid-channel point 2 km upstream. Figure 4 shows a comparison between observed and computed tidal currents at the mouth during a spring tide. Note that the peak tidal currents are strong, being greater than 1 m s^{-1}. Also there is a large vertical shear of the currents. Since the waters are vertically well-mixed in salinity and temperature, this shear can be attributed to bottom friction. The model cannot reproduce this vertical shear since it is one-dimensional (vertically averaged). Nevertheless the model appears to be successful in reproducing the main features

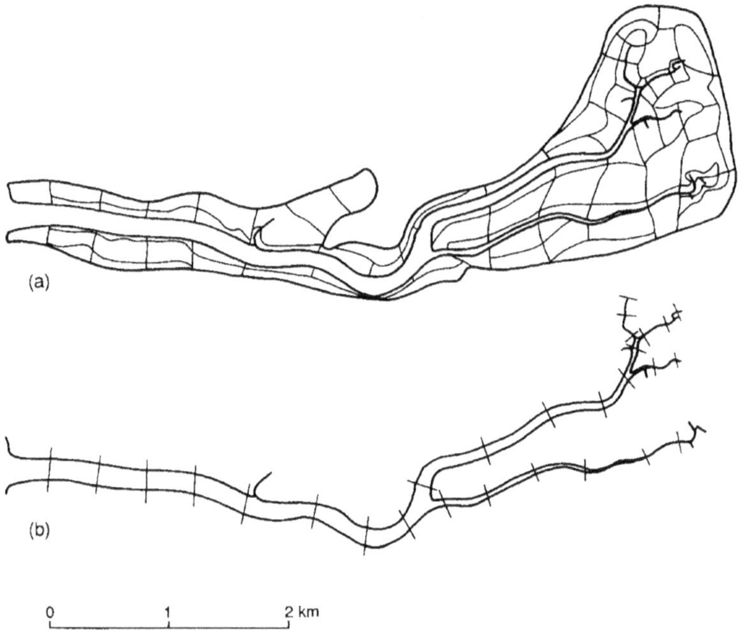

Fig. 3. (*a*) shows the distribution of cells in the mangrove swamp. (*b*) shows the location of the cross-sections.

of the depth-averaged currents, including the tidal asymmetry, the peak ebb tidal currents being much stronger than peak flood currents. The latter peak at 1.2 m s^{-1}, the former at 1.6 m s^{-1}.

This tidal asymmetry results in self-scouring of the tidal creek (Wolanski *et al.*, 1990). Numerical experiments suggest that destruction of the vegetation or reclamation of mangrove land, reduces both the peak tidal currents and the asymmetry of the tidal currents, thereby resulting in siltation of the channel.

Figure 5 shows the distribution of the predicted peak ebb and flood tidal currents in the mangrove swamp. This figures illustrates another important asymmetry of the tidal currents. The flood tidal currents are small, typically of order 0.03 m s^{-1}, and are oriented at about 90° to the banks of the creek. On the other hand, the ebb tidal currents are up to 3 times faster and are oriented at typically 20 to 30° from the creek. This asymmetry in the tidal currents was also confirmed in the field through measurements of currents and current directions at one point in the swamp. Hence, over a tidal cycle there exists in the fringing mangroves a net circulation towards the mouth. The implications of this circulation in the aggregation of mangrove litter are discussed later.

Groundwater flow

In the presence of strong tidal currents such as at Coral Creek, groundwater flow probably does not contribute much to the hydrodynamics of the system. In other cases, groundwater flow may be important. Ovalle *et al.* (1990) have suggested that the net outflow from the small Itacuruca mangrove swamp in Brazil is due to groundwater inflow in the swamp.

Wolanski (unpublished data) and Mazda *et al.* (1990a) have observed the initial stages of the flooding of the swamp at rising tide when a thin sheet of water advances in the swamp from the creek. This sheet is blocked by the high vegetation density. The first few mm of flooding water at a given point are not due to water from that sheet

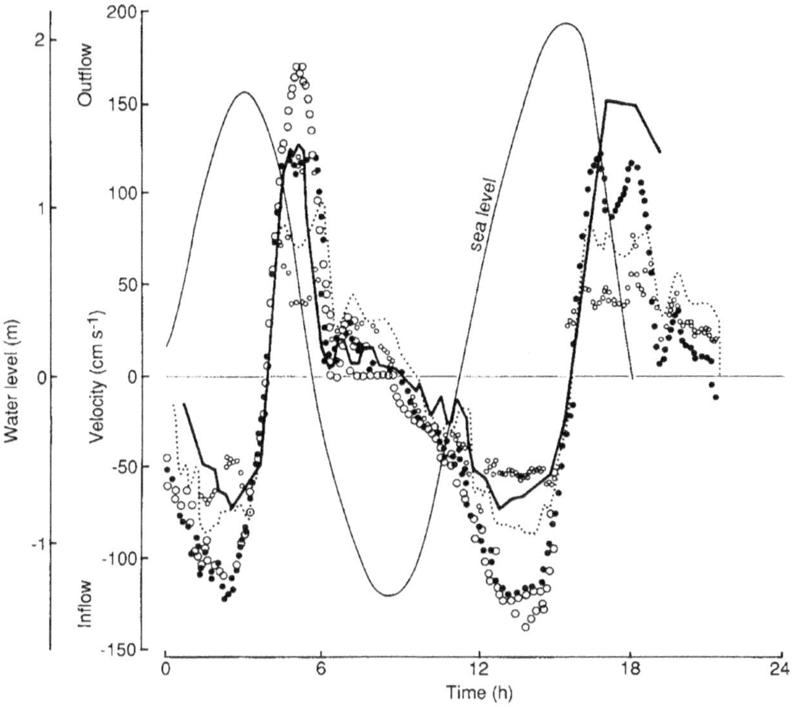

Fig. 4. Time series plot over one tidal cycle of the observed currents at different depths below the surface (from near the surface to near the bottom) at the mouth of Coral Creek; the thick line shows the depth-averaged predicted velocity time series. The thin line is the sea level time series.

spilling over obstructions on the ground, but are observed to come from the ground through the numerous holes due to crabs and rotting vegetation.

Wolanski & Gardiner (1981) have proposed that this groundwater flow is an important flushing mechanism preventing the accumulation of salt left behind after evapotranspiration.

Mazda *et al.* (1990) have carried out a study of groundwater flow in the 200 m long Bashita-Minato mangrove swamp in Japan. After storms, the tidal creek is occasionally ponded by a sill at the mouth. They found that when the creek is ponded the water level in the creek decreases by up to 10 cm day^{-1}, about 15 times larger than the evaporation rate. This observation suggests that the creek was draining out through groundwater flow. The flow appeared to be limited to the upper 90 cm of the substrate, i.e. the maximum depth of crab burrows, since below that point the groundwater was anaerobic.

The importance of crab burrows in enhancing groundwater flow is apparent in Coral Creek and in the Itacuruca swamp, where rainfall infiltrates readily in the soil and produces no local runoff, and in the Klong Ngao mangrove swamp in Thailand where crab burrows are few and direct runoff is observed (Wattayakorn *et al.*, 1990).

Longitudinal diffusion in mangrove creeks

Salinity in mangrove creeks increases with increasing distance from the mouth in the hot dry season when runoff is negligible. The salinity increase is due to evapotranspiration and is typically of order 0.5 to 1 p.p.t. for 5 km long mangrove creeks such as Coral Creek and Dickson

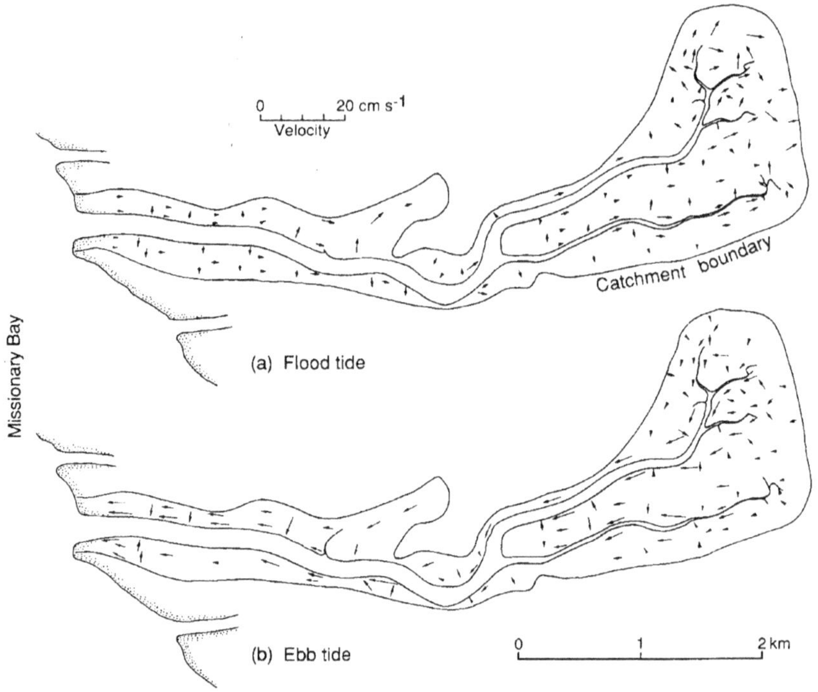

Fig. 5. Predicted peak tidal currents in the mangrove swamp of Coral Creek at flood and ebb tides.

Inlet in Australia (Wolanski *et al.*, 1980; Ridd *et al.*, 1990) and the Klong Ngao mangrove creek in Thailand (Wattayakorn *et al.*, 1990). This salinity gradient can introduce an inverse estuarine circulation as is sketched in Fig. 6a. This circulation becomes hydrodynamically important and measurable when both vertical and horizontal salinity gradients are large. the existence of a vertical gradient is largely dependent on the strength of the tidal currents. In Coral Creek for instance the tidal currents are particularly strong and maintain vertical homogeneity (Fig. 7), except possibly at neap tides. In Dickson Inlet, vertical gradients of salinity are larger, particularly in the upper reaches of the tidal creek (Fig. 8a). The resulting buoyancy effects inhibit vertical turbulent mixing.

In the presence of a small runoff, a salinity maximum zone can develop, which, as sketched in Fig. 6b, essentially isolates the upper reaches of the estuary from the coastal waters. An inverse estuarine circulation exists on either side of the salinity maximum zone. In tropical Australia,

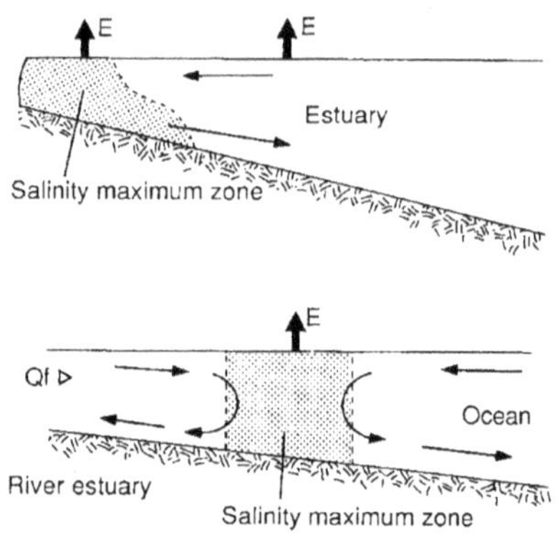

Fig. 6. (*a*) sketches the inverse estuarine circulation generated in a mangrove creek by evapotranspiration in the fringing swamp in the absence of runoff. (*b*) sketches the internal circulation generated by the 'presence of a salinity maximum zone in the presence of a vast evaporation area, such as a vast mangrove swamp, and of a small freshwater runoff.

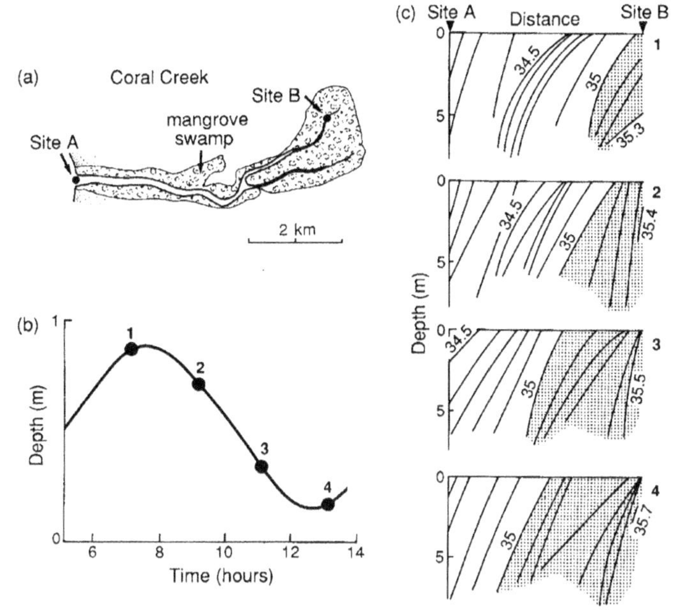

Fig. 7. Longitudinal distribution of salinity in Coral Creek in the hot dry season.

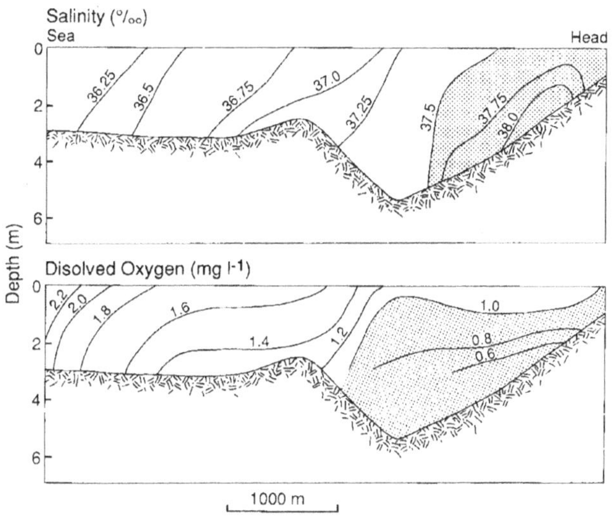

Fig. 8. (*a*) shows the salinity distribution in the Dickson Inlet mangrove creek in the hot dry season, at low tide. (b) shows the corresponding distribution of dissolved oxygen, also at low tide.

such an unusual estuarine circulation can last several months (Wolanski, 1986; Ridd *et al.*, 1988).

In the presence of an inverse estuarine circulation, high salinity mangrove water is then trapped in the creek at the bottom and is not aerated. The decomposition of the mangrove detritus in that layer and the biological oxygen demand of the mud, can result in low values of the dissolved oxygen concentration (less than 1 mg l^{-1} in Dickson Inlet, Fig. 8b). In some cases anoxic conditions can result, such as occasionally in Coral Creek (Boto, personal communication) and in the Bashito-Minato mangrove swamp (Mazda *et al.*, 1990).

The value of the longitudinal diffusion coefficient, B, hence also the flushing rates, may be estimated from the longitudinal salinity gradient. Since the gradient reaches a steady state, the salt accumulated in the swamp/creek system is exported by tidal diffusion, so that

$$A_t E_t S = BA \frac{dS}{dx} \tag{5}$$

where S is the salinity, E_t the evapotranspiration rate (0.2 to 0.6 cm d^{-1}), A_t the surface area of the swamp, and A is the creek cross-sectional area. Using observed salinity gradients in Coral Creek and the Klong Ngao estuary, equation (5) yields for these two systems

$$B = 20 - 50 \text{ m}^2 \text{ s}^{-1} \tag{6}$$

This value of B is surprisingly large for such small tidal creeks. Indeed, if diffusion was due to vertical turbulent mixing in the creek, then (Bowden & Hamilton, 1975; Fischer *et al.*, 1979) the turbulent eddy diffusivity (B, now called K to differentiate observed from predicted values) would be

$$K = C_d hU \tag{7}$$

where C_d is a drag coefficient, h is the depth and U a characteristic tidal velocity. If vertical motions were dominant, then $Cd = 0.003$. It results $K = 0.03$ m^2 s^{-1} for Coral Creek and the Klong Ngao mangrove creek, a result 1000 times too small. This implies that verticle turbulent motions are not the leading mixing mechanism in mangrove creeks.

Alternatively, following Fisher *et al.* (1979)

$$K = 0.15\, hu_* \tag{8}$$

where u_* is the shear velocity. Nokes (1986) found that

$$u_* = 0.05\, U \tag{9}$$

so that equation (8) becomes

$$K = 0.0075\, hU \tag{10}$$

The value of K from equation (10) is twice that from equation (7). This results in $K = 0.06$ m^2 s^{-1} for Coral Creek and the Klong Ngao mangrove creek, a result still 500 times too small.

Another mixing mechanism may be vertical shear dispersion. This process was first reported by Taylor (1921) for flows in pipes, and has been applied to open-channel flow by Elder (1959) and Fischer *et al.* (1979) who proposed two alternative formulae for diffusion in oscillatory currents,

$$K = 0.1\, Uh$$

or

$$K = 5.93\, hu_* \simeq 0.3\, Uh \tag{11}$$

which for Coral Creek and the Klong Ngao mangrove creek yields $K = 0.5–1.0$ m^2 s^{-1} a value still at least 20 times too small. This suggests that longitudinal diffusion in mangrove-fringed tidal creeks is controlled by another mechanism. This mechanism may be lateral trapping due to lateral shear. Fischer *et al.* (1979) proposed that in a wide, shallow, tidal estuary lateral shear is much more efficient than vertical shear at generating a large longitudinal diffusion. However, their model is not readily applicable to mangrove creeks as it requires a spatially homogeneous lateral shear. More detailed studies of this mechanism are discussed below.

Lateral trapping in mangrove swamps

The previous analytical models of the longitudinal diffusion coefficient in a mangrove creek neglect the presence of the fringing mangrove swamp. In fact this mangrove system is the reason for the existence of a large longitudinal eddy diffusion coefficient in mangrove creeks. The prevailing mechanism is lateral trapping, sketched in Fig. 9. This mechanism was first studied by Okubo (1973) who modeled dispersion in a tidal estuary with a lateral embayment. A tagged water mass moves upstream at flood tide, but a fraction of that mass ends trapped in the embayment. At ebb tide, this mass returns to the tidal creek and mixes with untagged water. Mixing is thus greatly enhanced and, following Okubo (1973),

$$B = \frac{A}{1+\varepsilon} + \frac{\varepsilon u_0^2}{2k(1+\varepsilon)^2(1+\varepsilon+\sigma/k)} \quad (12)$$

where u_0 is the peak tidal current, k^{-1} is the characteristic exchange time between the embayment and the estuary, σ is the tidal frequency and ε is the ratio of volume in the embayment to that in the estuary.

Wolanski & Ridd (1986) modified the Okubo (1973) model to remove two assumptions that are invalid in a mangrove creek, namely that the depths of water in the estuary and in the swamp are the same and are much larger than the tidal range, and that the exchange rate between water in the swamp and in the estuary are constant in time. They divided the mangrove creek in a rectangular main channel where tidal currents occur, and a swamp seen as the lateral embayment. The elevation of the swamp substrate was taken to be higher than mean sea level so that the swamp has

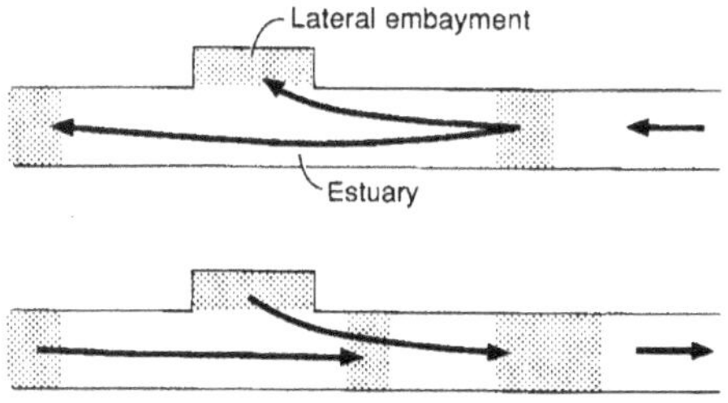

Fig. 9. Sketch of the lateral trapping phenomenon in a tidal estuary with an embayment, adapted from Okubo (1973).

no surface water for a fraction (1-a) of the tidal cycle of period 2T. They predicted

$$B = \frac{A}{(1+\varepsilon)} + \frac{\varepsilon u^2 a^2 T}{48\,(1+\varepsilon)} \quad (13)$$

Equation (13) yield $B = 10\text{–}40\ m^2\,s^{-1}$ for Coral Creek, Dickson Inlet and the Klong Ngao estuary, values that agree qualitatively well with observations. Equation (13) was recently modified by Ridd *et al.* (1990) to take into account the decrease in peak tidal velocity from the mouth to the upper reaches of the mangrove creek. Smaller values of the eddy diffusion coefficient are predicted in the upper reaches of the creek particularly in the uppermost 200 m or so.

Lateral trapping in mangrove swamps thus appear to be the dominant effect determining their flushing rate. The residence time in mangrove creeks is of order

$$T_0 = L^2/B \quad (14)$$

where L is the length of the mangrove-fringed creek. For Coral Creek and the Klong Ngao mangrove creek, it results $T_0 = 6$ days, suggesting that the upper reaches of mangrove creeks are flushed slowly.

Wolanski *et al.* (1990) also applied equations (13) and (14) to estimate the values of B and T_0 in Hinchinbrook Channel. They found $T_0 = 50$ days. This implies long-term trapping in this giant mangrove swamp. In fact they found experimental evidence for this long-term trapping in the flushing of brackish water from the channel following cessation of runoff. The salinity deficit followed an exponential decay in time, with an e-fold time scale of 54 days, in surprisingly good agreement with the predictions.

Buoyancy effects in mangrove swamps

Uncles *et al.* (1990) studied the dynamics of a large, mangrove-fringed, estuary in Malaysia. This estuary receives a large fairly steady fresh-

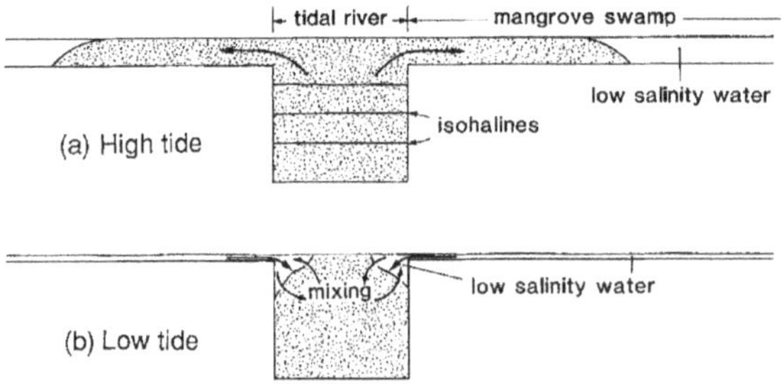

Fig. 10. Cross-channel distribution of salinity in a mangrove-fringed estuary at high and low tide following recovery of the system from a major river flood.

water inflow. They found that the salinity stratification is strong during neap tides but that the system destratifies during spring tides. Water flooding the mangroves is thus fresher at neap tides and saltier at spring tides. This may introduce a cycle, with a period of 2 weeks, in the productivity of the fringing mangrove forest.

Wattayakorn *et al.* (1990) showed that the small mangrove creek of Klong Ngao in Thailand stratifies and destratifies at a few hours period closely following fluctuations in both the tidal currents and in the freshwater inflow which varied rapidly with local short-lived storms in the small, steep watershed draining into that swamp. This unsteadiness precluded a reliable estimate of the outwelling, if any, in the wet season.

In tropical Australia, river runoff can be very intense but short-lived. Salt water can be totally flushed out of the estuary during floods. As soon as the flood is over, salt water creeps back in the estuary. If the estuary is mangrove-fringed, such as the Wenlock River in Australia (Wolanski & Ridd, 1986), the intruding salt water intrudes under the fresh water remaining near the surface. As the tide rises and salt water fills the estuary, the freshwater is pushed back laterally into the fringing mangrove swamps (Fig. 10). This freshwater is then trapped there, with minimal mixing

Fig. 11. Photograph showing an oceanographic front separating swamp water from creek waters, and made visible by a foam line.

with salt water in the estuary. As the tide falls, this freshwater returns to the estuary but stays trapped as a narrow river plume in shallow waters along the length of the estuary banks (Fig. 10). In the Wenlock River estuary, this front maintained its integrity throughout the ebb tide. Hence there was minimal mixing between fresh water in the mangrove swamp and sea water in the estuary.

The lateral trapping mechanism prevails also in mangrove creeks and is occasionally made readily visible by the presence of an oceanographic front forming at the start of the ebb tide. This front, shown in Fig. 11, separates swamp water from creek water. In small mangrove creeks, this front disappears in less than an hour but it indicates that the creek is not laterally well-mixed.

The front forming between creek and swamp water is important as it aggregates floating detritus, such as mangrove leaves. Visual observations reveal that at Coral Creek the flood tidal currents are too weak to break the organic film at the surface of the water in the mangrove swamp. Floating mangrove leaves are often trapped in this film and the flood tide does not carry the floating mangrove leaves further inland in the swamp. At ebb tide, the currents are stronger, break the surface film and export floating leaves to the creek. The leaves are then aggregated in long straight lines (foam lines) marking the small-scale oceanographic fronts separating mangrove swamp water from creek water (Fig. 11). These leaves can be outwelled from the mangrove creek to coastal waters where they can sometimes be found, still aligned in a straight line, several km from the swamp in calm weather.

Hence, the fate of mangrove detritus cannot be estimated using classical advection-diffusion models since these models predict that the detritus would be dispersed in a cloud of dimension increasing in time. The formation of small scale barotropic and baroclinic fronts in mangrove swamps has not been studied. These front are important since they result for a depth-averaged model in an apparent negative eddy diffusion as the mangrove seeds and leaves are not dispersed but are instead aggregated. This may have important biological consequences. For instance one often sees small fish aggregating under the leaves

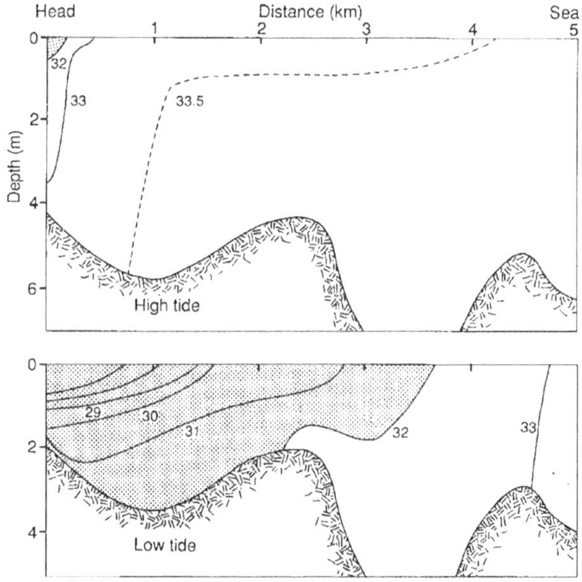

Fig. 12. Longitudinal distribution in Dickson Inlet at high and low tide about 10 days after cessation of runoff.

floating in a small-scale front in a mangrove creek. Wolanski & Hamner (1988) documented several small-scale mechanisms aggregating floating coral eggs. These mechanisms are due to the three-dimensional circulation in topographically controlled fronts. Most of the examples illustrated for reefs can be applied to mangrove creeks. For instance, leaves aggregate downstream of a sharp bend in the mangrove creek even in barotropic conditions as a result of the secondary circulation in a zone of strong flow curvature, in the same manner as coral eggs are aggregated in zones of strong flow curvatures near coral reefs. Axial convergence (Simpson & Turrell, 1986), due to the transverse transport in estuaries, also appears to be a process maintaining the floating leaves and mangrove seeds in a foamline, especially near the mouth of mangrove creeks.

A combination of lateral trapping and buoyancy effects can result in very long residence time of brackish water in mangrove creeks. Figure 12 shows the longitudinal distribution of salinity in mangrove-fringed Dickson Inlet at both low and high tide. Brackish water, a remnant of the runoff 10 days earlier, was still present near the surface in the creek at low tide. At high tide, the bulk of that water was trapped in the swamp. Because buoyancy effects inhibit vertical mixing, the brackish water is basically simply moving back and forth between the swamp at high tide and the creek and low tide, but mixes very little with creek water and is thus only very slowly exported from the system.

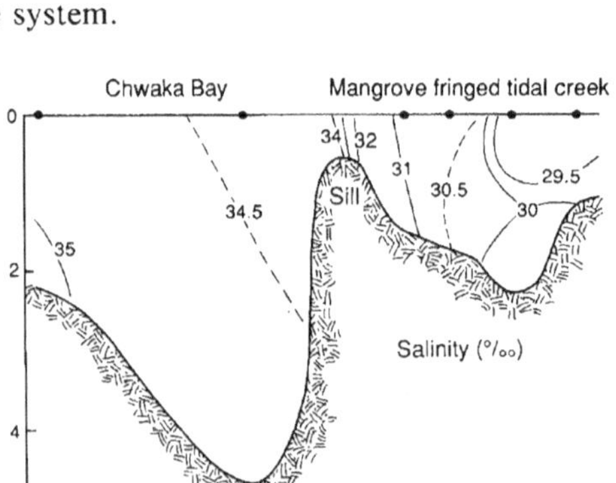

Fig. 13. Longitudinal salinity distribution in the Chwaka bay mangrove swamp in Zanzibar, in the dry season, July 1988.

The presence of a sill or narrow opening increases trapping. For instance, the mangrove creek of Chwaka Bay in Zanzibar is nearly ponded for a measurable portion of the tidal cycle by a small coral limestone sill at the mouth. Strong tidal mixing over shallow water of the sill results in the formation of a front separating creek waters from oceanic waters (Fig. 13). The sill thus inhibits mixing between creek waters and oceanic waters, and ensures long-term trapping in the mangrove-fringed tidal creek (Wolanski, 1989).

Trapping in the coastal zone

Straight coastline

Material outwelled from mangrove swamps may support offshore productivity but, if the coastal waters are shallow, this enrichment is restricted to either a narrow coastal zone or specific points near headlands.

When runoff is important, the outwelled material remains in the river plumes. In areas where river runoff is always important, such as on the mangrove-fringed coast of Thailand facing the Andaman Sea near Burma, longshore currents ensure that these river plumes usually remain coastally trapped together with the outwelled mangrove detritus and nutrients (e.g. Wattayakorn *et al.*, 1990). Any nutrient enrichment from the mangrove is then limited to coastal waters. Similar coastal trapping of steady river plumes for hundreds of km has been reported in Nicaragua (Murray & Young, 1985) and in South America (Bouyne & Roux, 1978).

In tropical Australia, river runoff is short-lived. Freshwater is injected into coastal waters in a pulse-like event. Once runoff ceases, tidal currents maintain vertical homogeneity in shallow coastal waters, and the brackish waters behave more or less like a passive tracer. The fate of this tracer, i.e. the time it takes to be flushed out from coastal waters, provides a good indication of the dynamics of the coastal zone. Two examples are illustrated below: the Gulf of Carpentaria and the coast of the Great Barrier Reef.

Figure 14 shows the distribution of salinity in the Gulf of Carpentaria at 3 monthly interval in 1976. The survey in April–May corresponds to the end of the wet season. It shows that the brackish water was coastally trapped and did not mix readily with offshore waters. Five months after, in August–September, runoff was negligible but brackish water was still found in the shallow coastal waters (depth less than 15 m or so). This finding suggests that brackish waters is trapped for several months in a coastal boundary layer, and that cross-shelf mixing between offshore waters and coastal boundary layer water is very slow (Wolanski & Ridd, 1990). A satellite image in the visible bands taken in calm weather (no wind stirring of bottom sediment) in the very dry season of November, 1986 (zero runoff hence no river plume) also shows a band of water of different colour in the dry season (negligible runoff) along the coast of the Gulf of Carpentaria, with a front running along the 15–20 m isobath (Fig. 15). The presence of a coastal boundary layer may be explained by the effects of friction in shallow water. Figure 16 shows the predicted mean Eulerian circulation under the action of both the tides and wind in the tradewind season. This shows a narrow coastal jet extending to the 15 to 20 m isobath with a return flow in deeper water. The tides alone drive a negligible net circulation even in shallow coastal waters, while wind alone in the absence of tides drives a coastal jet about 3 times as wide as that shown in Fig. 16 (Wolanski & Ridd, 1990). The non-linear interaction between tidal currents and wind-driven currents is an essential factor in the dynamics of the coastal boundary layer.

Rugged coastline

The east coast of the Gulf of Carpentaria is fairly straight, with no major headlands, and extensive areas of shallow waters, the 20 m isobath being located 20 km offshore. In contrast, the Great Barrier Reef continental shelf has numerous headlands (Fig. 17) and the waters are deeper, the 20 m contour line nearly touching these headlands. The salinity distribution in coastal waters was measured after the flooding of the Burdekin River and other smaller streams in January 1981. This distribution is shown in Fig. 17 for 26–27 January 1981, about 3 days past the peak of the flood and for 3–4 February 1981, 10 days after the flood. Note that in the first survey the plume moved northward but was continuous over 200 km and coastally trapped with the coast on its left-hand-side, as dictated by Coriolis forces, though offshore the current was southward (Wolanski & van Senden, 1983). In the second survey, the plume was broken up in patches. Brackish water remained preferentially in the wide bays sheltered by headlands.

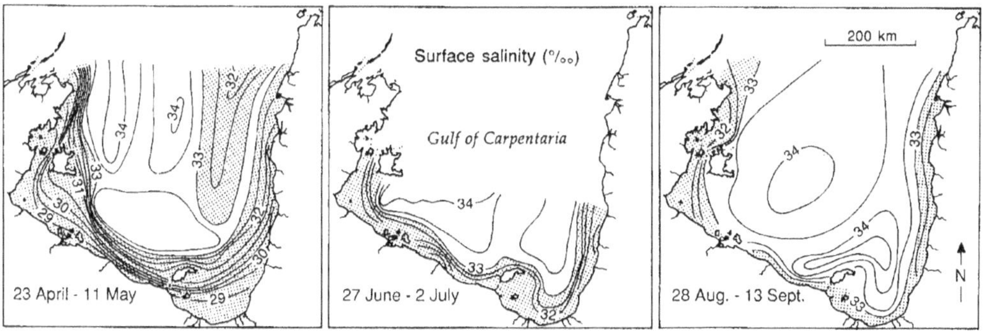

Fig. 14. Salinity distribution in the Gulf of Carpentaria at about three-monthly interval in 1976. These data were kindly provided by P. Rothlisberg.

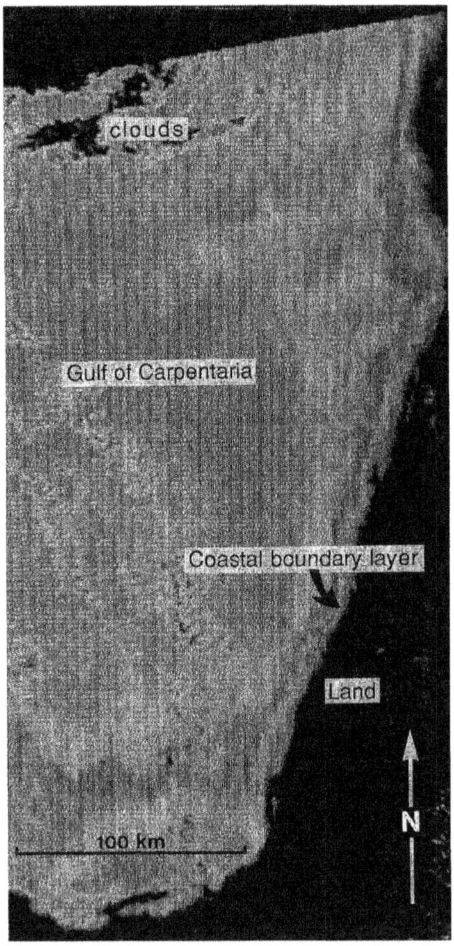

Fig. 15. NOAA satellite view in the visible bands of the east and south coasts of the Gulf of Carpentaria on 23 November 1986, during the hot dry season, in calm weather.

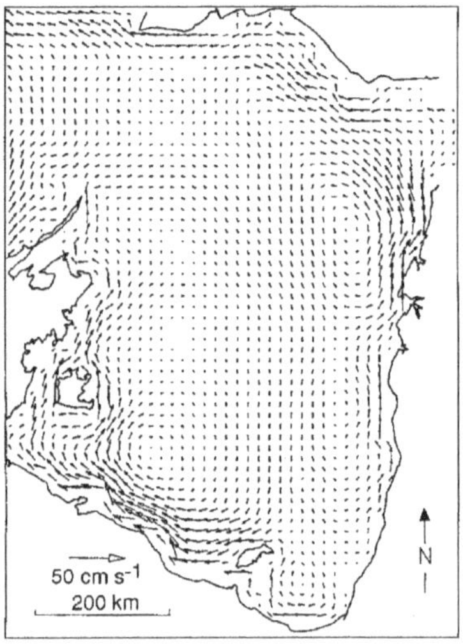

Fig. 16. Predicted mean circulation in the Gulf of Carpentaria under the combined influence of all tides and a $15\ \mathrm{m\ s^{-1}}$ southeasterly trade-wind.

Mixing and dispersion near headlands

These data imply that coastal trapping is much less efficient on the Great Barrier Reef coast than in the Gulf of Carpentaria. This difference may be attributed to the headlands and reefs on the Great Barrier Reef continental shelf, these obstructions to flow being absent in the Gulf of Carpentaria. Flow around headlands and islands generate eddies (Pingree & Maddock, 1979; Wolanski *et al.*, 1984; Falconer *et al.*, 1986; Pattiaratchi *et al.*, 1987; Geyer & Signell, 1990). The two most intensive field studies of currents around a headland or an island in shallow waters have been those of Wolanski *et al.* (1984) and Geyer & Signell (1990). Wolanski *et al.* (1984) documented the presence of energetic island wakes from the data from 24 current meters, float measurements, CTD studies and aerial photography. Geyer & Signell (1989) measured the circulation around a headland using a ship-mounted acoustic Doppler current profiler. Figure 18 shows an example of a fully developed, energetic headland eddy as measured with this latter technique.

Wolanski *et al.* (1984) and Falconer *et al.* (1986) have proposed respectively analytical and numerical models that attempt to explain these eddies in terms of vorticity dynamics at respectively steady and unsteady cases. Their models rely strongly on flow separation dynamics at the separation points at the island's tips. Black & Gay (1977) criticised the vorticity model and claimed that the eddy strength was not primarily governed by flow separation, but is a 'phase eddy' formed as a result of the oscillatory pressure gradient, which near the end of the half cycle, reverses the frictionally controlled flow near the

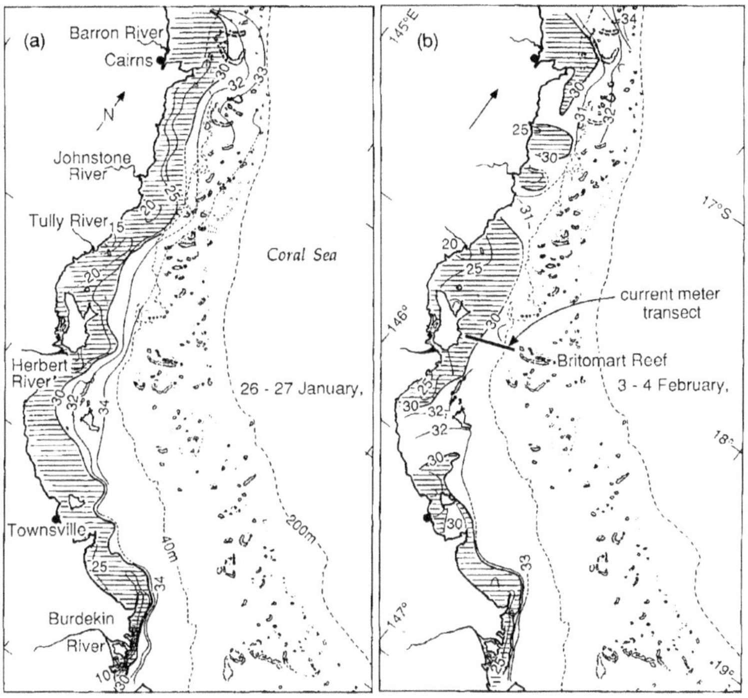

Fig. 17. Surface salinity distribution on the Great Barrier Reef continental shelf between the mouth of the Burdekin River and Cairns, following flooding of the Burdekin River and other smaller streams, in 1981.

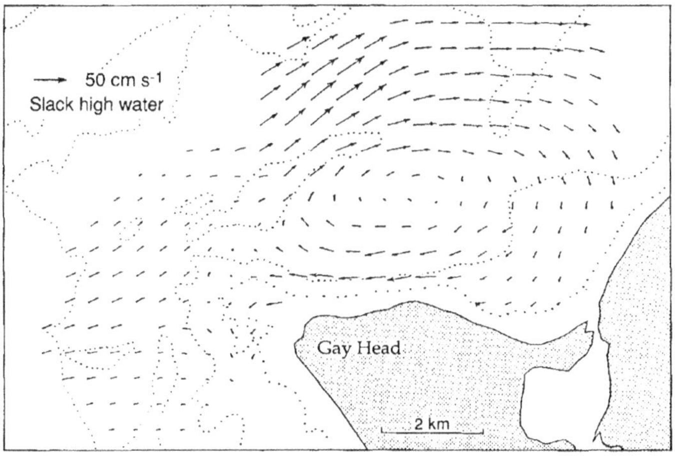

Fig. 18. Observations of the velocity field and evidence for a strong eddy behind Gay Head, Massachusetts, USA, adapted from Geyer & Signell (1990).

coast before the freestream flow. However, Signell (1989) carried out the most detailed, small-scale, numerical studies so far of headland wakes, and showed that the vorticity which forms transient eddies is produced near the headland tip and that flow separation is indeed critical to transport this vorticity into the interior of the eddy.

Deleersnijder *et al.* (1989) have developed a three-dimensional model of the tidal wakes and showed that, though the eddy dynamics are essentially two-dimensional (independent of depth), there exist nevertheless vertical motions, with upwelling near the center of the eddy and downwelling along the island sloping boundaries, as shown in Fig. 19.

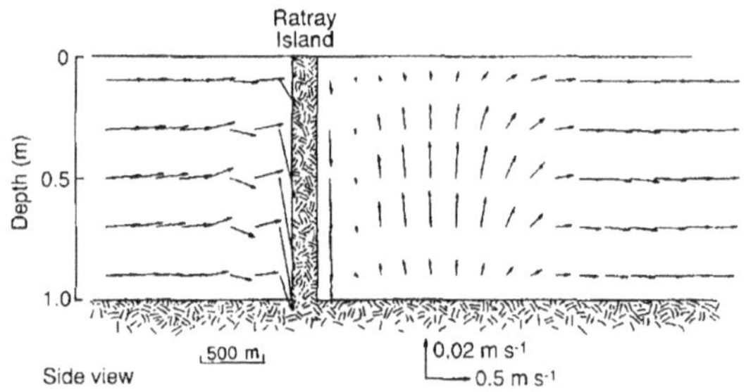

Fig. 19. Numerical predicted three-dimensional circulation around Rattray Island, Australia, after eddy spin-up. This figure shows the velocity field in a vertical plane oriented along the mean flow direction and intersecting the middle of the island. Adapted from Deleersnijder *et al.* (1989).

Signell (1989) also showed the importance of adequate grid resolution near the headland tip. He showed that a lack of small-scale resolution in that area leads to an underestimate of the wake strength (Tee, 1976). This is not simply of academic interest, but has profound implications for modeling the fate of water-borne material. The inclusion, or the neglect, of flow separation effects in numerical models lead to quite different results. When flow separation is included, patches of contaminants in the vicinity of the headlands are strongly deformed by straining associated with the separated shear layer and transient vortices (Signell, 1989). As is shown in Fig. 20, the distribution of material is extremely patchy and streaky for those contaminants entrained in the free shear layers, while those contaminants not entrained in the free shear layers are essentially carried through with minimal mixing. Figure 20 demonstrates that the large strain rate in the region of flow separation tends to stretch particles of fluid into long filaments, which are subsequently rolled and distorted by the transient eddy field (Geyer & Signell, 1990; Signell & Geyer, 1990).

The intense shear associated with the flow separation at the tip of the headland, is the agent responsible for the nearly explosive dispersion of the patch that started close to the headland in Fig. 20. Evidence for such patchiness and streakiness is found in field experiments of dispersion around coral reefs (e.g. Wolanski & Hamner, 1988; Wolanski *et al.*, 1988; Wolanski & King, 1990). On the other hand, models that neglect

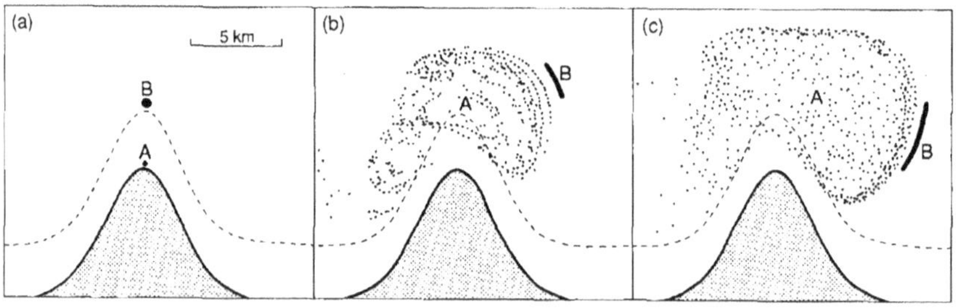

Fig. 20. Numerically predicted dispersion around a headland with reversing tidal currents in the absence of any net currents. Two patches of particles are released at the beginning of the eastward flowing tide, one just off the tip of the headland and the other some distance offshore (*a*). (*b*) shows the distribution after 6 tidal cycles, and (*c*) after 12 tidal cycles. Adapted from Signell & Geyer (1990).

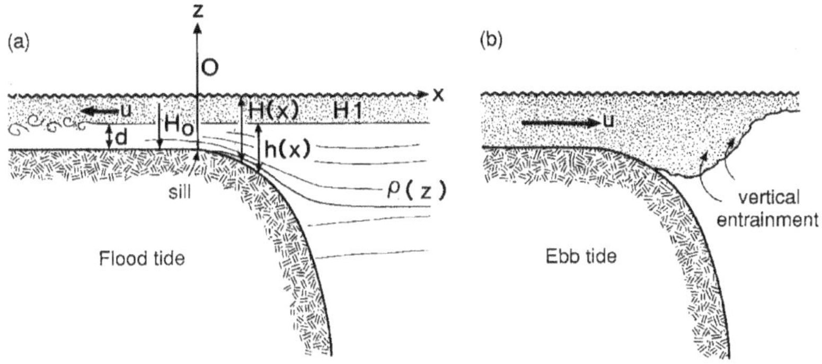

Fig. 21. Upwelling in a tidal jet at (*a*) flood tide and (*b*) ebb tide.

these flow separation dynamics, such as the 'phase eddy' model, are unreliable because they result in smooth distributions from gradient-type diffusion since they do not simulate straining and folding of vortex lines.

The inclusion, or no inclusion, of separation flow effects in numerical models, is not just a question of academic interest only. Indeed, the numerical models used at present to simulate the dispersal of crown-of-thorn starfish in the Great

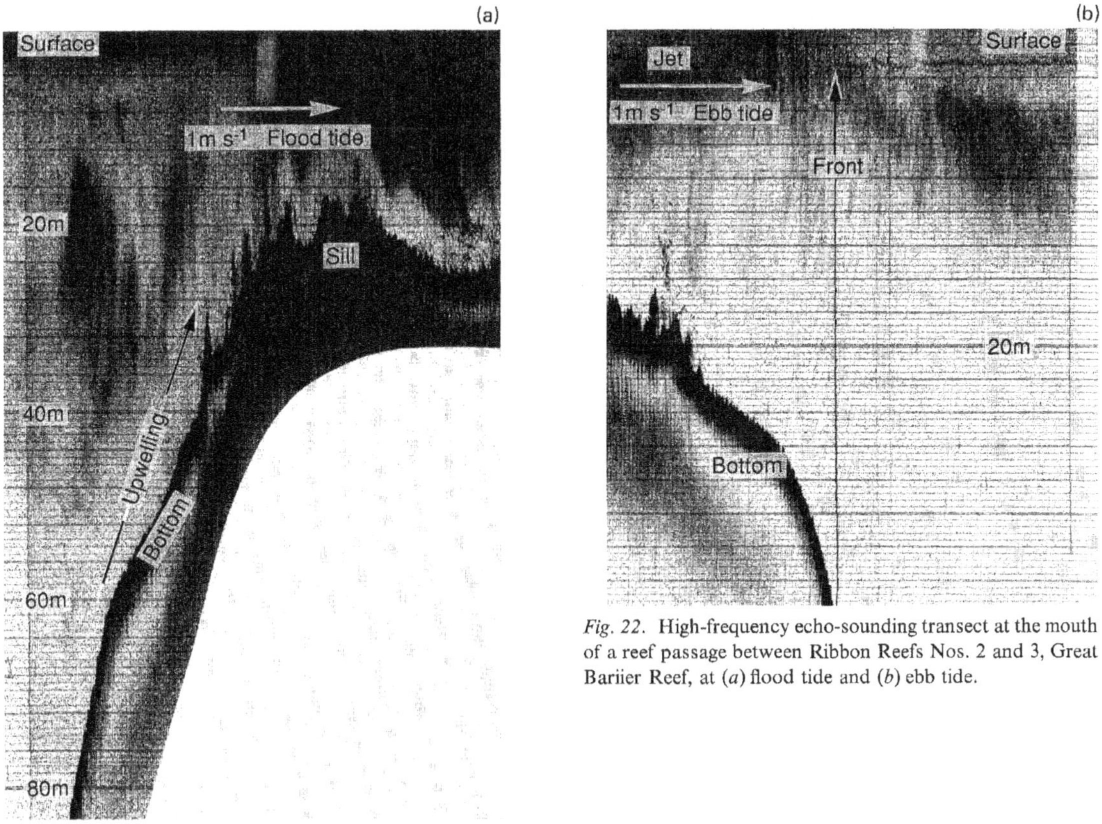

Fig. 22. High-frequency echo-sounding transect at the mouth of a reef passage between Ribbon Reefs Nos. 2 and 3, Great Bariier Reef, at (*a*) flood tide and (*b*) ebb tide.

Barrier Reef, a problem of great practical and management importance, neglect these dynamics and predict smooth distributions and long retention times near reefs (e.g. Black, 1988). Such model predictions are contrary to expectations if straining and folding of vortex lines were included in the models.

Another important mixing mechanism is that of tidal jet-vortex pair system, the dynamics of which is also intimately linked with flow separation effects (Awaji *et al.*, 1980; Awaji, 1982; van Senden, 1985; Wolanski *et al.*, 1988). Awaji *et al.* (1980) and Awaji (1982) found that the pattern of transport and dispersion varies greatly over short length scales and, exactly as in the case of island wakes, does not display the smearing tendency of a typical diffusive process.

Usually mangrove-fringed coastal waters are sufficiently shallow that, through friction (Ozsoy, 1977), tidal jets do not prevail at ebb tides at the mouth of tidal creeks (Wolanski & Ridd, 1990). The frictional slope term balances the surface slope term in the continuity equation, and thus the flow resembles potential flow. However, there exist some mangrove creeks, e.g. Tudor Creek in Mombasa, Kenya, that face deep water. Complex and biologically important flow patterns probably exist at such places and may resemble that observed in front of reef passages through the Ribbon Reefs, Great Barrier Reef of Australia (Thompson & Golding, 1981; Wolanski *et al.*, 1988). At flood tide, forced upwelling may occur as a result of the strong tidal currents at the mouth, as is sketched in Fig. 21a. This upwelling is readily visible in high-frequency echo-soundings, the scatterer probably being plankton (Fig. 22a). At ebb tide, a tidal jet may also develop, lifting off the bottom as a result of buoyancy effects, and resulting in upwards entrainment of deep water in the jet (Fig. 21b). High-frequency echo-soundings (Fig. 22b) also confirm the existence of such a mechanism.

Though a rugged coastline prevents the formation of a long, stable coastal boundary layer, trapping can occur in the shallow embayments sheltered by headlands such as Bowling Green bay, a wide, shallow embayment near Townsville, Australia (Fig. 23). The primary reason for this is that, because of friction, neither headland eddies nor the longshore currents found further offshore are generally able to propagate in these very shallow, sheltered, waters (Wolanski & Ridd, 1990).

Trapping is reinforced when the embayments are mangrove-fringed. Such is the case for instance at the northern region of Hinchinbrook Channel where the west coast is mangrove-fringed and very shallow (Fig. 24a). Water moves back and forth with the tides between the mangrove swamp and the coastal boundary layer in shallow

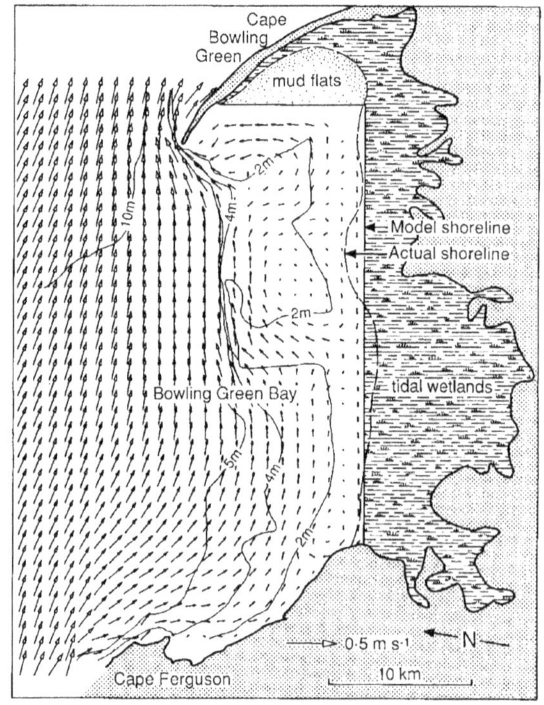

Fig. 23. Net circulation in Bowling Green Bay, a shallow mangrove-fringed embayment near Townsville, Australia, in calm weather. Tides and the southward flowing East Australian Current generate strong currents offshore. The bathymetry is in metres. The bay has extensive tidal wetlands. To make the very small nearshore currents visible on this graph together with the strong offshore currents, the length of the velocity arrows is proportional to the logarithm of the water currents. As a result a factor of 10 difference in the water velocity implies a difference by a factor of 2 in the length of the arrows.

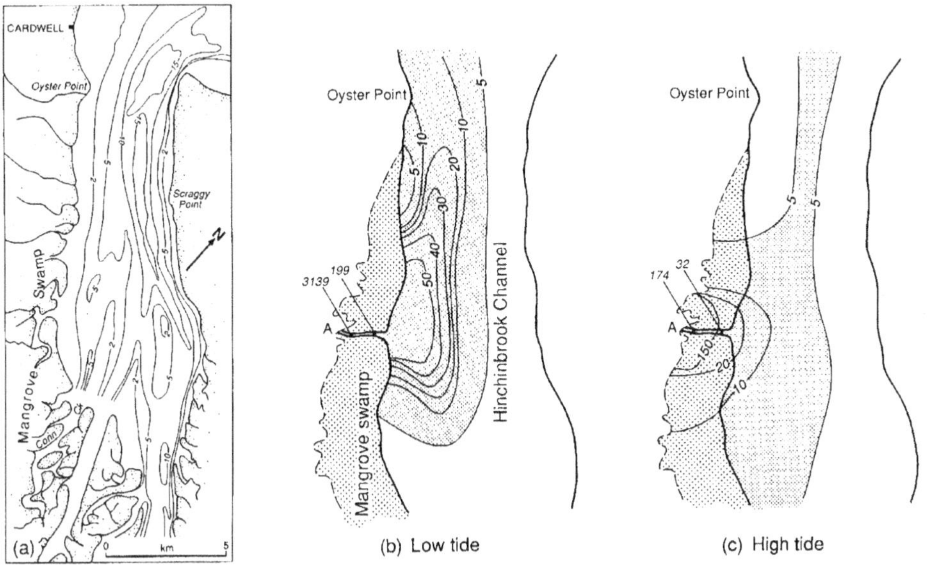

Fig. 24. (*a*) Map of the northern region of Hinchinbrook Channel with depths in m. (*b*) and (*c*) show the plume, at low and high tide, of a tracer after 15 days of continuous discharge at point A in the upper reaches of a mangrove creek in calm weather.

waters on the west coast. The coastal boundary layer waters being very shallow, there is little cross-shelf mixing between coastal boundary layer waters and waters further offshore in the deeper parts of the channel. Wolanski *et al.* (1990) have modeled the water circulation in this area. The mangrove swamps were considered to be inter-tidal storage areas. The model results were verified against field observations of currents. A tracer was released continuously for 15 days at point A in the model, in the upper reaches of a mangrove creek, and followed through time. Figure 24 (b and c) shows the tracer's plume at both low tide and high tide after 15 days in calm weather. The contaminant is found in shallow coastal waters at low tide, and the bulk of this contaminant is found trapped in the mangrove swamp at high tide.

The presence of a coastal boundary layer in shallow, mangrove-fringed, coastal waters, presents problems in interpreting outwelling studies based on a budget of material exported over a tidal cycle by a mangrove creek (e.g. Boto & Bunt, 1981) since water entering a mangrove creek in such areas may simply have left another nearby creek at the previous tidal cycle and is chemically different from offshore water. Thus a mangrove creek at the upstream side of a coastal boundary layer will have an apparent lower bulk return coefficient than one at the downstream end.

Under the action of wind and tides, coastal boundary layer waters, enriched in nutrient by outwelling from the tidal wetlands, can be ejected offshore as tidal jets peeling off capes and headlands (Bowman, 1988; Wolanski & Ridd, 1990). This process may explain the presence of baitfish and billfish schools found a few km offshore from Cape Bowling Green (Williams, 1988).

References

Awaji, T. A., N. Imasato & H. Kunishi, 1980. Tidal exchange through a strait: A numerical experiment using a simple model basin. J. Phys. Oceanogr. 10: 1499–1508.

Awaji, T., 1982. Water mixing in a tidal current and the effect ot turbulence on tidal exchange through a strait. J. Phys. Oceanogr. 12: 501–514.

Black, K. P., 1988. The relationship of reef hydrodynamics to variation in number of planktonic larvae on and around

coral reefs. Proc. 6th. Int. Coral Reef Symposium, Townsville, Vol. 2.
Black, K. P. & S. L. Gay, 1987. Eddy formation in unsteady flows. J. geophys. Res. 92: 9514–9522.
Boto, K. & J. S. Bunt, 1981. Tidal export of particulate organic matter from a northern Australian mangrove system. Estuar. coast. Shelf Sci. 13: 247–255.
Bowden, K. F. & P. Hamilton, 1975. Some experiments with a numerical model of circulation and mixing in a tidal estuary. Estuar. coast. Mar. Sci. 3: 281–301.
Bowman, M. J., 1988. Estuarine fronts. In B. Kjerfve (ed.), Hydrodynamics of estuaries, Vol. 1, CRC Press, Boca Raton, Florida, 85–132.
Bouyne, P. & J. D. Roux, 1978. Remarques sur la circulation d'eaux turbides d'origine amazonienne le long de la cote des Guyanes. C.r. Acad. Sci. Paris, 287, Serie D.: 203–205.
Burke, R. W. & K. H. Stolzenbach, 1983. Free surface flow through salt marsh grass. Massachusetts Institute of Technology Sea Grant College Program, Publication No. MITSG 83-16, Cambridge, MA., 252 pp.
Deleersnijder, E., E. Wolanski & A. Norro, 1989. Numerical simulation of the three-dimensional tidal circulation in an island's wake. In G. M. Carlomagno & C. A. Brebbia (eds), Computer and experiments in fluid flow. Computational Publications, Springer-Verlag, Berlin: 355–381.
Elder, J. W., 1959. The dispersion of marked fluid in turbulent shear flow. J. Fluid Mech. 5: 544–560.
Falconer, R. A., E. Wolanski & L. Mardapitta-Hadjipandeli, 1986. Modelling tidal circulation in an island's wake. J. Waterway, Port, Coastal and Ocean Engineering, ASCE, 112: 234–254.
Fischer, H. B., E. J. List, R. C. Y. Koh, J. Imberger & N. H. Brooks, 1979. Mixing in inland and coastal waters. Academic Press, New York, 484 pp.
Geyer, W. R. & R. P. Signell, 1990. A reassessment of the role of tidal dispersion in estuaries and bays. Estuaries, in press.
Geyer, W. R. & R. P. Signell, 1990. Measurements of tidal flow around a headland with a shipboard acoustic Doppler current profiler. J. geophys. Res. 95: 3159–3197.
Mazda, Y., Y. Sato, S. Swamoto, H. Yokochi & E. Wolanski, 1990. Links between physical, chemical and biological processes in Bashita-Minato, a mangrove swamp in Japan. Estuar. coast. Shelf Sci., in press.
Mazda, Y., H. Yokochi & Y. Sato, 1990. The behaviour of groundwater in a mangrove area and the influence on the properties of water and bottom mud. Estuar. coast. Shelf Sci. 31: 621–638.
Murray, S. P. & M. Young, 1985. The nearshore current along a high-rainfall, tradewind coast – Nicaragua. Estuar. Coast. Shelf Sci. 21: 687–699.
Nokes, R. I., 1986. Problems in turbulent dispersion. Ph.D. Thesis, Dept. Civil Eng., Univ. of Canterbury, Christchurch, New Zealand, 229 pp.
Okubo, A., 1973. Effect of shoreline irregularities on streamwise dispersion in estuaries and other embayments. Neth. J. Sea Res. 6: 213–224.
Ovalle, A. R. C., C. E. Rezende, L. D. Lacerda & C. A. R. Silva, 1990. Factors affecting the hydrochemistry of a mangrove tidal creek, Sepetiba Bay, Brazil. Estuar. coast. Shelf Sci.: 639–650.
Ozsoy, E., 1977. Flow and mass transport in the vicinity of tidal inlets. Technical Rep. UFL/COEL/TR-036, Coastal and Oceanographic Engineering Department, University of Florida.
Pattiaratchi, C., A. James & M. Collins, 1987. Island wakes and headland eddies: a comparison between remotely sensed data and laboratory experiments. J. geophys. Res. 92: 783–794.
Petryk, S. & G. Bosmajan, 1975. An analysis a flow through vegetation. J. Hydraul. Div., Am. Soc. Civil Eng. 101: 871–874.
Pingree, R. D. & L. Maddock, 1979. The tidal physics of headland flows and offshore tidal bank formation. Mar. Geol. 32: 269–289.
Ridd, P. V., M. W. Sandstrom & E. Wolanski, 1988. Outwelling from tropical tidal salt flats. Estuar. coast. Shelf Sci. 26: 243–253.
Ridd, P. V., E. Wolanski & Y. Mazda, 1990. Longitudinal diffusion in mangrove-fringed tidal creeks. Estuar. coast. Shelf Sci. 31: 541–554.
Robertson, A. I., D. M. Alongi, P. Daniel & K. G. Boto, 1988. How much mangrove detritus enters the Great Barrier reef lagoon? Proc. 6th. Inter. Symp. Coral Reefs, Townsville, Aug. 1988.
Signell, R. P., 1989. Tidal dynamics and dispersion around coastal headlands. Ph.D. thesis, MIT/WHOI-89-38.
Signell, R. P. & W. R. Geyer, 1990. Numerical simulation of tidal dispersion around a coastal headland. In R. T. Cheng (ed.), Residual currents and long-term transport in estuaries and bays, Lecture notes on Coastal and Estuarine Studies, Springer-Verlag: 210–222.
Simpson, J. H. & I. D. James, 1986. Convergent fronts in the circulation on tidal estuaries. In D. A. Wolfe (ed.), Estuarine variability. Academic Press, London.
Taylor, G. I., 1921. Diffusion by continuous movements. Proc. London Math. Soc., Ser. A, 20: 196–211.
Tee, K. T., 1976. Tide-induced residual current, a 2-D nonlinear numerical tidal model. J. Mar. Res. 34: 603–628.
Thompson, R. O. R. Y. & T. J. Golding, 1981. Tidally induced upwelling by the Great Barrier Reef. J. geophys. Res. 86: 6517–6521.
Uncles, R., J. E. Ong & W. K. Gong, 1990. Observations and analysis of a stratification-destratification event in a tropical estuary. Estuar. coast. Shelf Sci. 31: 651–666.
van Senden, D., 1985. Structure of tidal jets under variable source conditions. Proc. 21st Congress I.A.H.R., Melbourne, 3C, pp. AI/3–AI/8.
Wattayakorn, G., E. Wolanski & B. Kjerfve, 1990. Mixing, trapping and outwelling in the Klong Ngao mangrove swamp, Thailand. Estuar. coast. Shelf Sci., in press.
Williams, D. McB., 1988. Significance of coastal resources to sailfish and blackmarlin in NE Australia: an on-going re-

search program. Proc. International Billfish Symposium II, Kailua-Kona, Hawaii, Aug. 1–5, 1988.
Wolanski, E., 1986. An evaporation-driven salinity maximum zone in Australian tropical estuaries. Estuar. coast. Shelf Sci. 22: 415–424.
Wolanski, E., 1989. Measurements and modeling of the water circulation in mangrove swamps. UNESCO-COMARAF Serie Documentaire No. 3, 43 pp.
Wolanski, E., 1989. Circulation anomalies in tropical Australian estuaries. In B. J. Kjerfve (ed.), Hydrodynamics of estuaries. Vol. II, CRC Press, Boca Raton: 53–59.
Wolanski, E. & R. Gardiner, 1981. Flushing of salt from mangroves. Aust. J. mar. Freshwat. Res. 32: 681–683.
Wolanski, E. & W. M. Hamner, 1988. Topographically controlled fronts in the ocean and their biological influence. Science 241: 177–181.
Wolanski, E. & B. King, 1990. Flushing of Bowden Reef lagoon. Estuar. coast. Shelf Sci. 31: 789–804.
Wolanski, E. & P. Ridd, 1986. Tidal mixing and trapping in mangrove swamps. Estuar. coast. Shelf Sci. 23: 759–771.
Wolanski, E. & P. Ridd, 1990. Mixing and trapping in Australian tropical coastal waters. In R. T. Cheng (ed.), Residual currents and long-term transport in estuaries and bays. Lecture notes on Coastal and Estuarine Studies, Springer-Verlag: 165–183.
Wolanski, E. & D. van Senden, 1983. Mixing of Burdekin River flood waters in the Great Barrier Reef. Austr. J. mar. Freshwat. Res. 34: 49–63.
Wolanski, E., D. Burrage & B. King, 1989. Trapping and dispersion of coral eggs around Bowden Reef, Great Barrier Reef, following mass coral spawning. Contin. Shelf Res. 9: 479–496.
Wolanski, E., J. Imberger & M. L. Heron, 1984. Island wakes in shallow coastal waters. J. geophys. Res. 89: 10553–10569.
Wolanski, E., M. Jones & J. S. Bunt, 1980. Hydrodynamics of a tidal creek-mangrove swamp system. Aust. J. mar. Freshwat. Res. 31: 431–450.
Wolanski, E., E. Drew, K. Abel & J. O'Brien, 1988. Tidal jets, nutrient upwelling, and their influence on the productivity of the alga Halimeda in the Ribbon Reefs, Great Barrier Reef. Estuar. coast. Shelf Sci. 26: 169–201.
Wolanski, E., Y. Mazda, B. King & S. Gay, 1990. Dynamics, flushing and trapping in Hinchinbrook Channel, a giant mangrove swamp, Australia. Estuar. coast. Shelf Sci. 31: 555–580.

Hydrobiologia **247**: 163–171, 1992.
V. Jaccarini & E. Martens (eds), The Ecology of Mangrove and Related Ecosystems.

Intratidal fluctuations in stratification within a mangrove estuary

R. J. Uncles,[1] W.-K. Gong[2] & J.-E. Ong[2]
[1] *Plymouth Marine Laboratory, Prospect Place, Plymouth, PL1 3DH, UK;* [2] *School of Biological Sciences, Universiti Sains Malaysia, Penang, Malaysia*

Key words: mangrove estuary, salt transport, estuarine stratification, dispersion, vertical shear transport

Abstract

The dependence of salinity stratification on the vertical structure of the longitudinal salinity gradient is investigated by solving the equations of longitudinal momentum and salt conservation. The results are used to interpret measurements made throughout 31 tidal cycles in the Merbok Estuary, Malaysia, as part of an ecological study of a tropical mangrove estuary. The solutions show that a substantial part of the salinity stratification was caused by vertical shear in the currents coupled with advection of the longitudinal salinity gradient. This mechanism led to maximum stratification around low water and minimum stratification around high water. Observed intratidal variations in stratification at neap tides greatly exceeded modelled values when a depth-independent, longitudinal salinity gradient was assumed. Simulations made using a longitudinal salinity gradient of simple depth-dependent, power-law form implied that this difference arose mainly because the near-surface, longitudinal salinity gradient in the field was significantly larger than that near the bed during neap tides and at times of high run-off. The longitudinal, dispersive transport of salt, fresh water and solutes due to vertical shear is shown to be important during periods of stratification at neap tides and during high run-off. The relevance of stratification to mangrove estuaries is discussed.

Introduction

Mangroves are thought to be important in the nutrient budgets of many tropical coastal areas. Boto and Bunt (1981), Boto *et al.* (1984) and Ridd *et al.* (1988) have shown that large quantities of organic and inorganic nutrients are exported from mangrove swamps to coastal waters. Such nutrient outwelling may affect the productivity of coastal fisheries. The longitudinal, tidally-induced dispersive transport of nutrients, fresh water and detritus from mangrove to coastal areas has been investigated by Wolanski and Ridd (1986), Ridd *et al.* (1990) and Wolanski *et al.* (1990). Dispersive transport due to vertical shear in the currents, coupled with density stratification (Dyer, 1974; Rattray & Dworski, 1980) has received less attention, but is likely to be important for these systems during periods of stratification at neap tides and during times of high freshwater runoff.

Uncles *et al.* (1990) showed that salinity stratification and longitudinal transport due to vertical shear dispersion of fresh water (and presumably nutrients and other solutes) were large during neap tides. The stratification and transport were considerably enhanced in the presence of high freshwater runoff. A model of the stratification, in which stratification was related to the longitudinal advection of the horizontal density gradient and the effects of vertical mixing, generated the correct trends and tidal averages but failed to reproduce sufficiently strong intratidal fluctuations in stratification. In this paper we consider intratidal processes in more detail and show that it is necessary to specify the vertical profile of the longitudinal density gradient if the intratidal fluctuations are to be reproduced correctly.

This article briefly reviews observed data for the Merbok Estuary, followed by an outline of the mathematical model and its solutions. The original aspects are the treatment of a schematic form for the vertical profile of longitudinal density gradient and the analysis of intratidal fluctuations in stratification. It is not our objective to derive 'best fits' between model and data, but rather to emphasize the importance of stratification and the physical processes affecting it, as well as complement recent work on this topic (Nunes *et al.*, 1989; Simpson *et al.*, 1990; Uncles *et al.*, 1990; Simpson & Sharples (in press)).

Observations

Study area and measurements

The Merbok Estuary (Fig. 1) is situated in the state of Kedah, Malaysia (5° 40′ N, 100° 25′ E). The tidal influence extends for 35 km. The width ranges from about 2 km at the mouth to about

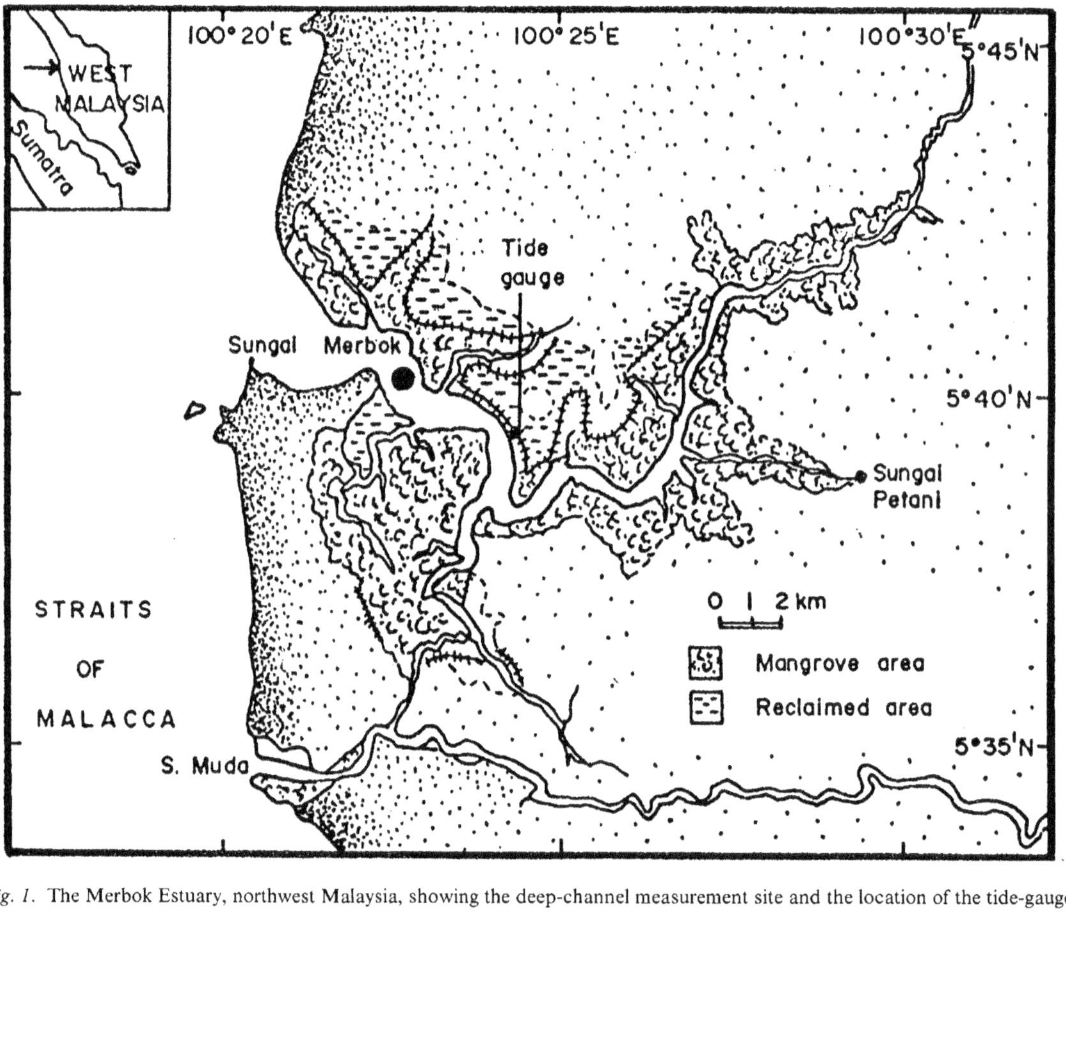

Fig. 1. The Merbok Estuary, northwest Malaysia, showing the deep-channel measurement site and the location of the tide-gauge.

20 m towards the upper reaches of the estuary, with depths generally varying from 15 to 3 m. The mean annual freshwater discharge is estimated to be 20 $m^3 s^{-1}$.

Measurements were made throughout 31 tidal cycles during June 1987 as part of an ecological study of a tropical mangrove estuary. Four stations were worked near the mouth of the Merbok from 0730 h on 2 June until 1100 h on 18 June 1987. Current velocities were measured every lunar hour at five depths using calibrated current vanes (Kjerfve & Medeiros, 1989). Water samples were collected and analysed for salinity at three depths (surface, mid-depth and bed) every alternate lunar hour and at one depth (mid-depth) every other lunar hour. We will present data from the deep-channel (15 m) station in order to extend the analyses presented in Uncles *et al.* (1990) and to provide additional insights into the physical processes at work in both mangrove and temperate estuaries.

Tides, salinity and runoff

The amplitudes of the water-surface elevations were small; root-mean-square (RMS) amplitudes varied from about 0.3 m during the neap tide to about 0.8 m during the following spring tide. These amplitudes were less than about 5% of the mean depth. The depth-averaged tidal currents (ebb positive) are shown in Fig. 2A against time in lunar hours. The RMS current speeds ranged from about 0.2 m s^{-1} during the neap tide to 0.7 m s^{-1} during springs. Freshwater runoff data are also shown in Fig. 2A (plotted as the dashed line at 10 times actual value for clarity). Runoff reached a maximum of about 15 $m^3 s^{-1}$ during the neap tides.

The surface and bed salinity data are shown in Fig. 2B. A considerable reduction in surface salinity occurred during the neap tide, whereas the mid-depth salinity (not shown) and bed salinity remained high. This reduction was due to weak vertical mixing during the neap-tide period and increased runoff (Fig. 2A). The stratification (the difference in salinity between bottom and surface waters) became very small over the spring tide due to increased mixing and showed pronounced intratidal variability over the period.

Longitudinal salinity gradient

The observations permit only the depth-averaged, longitudinal salinity gradient to be estimated (Uncles *et al.*, 1990). A large peak occurred at 100 h when salinity minimized and maximum effects of the runoff were experienced (Fig. 3A). However, because the salinity of waters at the bed (and

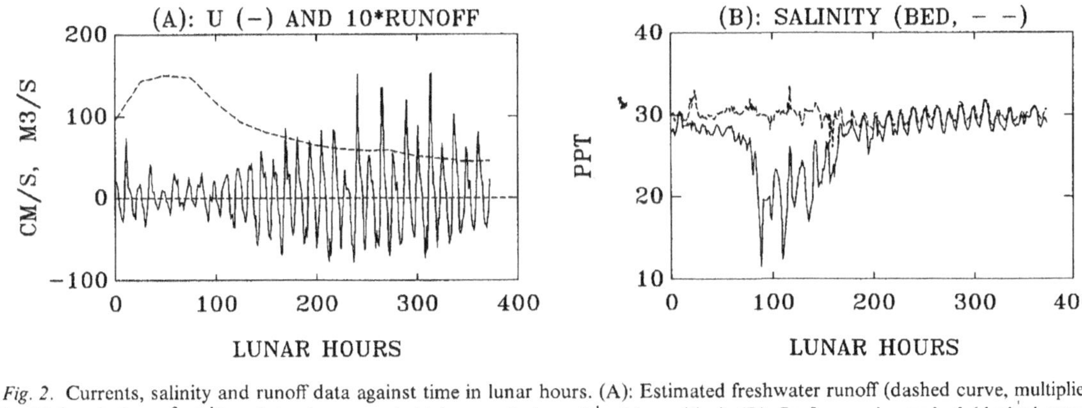

Fig. 2. Currents, salinity and runoff data against time in lunar hours. (A): Estimated freshwater runoff (dashed curve, multiplied by 10 for clarity, $m^3 s^{-1}$) and depth-averaged tidal currents (cm s^{-1}, ebb positive); (B): Surface and near-bed (dashed curve) salinity (ppt).

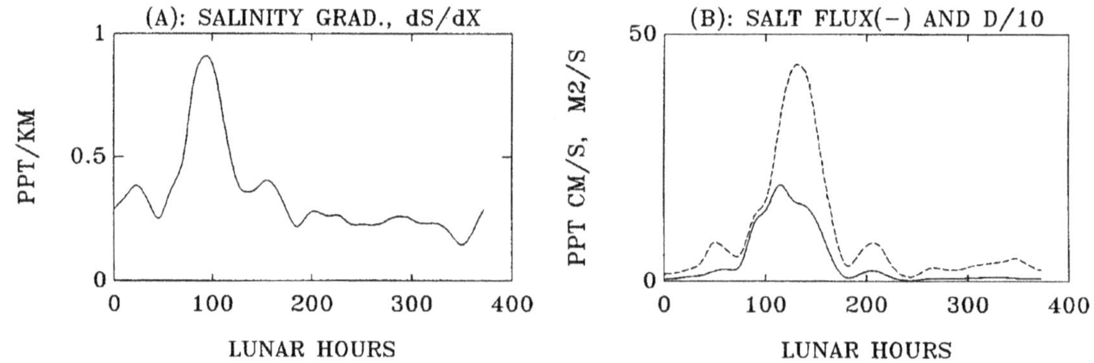

Fig. 3. Tidally averaged variables against time in lunar hours. (A): Estimated, depth-averaged, longitudinal salinity gradient (ppt km^{-1}); (B): Magnitude of the up-estuary salt flux due to vertical shear dispersion (ppt cm s^{-1}) and the associated dispersion coefficient, D (dashed line, plotted one-tenth actual value for clarity, $m^2 s^{-1}$).

mid-depth) were little affected by runoff and tidal state over the period (Fig. 2B), much of the strong longitudinal salinity gradient at 100 h must have been confined to the upper half of the water column.

Vertical shear salt transport

The vertical shear dispersive flux of salt is denoted by F and is defined as (Uncles *et al.*, 1985; Murray & Siripong, 1978; Dronkers & van de Kreeke, 1986; Hunkins, 1981):

$$F = \overline{(U'S')}$$

with $U' = U - \overline{U}$ and $S' = S - \overline{S}$ and where the overbar is a depth average. An estimate of the dispersion coefficient due to vertical shear is:

$$D = -\langle F \rangle / \partial_x \langle \overline{S} \rangle$$

where $\langle\ \rangle$ denotes a tidal average.

The tidally-averaged salt flux due to vertical shear, $\langle F \rangle$, is always directed up-estuary but is small during spring tides (Fig. 3B). Dividing the tidally averaged salt flux by the depth-averaged longitudinal salinity gradient (Fig. 3A) gives an estimate of the dispersion coefficient, D, due to vertical shear (plotted as the dashed line at one tenth actual value for clarity in Fig. 3B). The dispersion coefficient is about 450 $m^2 s^{-1}$ during strong stratification at neap tides and less than about 50 $m^2 s^{-1}$ during spring tides. Therefore, during neap tides, the vertical shear flux is important for flushing fresh water and solutes from the Merbok Estuary to its coastal waters and for importing salt water.

Intratidal stratification model

Basic equations

A vertical mixing model has been presented in Uncles *et al.* (1990). The model includes the competing influences of tidal mixing and buoyancy. The longitudinal acceleration of currents is the result of an imbalance between vertical eddy viscosity and horizontal pressure gradients. There is no wind stress at the surface and a quadratic friction law applies at 1 m above the bed. Longitudinal advection is ignored, so that the only nonlinear aspects of the model are the shear and stratification dependence of the eddy viscosity, and the quadratic dependence of frictional drag on near-bed velocity.

The momentum equation can be written (with, for example, $\partial_t = \partial / \partial t$):

$$\partial_t U = -g\bar{\rho}^{-1}\int_z^{\zeta} \partial_x \rho dz - g[\rho(\zeta)/\bar{\rho}]\partial_x \zeta + \partial_z(N\partial_z U) \quad (1)$$

where longitudinal current is denoted by U, density, ρ, is equal to its depth-averaged value $\bar{\rho}$, except when associated with buoyancy forcing. Eddy viscosity, N (a function of the Richardson number) is determined using a mixing length approach. The surface is at $z = \zeta$, where ζ is surface elevation relative to datum, and the bed is at $z = -h$. The surface elevation displacements within a tidal cycle are ignored ($\zeta \ll h$).

We will explore the effects of depth variations in the longitudinal density gradient ($\partial_x \rho$) of the form:

$$\partial_x \rho = [1 + \beta\{1 - (n + 1)\sigma^n\}]\partial_x \bar{\rho} \quad (2)$$

where β and n are constants, such that $n \geq 0$ and $\beta = 0$ when $n = 0$, and where σ is defined by:

$$\sigma = (z + h)/(\zeta + h) = (z + h)/H$$

such that $\sigma = 0$ is the bed and $\sigma = 1$ is the surface with $H = h + \zeta$ used to denote total depth. If the longitudinal density gradient is directed down-estuary throughout the water column (that is, the density always increases towards the sea) and is never zero, then:

$$-1 < \beta < 1/n$$

Further, if the longitudinal density gradient monotonically increases in the fresher layers towards the surface, then:

$$-1 < \beta < 0$$

In homogeneous conditions, $\beta = 0$.

Averaging Equation (1) over depth and subtracting the depth-averaged version from Equation (1) eliminates the barotropic pressure gradient and gives:

$$\partial_t U' = \bar{G}H[(1 - 2\sigma) + \beta(1 - 2\sigma) - 2\beta\{(n + 2)^{-1} - \sigma^{n+1}\}] + \partial_z(N\partial_z U') + C_D \cdot U_1|U_1|/H \quad (3)$$

where a bottom drag coefficient of $C_D = 2.5 \times 10^{-3}$ is used and where U' is a deviation from the depth-averaged velocity:

$$U' = U - \bar{U}$$

and

$$\bar{G} = -g\partial_x\bar{\rho}/2\bar{\rho} = -3.9 \times 10^{-4} g\partial_x \bar{S} \quad (4)$$

(Bowden & Hamilton, 1975). Boundary conditions for Equation (3) are: At the surface ($z = \zeta$):

$$N\partial_z U' = 0$$

At the bed ($z = -h$):

$$N\partial_z U' = C_D \cdot U_1|U_1|$$

where U_1 is the current velocity at 1 m above the bed. Standard notation is used. Equation (3) is solved by specifying the depth-averaged tidal velocity from observations (Uncles *et al.*, 1990) together with the longitudinal density gradient. This approach avoids the need to solve a continuity equation in order to specify the barotropic pressure gradient (which varies in a complicated way with both intratidal and spring-neap tidal state and freshwater flow).

The salt-balance equation (in which lateral and vertical flows, longitudinal variations in U, horizontal mixing and temporal variations in surface elevation are ignored) may be written:

$$\partial_t S = -U\partial_x S + \partial_z(K\partial_z S) \quad (5)$$

Subject to zero salt flux, $K\partial_z S = 0$, at the surface and bed.

Averaging Equation (5) over the depth and subtracting the depth-averaged version from Equation (5) gives:

$$\partial_t S' = -[U\partial_x S - \overline{U\partial_x S}] + \partial_z(K\partial_z S') \quad (6)$$

In which the advection term can be computed having solved for U' and knowing $\bar{U}$ from the observations.

The eddy coefficients, N and K are functions of the local Richardson Number, velocity shear and mixing length and possess the classical Munk-Anderson damping dependence on water-column stability (Uncles *et al.*, 1990).

Results of model

Longitudinal density-gradient profiles

Having estimated the depth-averaged, longitudinal density (salinity) gradient (Fig. 3A), the stratification, vertical profile of longitudinal current and the longitudinal salt flux due to vertical shear, F, can be computed using Equations 1–6. The results depend upon the way in which the longitudinal density gradient is assumed to vary with depth (that is, the values assigned to β and n in Equation 2).

Figures 4(A,B) show hypothetical, vertical profiles of the longitudinal density (or salinity) gradient, relative to its depth-averaged values, for $\beta = -1/3$ and $\beta = -1$ in Equation 2, respectively. The case $\beta = -1$ corresponds to zero longitudinal gradient at the bed; $\beta = 0$ (equivalent to $n = 0$) corresponds to a longitudinal gradient which is everywhere equal to its depth-averaged value, and $\beta = -1/3$ is an intermediate and more realistic case. Shown in Figs 4(A,B) are the vertical profiles corresponding to $n = 5$, 10 and 20, (Equation 2) together with vertical, dashed lines corresponding to the depth averages (or $n = 0$).

Vertical profiles having values of n greater than

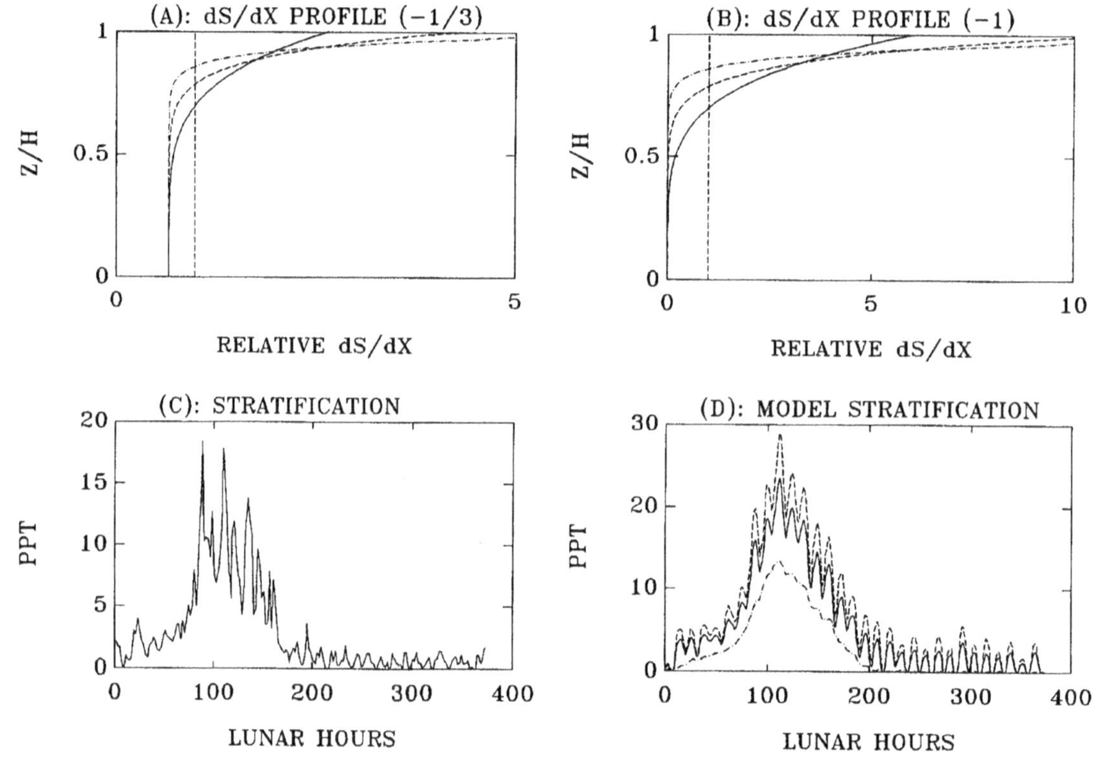

Fig. 4. The effects of a hypothetical, depth-dependent, longitudinal density (salinity) gradient. (A): Vertical profiles of $\partial_x S$ for $\beta = -1/3$ in Equation 2 and $n = 5$ (–), 10 (– –) and 20 (– · –); (B): The same except that $\beta = -1$; (C): Observed stratification (ppt) against time; (D): Modelled stratification (ppt) for $\beta = -1/3$ in Equation 2 and for $n = 0$ (– · –), 5 (–) and 10 (– –) against time. Note that $Z = z + h$ in (A) and (B).

about 5 demonstrate an almost uniform value of the longitudinal density gradient in the bottom half of the water column (Figs 4(A,B)). This near-uniformity of the hypothetical profiles is consistent with the small intratidal variations in salinity which were observed to occur during neap tides in the bottom half of the water column in the Merbok (Fig. 2B).

Calculations of intratidal variability

Results are presented of a calculation in which the depth-averaged, longitudinal density gradient corresponds to the estimated values shown in Fig. 3A. By using the time-varying, longitudinal density gradient we are also effectively taking into account the time-varying freshwater flow. A value of $\beta = -1/3$ is used in Equation 2 as an example of the effects of a hypothetical, vertical profile in the longitudinal density gradient and calculations are made for values of $n = 0$, 5, 10 and 20. The case $n = 0$ corresponds to a constant value of the longitudinal density gradient throughout the water column (equivalent to $\beta = 0$). In nature, one would expect both β and n to vary with time and especially with the state of the spring-neap cycle.

The observed stratification (bed minus surface salinity) is shown in Fig. 4C. The intratidal fluctuations are very pronounced. Maximum stratification tends to occur at low water and minimum stratification at high water. During ebb currents, low salinity water is advected down-estuary in the faster surface layers, which tends to minimize salinity at the surface around low water and maximize stratification. During the flood, vertical shear in the tidal currents is reduced by the density-driven circulation, so that the tendency for stratification to occur as a result of differential advection of the longitudinal density gradient is less. The modelled stratification is shown in Fig. 4D for values of $n = 0$ (dash-dot line), 5 (continuous line) and 10 (dashed line). There is very little difference in the modelled stratification between the cases for $n = 10$ and $n = 20$, so that only the case for $n = 10$ is plotted in Fig. 4D.

It is clear from Figs 4(C,D) that the intratidal variations in stratification cannot be reproduced with $n = 0$ ($\beta = 0$), which corresponds to a depth-independent longitudinal density gradient, although the magnitude of the tidally-averaged stratification is reproduced quite well. With $n = 5$ and $n = 10$ (Fig. 4D) the intratidal fluctuations in stratification are strong and are such that maximum and minimum stratification tend to occur at low and high water, respectively, as observed in the Merbok.

The modelled stratification is greater with increasing n. Almost-fresh water is predicted to occur in the surface layers during minimum neap tides when $n = 10$. This result is a consequence of our assumption that the longitudinal density gradient is independent of distance along the estuary and is therefore not affected by the computed salt flux. The neglect of density-gradient variations along the estuary also precludes us from modelling the behaviour of a surface, freshwater lens. In some estuaries, such as the Hawkesbury, Parramatta and Port Hacking Estuaries in New South Wales, Australia, a short-lived (few hours to a day) freshwater spate can flow to sea as a surface lens and have little effect on the deeper waters (Wolanski, personal communication).

The modelled, intratidal fluctuations in stratification for $n = 5$ and 10 at neap tides are very similar to those observed, whereas at spring tides they are much larger. This is a result of our assumption that the vertical profile of longitudinal density gradient does not change with time. In reality, enhanced vertical mixing at spring tides will tend to reduce variations with depth and thus decrease the absolute magnitudes of β and n.

Relevance of stratification to mangrove systems

The presence of mangrove swamps is most likely to influence the water column stratification indirectly, through an influence on morphology and thus tidal flow and mixing and through an ability to trap fresher water and thus modify longitudinal salinity gradients. However, the longitudinal, dispersive transport of salt and freshwater due to vertical shear in both density-driven and tidal currents, coupled with stratification, will directly influence the export of nutrients, mangrove litter and detritus from the mangrove system to the coastal area and may thus affect the productivity of coastal fisheries.

Wolanski & Ridd (1986) have shown that tidal trapping in the swamps of mangrove estuaries can enhance longitudinal dispersion of nutrients and other materials by temporarily storing water in the swamps during the late flood and releasing it to a different part of the main flow during the ebb. It is likely that stratification will modify this effective longitudinal dispersion of fresh water and nutrients by supplying fresher and more nutrient-rich surface waters to the swamp. Because of stratification, it is possible that mangroves will experience less stress and therefore spend less energy dealing with salt in their effort to extract fresh water for physiological and metabolical processes. It is also possible, as a result of fresher water flooding the mangrove muds during the neap stratification, that more nutrient leaching and export may occur.

Conclusions

(1) A substantial part of the salinity stratification was caused by vertical shear in the currents coupled with advection of the longitudinal salinity gradient. This mechanism led to maximum stratification around low water and minimum stratification around high water.

(2) Observed intratidal variations at neap tides greatly exceeded modelled values when a depth-independent, longitudinal density gradient was assumed. Simulations made using a hypothetical, longitudinal density gradient of power-law form with depth demonstrated that this difference arose mainly because the near-surface, longitudinal density gradient in the field was significantly larger than values lower in the water column during neap tides and at times of high run-off.

(3) The tidally-averaged salt transport due to vertical shear was directed up-estuary and was much smaller during spring tides. The greatest longitudinal salt flux and dispersion coefficient due to vertical shear occurred approximately during maximum stratification, when vertical mixing was very small. This is consistent with theoretical analyses (Fischer *et al.*, 1979) and with the conclusion of Nunes *et al.* (1989) that the spring-neap cycle has an important influence on estuarine stratification and mass transport. This transport will influence the export of nutrients, mangrove litter and detritus from the mangrove system to the coastal area.

Acknowledgements

This project ('Mangrove Malaysia', Centre file 3-P-82-0192) was funded by the International Development and Research Centre (IRDC) of Canada with infrastructure support from the Universiti Sains Malaysia. We are particularly grateful to Dr Brian Davy, Associate Director (Fisheries) and his staff at the IRDC Asia Regional Centre in Singapore for their excellent support. We are also grateful to the British Council for providing a travel grant that allowed one of us (R J U) to participate in this aspect of the project. Our thanks go to Dr Wong Chee Hoong and Dr Zubir Din as well as the 60 or so laboratory staff and students of Universiti Sains Malaysia who cheerfully assisted in the very arduous field and laboratory work.

References

Boto, K. G. & J. S. Bunt, 1981. Tidal export of particulate organic matter from a northern Australian mangrove system. Estuar. coast. Shelf Sci. 13: 247–255.

Boto, K. G., J. S. Bunt & J. Wellington, 1984. Variations in mangrove forest productivity in northern Australia and Papua New Guinea. Estuar. coast. Shelf Sci. 19: 321–329.

Bowden, K. F. & P. Hamilton, 1975. Some experiments with a numerical model of circulation and mixing in a tidal estuary. Estuar. coast. mar. Sci. 3: 281–301.

Dronkers, J. & J. van de Kreeke, 1986. Experimental determination of salt intrusion mechanisms in the Volkerak Estuary. Neth. J. Sea Res. 20: 1–19.

Dyer, K. R., 1974. The salt balance in stratified estuaries. Estuar. coast. mar. Sci. 2: 273–281.

Fischer, H. B., E. J. List, R. C. Y. Koh, J. Imberger & N. H.

Brooks, 1979. Mixing in inland and coastal waters. Academic Press, New York, 483 pp.
Hunkins, K., 1981. Salt dispersion in the Hudson Estuary. J. Phys. Oceanogr. 11: 729–738.
Kjerfve, B. & C. Medeiros, 1989. Current vanes for measuring tidal currents in Estuaries. Estuar. coast. Shelf Sci. 28: 87–93.
Murray, S. P. & A. Siripong, 1978. Role of lateral gradients and longitudinal dispersion in the salt balance of a shallow well mixed estuary. In: B. Kjerfve (ed.), Estuarine Transport Processes. University of South Carolina Press: 113–124.
Nunes Vaz, R. A., G. W. Lennon & J. R. de Silva Samarasinghe, 1989. The negative role of turbulence in estuarine mass transport, Estuar. coast. Shelf Sci. 28: 361–377.
Rattray, M. & J. G. Dworski, 1980. Comparison of methods for analysis of the transverse and vertical circulation contributions to the longitudinal advective salt flux in estuaries. Estuar. Coast. Mar. Sci. 11: 515–536.
Ridd, P. V., M. W. Sandstrom & E. Wolanski, 1988. Outwelling from tropical tidal salt flats. Estuar. coast. Shelf Sci. 26: 243–253.
Ridd, P., E. Wolanski & Yoshihiro Mazda, 1990. Longitudinal diffusion in mangrove-fringed tidal creeks. Estuar. coast. Shelf Sci. 31: 541–554.
Simpson, J. H., J. Brown, J. Mathews & G. Allen, 1990. Tidal straining, density currents and stirring in the control of estuarine stratification. Estuaries 13: 125–132.
Simpson, J. H. & J. Sharples, In press. Dynamically-active models in the prediction of estuarine stratification. In D. Prandle (ed.), Physics of Estuaries and Coastal Seas. Springer-Verlag, Berlin/Heideberg.
Uncles, R. J., R. C. A. Elliott & S. A. Weston, 1985. Dispersion of salt and suspended sediment in a partly mixed estuary. Estuaries 8: 256–269.
Uncles, R. J., J. E. Ong & W. K. Gong, 1990. Observations and analysis of a stratification-destratification event in a tropical estuary. Estuar. coast. Shelf Sci. 31: 651–665.
Wolanski, E. & P. V. Ridd, 1986. Tidal mixing and trapping in mangrove swamps. Estuar. coast. Shelf Sci. 23: 759–771.
Wolanski, E., Yoshihiro Mazda, B. King & S. Gay, 1990. Dynamics, flushing and trapping in Hinchinbrook Channel, a giant mangrove swamp, Australia. Estuar. coast. Shelf Sci. 31: 555–579.
Wolanski, E. Personal Communication. Australian Institute of Marine Science, PMB No. 3, Townsville M.C.. OLD 4810, Australia.

Hydrobiologia **247**: 173–179, 1992.
V. Jaccarini & E. Martens (eds), The Ecology of Mangrove and Related Ecosystems.

Physical processes in the Rufiji delta and their possible implications on the mangrove ecosystem

Julius Francis
Institute of Marine Sciences, P.O. Box 668, Zanzibar, Tanzania

Key words: mangrove swamps, trapping, mixing, estuarine circulation, coastal currents, Rufiji delta

Abstract

The physical processes such as coastal currents, estuarine circulation and monsoon winds prevailing in the Rufiji delta are discussed. The relationship between these processes and the occurrence of long-term trapping of the river discharge and the outflow of waters from the mangrove swamps into the nearshore zone has been observed. The trapped waters in the nearshore zone significantly reduce the mixing between the estuarine and offshore waters, leading to the two waters having distinctive properties. The existence of the trapped waters in the nearshore zone is supported by evidence from a satellite image and aerial photograph and measurements of salinity and suspended sediment concentration. The trapping phenomenon is discussed in the light of its possible implications on the ecology of mangrove ecosystems. Trapping may explain the enhanced growth of the mangrove in the delta compared to other areas. This trapping effect may be providing more time for nutrient retention in the mangrove zone, incorporation of the decomposed leaf litter and fine sediments in the substrate, and settling of fruits and seedlings in the swamps, thereby enhancing the regeneration of the mangrove.

Introduction

The Rufiji delta, which is located in the southern Tanzania, is aligned approximately in a N–S direction. The N–S extent of the delta is about 65 km and extends for about 23 km inland. The delta is traversed by a number of channels among which seven are major (see Fig. 1). Apart from being the largest delta in East Africa, it is also contains the largest estuarine mangrove forest in the eastern part of the continent. The area covered by the mangrove forest is estimated to be about 1022 km^2 (McNae, 1968). The mangrove swamps are dissected by tidal channels and numerous tidal creeks. The flow between the mangrove swamps and the ocean through these channels and creeks is largely controlled by the semi-diurnal tide, with an average tidal range of 2–2.5 m.

The Rufiji river has a mean annual freshwater discharge of about 30×10^9 m^3. The maximum discharge occurs between March and May and the minimum in late October and November, VHL, (1979) classified the Simba Uranga branch as being a partially-mixed estuary during the low freshwater discharge, and stratified during the high discharge.

The northern part of the delta (where most of the observations were made), experiences strong winds during the NE monsoon and therefore, the sea is rough. During the SW monsoon due to relatively weaker winds prevailing, the sea is calm. This is because the northern part of the delta is more exposed to the NE monsoon winds and

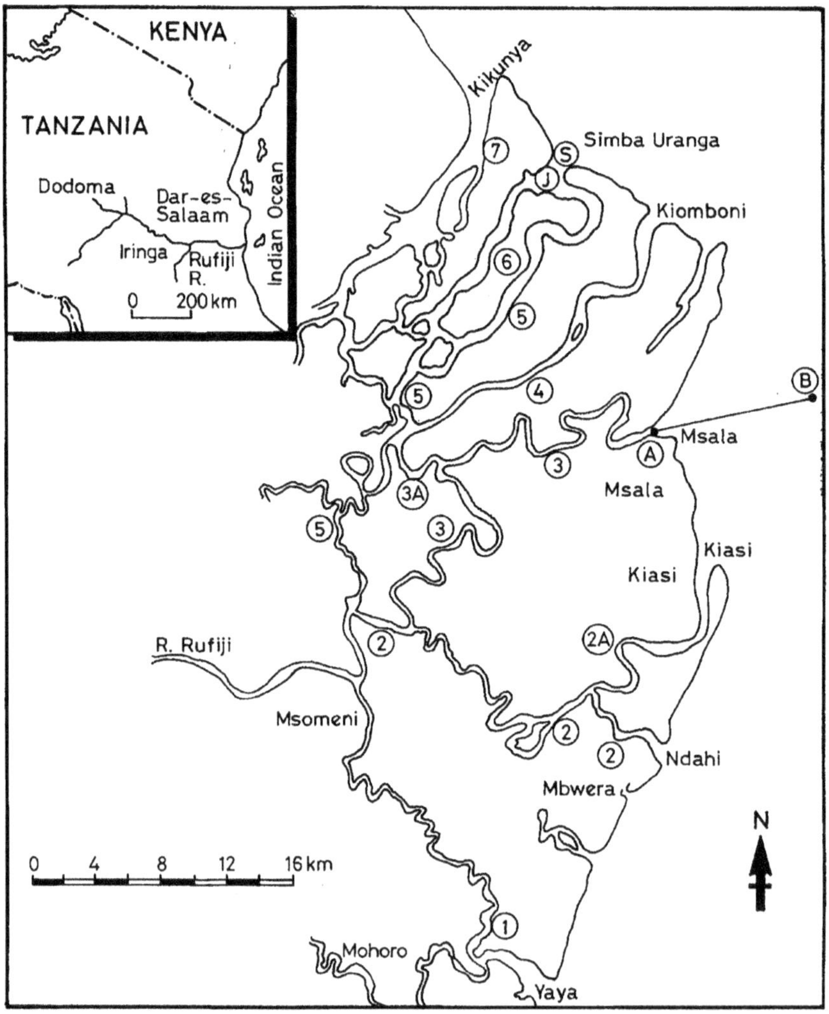

Fig. 1. Map of Rufiji delta area; A–B Suspended sediment concentration transect line, J- Station at the mouth of tidal creek J, and S- Station at the mouth of Simba Uranga channel.

protected to the SW monsoon winds. Generally however, Tanzanian coast experiences stronger winds during the SW monsoon compared to the NE monsoon (Bryceson, 1977).

The freshwater discharge from the delta channels of the Rufiji is trapped in the nearshore zone. The prevailing winds and current conditions are thought to be responsible for the existence of this trapping (Iversen *et al.*, 1984). The existence and the causes of the trapping effect in the nearshore zone and in the mangrove swamps have been established in several areas such as Chwaka Bay in Zanzibar and in the Gulf of Carpentaria and in its adjacent mangrove swamps in Australia (Wolanski & Ridd, 1986; Wolanski & Ridd, 1989; Wolanski, 1989).

Besides the report by Iversen *et al.* (1984), which has discussed briefly the existence of the trapping effect in the nearshore zone off the Rufiji delta, there are no other report that have discussed the subject. Knowledge of the existence of the trapping effect and its implications on the mangrove ecosystems is fundamental to our understanding of the processes that might be affecting the ecology of the mangrove ecosystems.

This paper intends to establish the existence and the causes of the trapping effect in the mangrove swamps and in the nearshore zone off the Rufiji delta, and to discuss the implications of the effect on the mangrove ecosystems.

Material and methods

The salinity measurements were done along the Simba Uranga channel (see Fig. 1). The channel was selected in this study because it is the main branch in the delta river system. The measurements of suspended sediment concentration were done along the Simba Uranga and Msala channels however, in this paper only data from the latter has been used.

A WPA cm 35 conductivity meter was used to obtain the conductivity of the water. The conductivity readings were later converted to salinity readings. The conductivity data were taken from two anchored stations. The first station at the middle of the mouth of the J creek (see Fig. 1). The creek is about 100 m wide at the mouth and 1 km long. This creek drains only the mangrove swamp around it. The second station was at the middle of the mouth of the Simba Uranga channel. The measurements at the two stations were taken for four tidal cycles in four consecutive days i.e. from 21 July, 1990 to 24 July, 1990.

Suspended sediment concentration measurements were done using the Partech Suspended Matter Monitor. At Msala channel the measurements were made in a transect running from the offshore to the mouth of the channel (see Fig. 1). This was done purposely to determine the variation of sediment concentration in the estuarine, nearshore, and offshore waters. The measurements were done on the ebb tide as the sediment concentration on the water is maximum during this period. At Msala channel, the data were collected on 5 September, 1989. Information for this paper have also been obtained from the interpretation of satellite image and aerial photograph.

Results and discussion

At the mouth of J creek, the salinity values for the surface waters were found to fluctuate slightly at tidal frequencies, being relatively high during flood tide and low during the ebb tide period. The salinity values ranged from 10.60‰ to 14.26‰ (see Fig. 2a). The data collected at a depth of more than 1 m below the surface showed that during the flood tide more saline water was flowing into the creek, as the values of salinity ranged from 18.12‰ to 30.70‰. The small variation and small range in salinity values at the J creek over tidal cycle suggest there is little mixing between the water from the creek and the water in the channel. Also the low salinity values obtained were probably due to freshwater discharged several weeks earlier during the rainy season, as for the period of about one month preceding the data collection there was no rain in the areas around the Simba Uranga Island and the floods of the Rufiji river had ceased about two months before. All these facts are implying the existence of the trapping of low salinity water in the mangrove swamps.

The surface salinity values obtained at the mouth of the Simba Uranga channel showed a relatively large variation between the peak flood tide values and the corresponding ebb tide values. The salinity values were low during the ebb tide and high during the flood tide period (see Fig. 2b). This large variation can be explained by the fact that the more saline nearshore waters move inland along the channel during the flood tide and back to the nearshore zone during the ebb tide.

Measurements of suspended sediment concentration taken during the ebb tide at the Msala channel showed that there was high concentration of sediment in the nearshore waters compared to either the estuarine or the offshore waters. The concentration ranged from less than 490 mgl^{-1} for the estuarine waters to about 1000 mgl^{-1} for the nearshore waters. For the off-

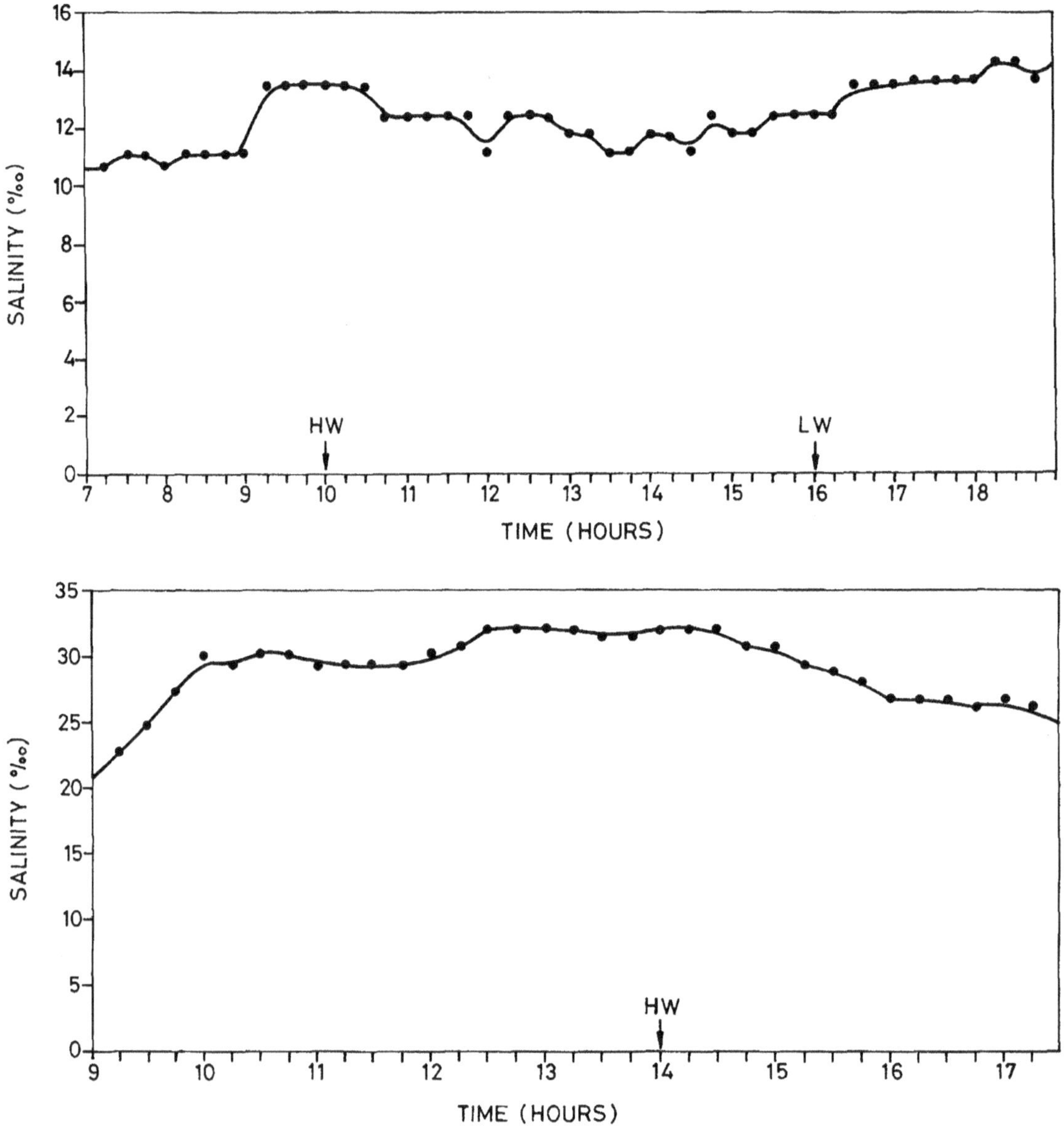

Fig. 2. Variation over a tidal cycle of surface salinity; (a) at the mouth of tidal creek J – 21 July, 1990 and (b) at the mouth of Simba Uranga channel – 24 July, 1990.

shore waters zero values were dominant (as shown in Table 1). A parallel picture is also drawn from the aerial photograph (Fig. 3), which indicates that nearshore waters contain more suspended sediment than either the estuarine or the offshore waters. It is clear that the separation between the nearshore and the offshore waters is well defined in many areas, suggesting the absence of intensive mixing between the two water masses. The suspended sediment concentration

Table 1. Suspended sediment concentration along the transect A–B.

Time (hour)	Depth (m)	Concentration (mg l^{-1})
1235	3.00	490
1235	1.00	480
1245	1.00	500
1251	1.00	490
1304	0.50	720
1312	0.50	890
1320	0.30	1000
1332	1.50	1000
1345	1.00	980
1256	1.00	1000
1405	1.00	50
1416	1.00	100
1428	1.00	150
1440	1.00	0
1456	1.00	0

measurements and the interpretation of the aerial photograph indicate a trapping effect in the nearshore waters off the delta.

The East African Coastal Current (EACC) system may be responsible for causing this trapping effect. They are generally flowing northwards throughout the year, being faster during the SW monsoon and slower during the NE monsoon (Newell, 1959). The aerial photography shows the existence of northward flowing currents, which could be the EACC system. These currents are inhibiting the flow of nearshore waters into the deeper offshore waters, consequently leading to the outflows from the delta channels being restricted in the nearshore zone.

The Simba Uranga channel being stratified during the period of high discharge and partially

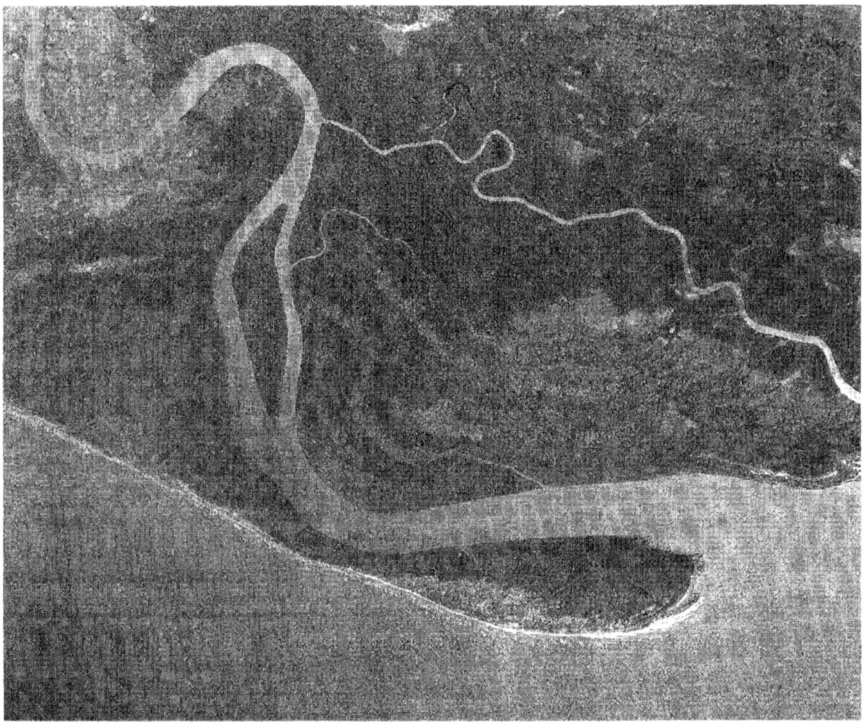

Fig. 3. Satellite image and aerial photograph; (a) Landsat image of the Rufiji delta with dates: 23 June, 1981, 23 June, 1981, 24 June, 1981, and 18 March, 1985 and (b) Aerial photograph of part of the delta taken on 10 June, 1965.

well mixed during the low freshwater discharge suggests the absence of strong mixing processes. It seems that the monsoon winds and/or the tidal currents are unable to cause intensive mixing between either the estuarine and the nearshore waters or the nearshore and the offshore waters.

The trapping effect discussed above may have several implications on the ecology of the mangrove ecosystems as will be indicated in the foregoing discussion. The trapping effect may be among factors controlling the amount of suspended particulate matter incorporated in the substrate in the mangrove swamps relative to the amount exported to the adjacent marine ecosystems. Other factors include the substrate characteristics, the frequency of tidal inundation, the energy of the tide, and the pattern of production and supply of mangrove litter (Woodroffe, 1985). This is due to the fact that trapping provide more time for the suspended matter to settle out. Also more time is provided for the fruits and seedlings of mangrove to settle in the swamp and regenerate instead of being washed away by the tidal action as it is in the case where trapping does not exist or exist for few days. Due to little mixing which exists between the estuarine and the nearshore waters, the dissolved matter is not flushed from the system by tidal action, and remains within the swamps for a long time. This provides more opportunity for absorption of the nutrients by the mangrove as practically the same water would be passing through them over and over again for prolonged period.

The mangroves develop best in areas which experiences periodic deposition of sediments (Cintron & Schaeffer-Novelli, 1983). In the Rufiji delta, the maximum sediment transport occurs during the period of maximum outflow (Euroconsult, 1980). Some of those sediments are deposited in the mangrove swamps and the rest are transported to the nearshore zone where as discussed, they are trapped. However, during the SW monsoon, there is also considerable deposition of sand in the mangrove swamps of Kiomboni and Simba Uranga islands (pers. obser.). The origin of these sands is thought to be the nearshore zone as they are well rounded and are found together with broken shells. In view of this it seems that the mangrove are receiving sediments from both terrestrial and marine environments at different times in a year.

Conclusion

Physical processes play a significant role in the existence of the long-term trapping effect in the mangrove swamps and in the nearshore zone. The presence of northward flowing EACC system and the slow mixing processes are probably the main factors responsible for the existence of the trapping effect in the swamps and in the nearshore zone. The intensity of trapping effect may be varying with the monsoon season, being more pronounced during SW monsoon than during the NE monsoon. This is due to the fact that the NE monsoon are characterized by weaker EACC system, lower freshwater discharge, and relatively strong monsoon winds.

The enhanced growth of the mangrove in the delta may be explained not only by the fact that the mangrove swamps receive significant amount of nutrients from freshwater discharge, but also from the fact that most of nutrients supplied by the river and from other sources are retained in the swamps or in the nearshore zone for a prolonged period due trapping effect.

Due to limited data available some of the issues raised in the discussion such as the ecological implications of the trapping effect on the mangrove and the possibility of the seasonal variations in the trapping effect etc. are not conclusive.

Acknowledgements

It is a pleasure to thank FRONTIER Expeditions for all the support they gave me during data collection. Thanks also should go to Messrs. J. P. Shunula, A. N. N. Muzuka, Y. W. Shaghude, and A. J. Mmochi all of whom helped me in various ways, especially for their valuable constructive comments and suggestions to the first draft of the manuscript.

References

Bryceson, I., 1977. An ecological study of the phytoplankton of the coastal waters of Dar-es-Salaam. Ph.D. Thesis, University of Dar-es-Salaam.

Cintron, G. & Y. Schaeffer-Novelli, 1983. Mangrove Forests: Ecology and response to natural and man-induced stressors. In J. C. Ogden & E. H. Gladfelter (eds), Coral reefs, seagrass beds, mangrove: Their interaction in the coastal zones of the Caribbean. UNESCO report in marine science 23: 87–113.

Euroconsult., 1980. Identification study on the ecological impacts of the Stiegler's Gorge Power and Flood Control Project Part 2: Methodology, Definition of impact area and Resources inventory.

Iversen, S. A., S. Myklevoll, K. Lwiza & Y. Yonazi., 1984. Tanzanian marine fish resources in the depth region 10–500 m investigated by R/V 'Dr. Fridtjof Nansen'. In S. A. Iversen & S. Myklevoll (eds), The Proc. of the NORAD-Tanzania Seminar to review the marine fish stocks and fisheries in Tanzania. Mbegani, Tanzania. March 6–8, 1984: 45–83.

MacNae, W., 1968. A general account of the fauna and flora of mangrove swamps and forest in the Indo-West Pacific region. Adv. Mar. Biol. 6: 73–270.

Newell, D. S., 1959. The hydrography of the British East African Coastal Waters. Fishery Publs. Colon. Off. No. 12. 18 pp.

VHL., 1979. Hydrological studies in the Lower Rufiji River. Trondheim, Norway. 100 pp.

Wolanski, E., 1989. Measurements and modelling of the water circulation in the mangroves swamps. Results of practical work at the COMARAF Regional Field Training Workshop on Mangrove Ecosystems and Estuarine Processes. Chwaka Bay, Zanzibar. June 6–16, 1989. Regional Project RAF/87/038.

Wolanski, E. & P. Ridd, 1986. Tidal mixing and trapping in mangrove swamps. Estuar. coast. mar. Sci. 23: 759–771.

Wolanski, E. & P. Ridd, 1989. Coastal mixing and trapping in tropical Australia. In R. T. Cheng (ed.), Dynamics of Shallow Bays and Estuaries. (in press).

Woodroffe, C. D., 1985. Studies of a mangrove basin, Tuff Crater, New Zealand: 1. Mangrove biomass and production of detritus. Estuar. coast. mar. Sci. 20: 265–280.

Hydrobiologia **247**: 181–186, 1992.
V. Jaccarini & E. Martens (eds), The Ecology of Mangrove and Related Ecosystems.

Diurnal variation in physical-chemical properties and primary production in the interconnected marine, mangrove and freshwater biotopes of Kakinada coast, Andhra Pradesh, India

V. Selvam, Jayapaul Azariah & Hilda Azariah*
*Department of Zoology, University of Madras, Guindy Campus, Madras – 600 025, India; *CAS in Botany, University of Madras, Guindy Campus, Madras – 600 025, India*

Key words: Diurnal variation, mangrove water, dissolved inorganic nutrients, primary production

Abstract

Diurnal variation in hydrological variables and dissolved inorganic nutrients such as PO_4^{3-}-P, N O_2^--N, NO_3^--N and NH_4^+-N were studied in three interconnected biotopes including freshwater, marine and mangrove brackish water of the Kakinada coastal zone, Andhra Pradesh. Samples were collected at intervals of 3 hours, for a period of 24 hours. In the marine environment salinity varied from 26‰ to 32‰ whereas in the mangrove waters it fluctuated from 12‰ to 20‰ and in both biotopes salinity showed bimodal type of oscillation. Dissolved oxygen content was high in the mangrove waters during day time but decreased rapidly during the night hours. In the marine environment PO_4^{3-}-P concentration varied from 0.345 to 1.195 μg at l^{-1}, NO_3^--N from 1.03 to 6.62 μg at l^{-1} and NO_2^--N from 0.086 to 0.506 μg at l^{-1}. The highest and the lowest concentrations of PO_4^{3-}-P, NO_3^--N, NO_2^--N recorded in the mangrove waters were 0.790 and 0.325 μg at l^{-1}, 7.10 and 1.60 μg at l^{-1} and 0.278 and 0.060 μg at l^{-1}, respectively. The concentration of PO_4^{3-}-P, NO_3^--N and NO_2^--N were high in the freshwater canal, the maximum and minimum values being 1.110 and 0.730 μg at l^{-1}, 26.40 and 9.98 μg at l^{-1} and 0.520 and 0.252 μg at l^{-1} respectively. The concentration of ammonia was relatively high in the mangrove water. Gross and net primary production in the mangrove water was 4 times higher than in the marine biotope. There was no 'export' of dissolved nutrients from the mangrove environment to the adjacent marine waters.

Introduction

Understanding the hydrography and nutrient cycling in the mangrove water is complicated by the continual mixing of water masses with different physical-chemical properties (Odum *et al.*, 1982). Saraya (1984) showed that the environmental conditions of mangrove waters are largely governed by two dominant factors, namely, short term changes resulting from tidal inundation and seasonal changes induced by the monsoonal cycles. Most of the investigations carried out on the hydrography and nutrients in mangrove waters are based on long term observations and are restricted to the brackish water surroundings of the mangroves. Hence, there is a lack of understanding on the influence of marine and freshwater discharge on the diurnal and tidal variations in the physico-chemical properties of the mangrove waters. In the present study diel and tidal changes in the hydrographical variables and nutrients were studied at three interconnected biotopes namely, freshwater, marine and mangrove brackishwater in order to obtain: (i) the pattern of the diurnal and tidal variations in the physical-chemical properties of the mangrove water and (ii) to understand the influence of marine and fresh water discharge on such variations, if any.

Materials and methods

A mangrove swamp covering an area of approximately 96 hectare is present near the Kakinada town (16° 45′ N and 82° 15′ E) in Andhra Pradesh. The vegetation of the swamp consists of *Avicennia officinalis* L., *A. marina* (Forsk.) Verh, *Excoecaria agallocha* L., *Acanthus ilicifolius* L. with *A. officinalis* as the dominant species. The swamp, on the southern side is bordered by the Commercial Canal which receives surplus freshwater from irrigated land and is characterised by the presence of the inflow of freshwater throughout the year. On the western side of the Kakinada Bay (average salinity 32‰) lies a mangrove swamp. The swamp, Commercial Canal and the Kakinada Bay are interconnected as shown in Fig. 1. When the freshwater is minimal in the Commercial Canal seawater from the Kakinada Bay enters it at high tide and mixes with the freshwater. The resulting brackish water enters through a channel branching off the left side of the Commercial Canal. During the monsoon seasons, heavy influx of freshwater in the Commercial Canal prevents the entry of seawater from the Kakinada Bay and during such a period freshwater from the Commercial Canal directly reaches the mangrove environment. In each biotope one sampling site was fixed and designated as Station I (Kakinada Bay), Station II (Mangrove Swamp) and Station III (Commercial Canal). Single surface samples were collected from these stations at 3 hour intervals from 0600 h, 20th August, 1988 to 0600 h, 21st August, 1988. Salinity was measured by an optical salinometer (American Opticals, Hong Kong, Cat. No. 10423, ± 1‰). For dissolved oxygen analysis, samples were fixed in the field using manganous sulphate and alkaline iodide solution. Winkler's method

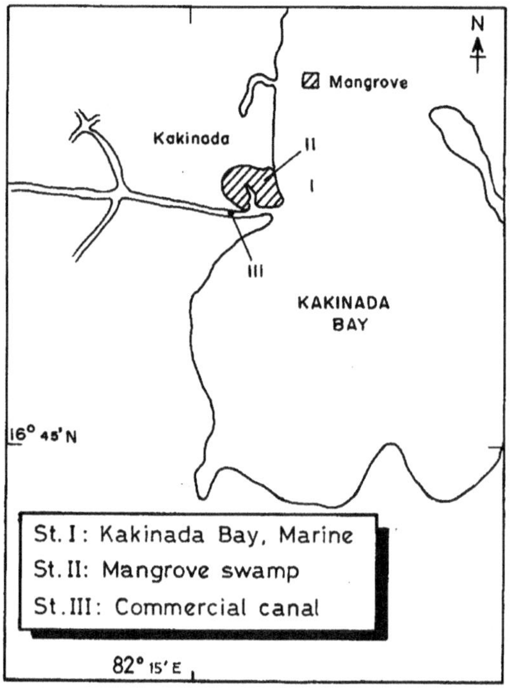

Fig. 1. Map of the Kakinada Bay showing the mangrove swamp and sampling stations.

given by Strickland & Parsons (1968) was followed for the estimation of dissolved oxygen. For the analysis of dissolved nutrients, water samples collected from surface were brought to the laboratory within 30 min of collection and filtered through a Whatman GF/C filter paper of 0.05 μm porosity. The filtered samples were deep frozen and the nutrients were analysed the next day. Analysis of PO_4^{3-}-P, NO_3^-N, NO_2^--N and N H_4^+-N was carried out according to the methods given by Parsons *et al.* (1984). The light and dark bottle method as described by Strickland & Parsons (1968) was followed for assessing the primary production. Water samples in duplicate were incubated *in situ* for 6 hrs from 0900 hrs to 1500 hrs. Due to logistical difficulty primary production was estimated only at Stations I and II. Atmospheric temperature was noted at the time of sampling.

Results

Tide

The tide of the study area is of semi-diurnal type (Fig. 2a). Tidal level increased gradually from 0600 hrs and reached the first peak at 1200 hrs. The second peak was recorded at 2400 hrs. The tidal amplitude was 1.52 m.

Water temperature

Water temperature at the three stations varied from 26.9 to 30.8 °C (St. I), 26.9 to 31.2 °C (St. II) and 27.4 to 30.4 °C (St. III). Water temperature at all stations increased gradually from 0600 hrs and reached its peak at 1500 hrs (Fig. 2b), closely following the atmospheric temperature. Water temperature in the mangrove water was slightly higher than in the other two biotopes during most of the period of observation.

Salinity

Salinity showed a semidiurnal oscillation at Sts. I and II. Salinity remained around 0‰ at St. III (Fig. 2c). At St. I salinity varied from 26‰ to 32‰ with two peaks one occurring at 1200 hrs and the other at 1800 hrs, indicating the semidiurnal nature of the tidal oscillation. At St. II salinity reached the peak values at the same time as at St. I but the magnitude was smaller, the maximum and minimum values being 20‰ and 12‰ respectively.

Dissolved oxygen

Diel variation in dissolved oxygen concentration at St. I and III was smaller than at St. II. It was within a range of 2.77 to 5.07 ml l^{-1} at St. I and 2.65 to 4.83 ml l^{-1} at St. III. At St. II dissolved oxygen concentration fluctuated from 2.41 to 7.24 ml l^{-1} and at this station dissolved oxygen content was high during the day time, and decreased rapidly during the night hours (Fig. 2d). High concentration of dissolved oxygen in the

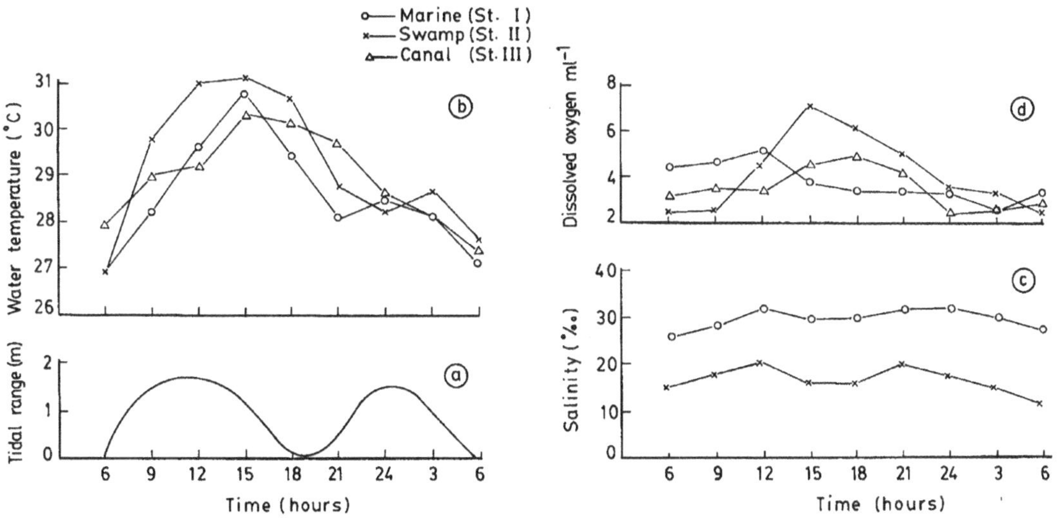

Fig. 2. Diurnal variations in (a) tidal range (m), (b) water temperature (°C), (c) salinity (‰), and (d) dissolved oxygen (ml l^{-1}).

mangrove water (St. II) might be due to a high rate of oxygen production by phytoplankton.

Nutrients

Dissolved inorganic phosphate

Dissolved inorganic phosphate at the three stations varied from 0.345 to 1.195 μg at l^{-1} (St. I), 0.325 to 0.790 μg at l^{-1} (St. II) and 0.730 to 1.110 μg at l^{-1} (St. III) (Fig. 3a). At Stations I and II PO_4^{3-}-P concentration, during the day time, was low and showed no relationship with tides whereas during the night hours PO_4^{3-}-P concentration increased at the time of high tide and decreased during the low tide regime. From Fig. 3a it is evident that there was significant variation in PO_4^{3-}-P concentration at all stations over the sampling period.

Nitrite-nitrogen

At Stations I and II, NO_2^--N concentration varied within the range of 0.086 to 0.506 μg at l^{-1} and 0.060 to 0.278 μg at l^{-1} respectively which is comparatively much lower than the levels recorded at St. III. At Sts. I and II, NO_2^--N was lower during the day time but rose at night. At St. I, NO_2^--N concentration during the night hours increased the flood tide whereas at St. II nitrite showed significant variation between day and night time but this did not correlate with the tide. At St. III, NO_2^--N fluctuated from 0.252 to

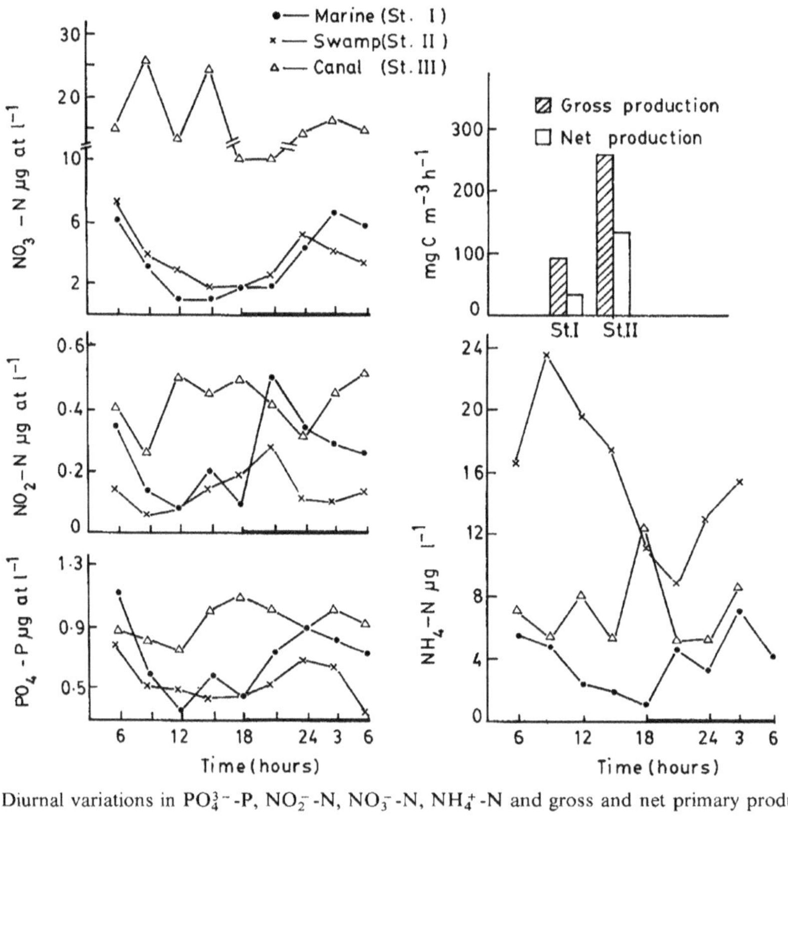

Fig. 3. Diurnal variations in PO_4^{3-}-P, NO_2^--N, NO_3^--N, NH_4^+-N and gross and net primary production.

0.520 μg at l^{-1} and showed only a minor variations from 1500 hrs to 2100 hrs.

Nitrate-nitrogen

In the Kakinada Bay (St. I) and mangrove waters (St. II) NO_3^--N showed a more or less similar range of variation, the value being 1.03 to 6.62 μg at l^{-1} and 1.60 to 7.10 μg at l^{-1} respectively. At these two stations NO_3^--N concentration decreased gradually from 0600 hrs and reached the lowest value at 1500 hrs and thereafter an increasing trend was noticed till 0300 hrs. Like PO_4^{3-}-P and NO_2^--N, NO_3^--N concentration at these two stations, was independent of the oscillations in tides during day time. However, during night hours NO_3^--N concentration increased during the high tide period and decreased during the low tide regime. At St. III, NO_3^--N concentration during day time fluctuated erratically with the following values: 0600 hrs: 14.86 μg at l^{-1}; 0900 hrs: 26.40 μg at l^{-1}; 1200 hrs: 13.88 μg at l^{-1}; 1500 hrs: 25.08 μg at l^{-1}; 1800 hrs: 9.98 μg at l^{-1} (Fig. 3c). At St. III there was intensive human activity, with the movement of mechanised fishing boats. These boats leave around 0500 hrs to 0900 hrs and return around 1500 hrs. It might be thought that boat activity after 0500 hrs and 1500 hrs explains the peaks in NO_3^--N in the freshwater canal. However, this is an unlikely explanation because one would expect similar peaks in the concentrations of the other nutrients which were not observed.

Ammonia-nitrogen

Ammonia concentration was low at St. I, the highest and lowest values being 1.120 and 7.128 μg at l^{-1} respectively. At St. II, NH_4^+-N concentration was high, the maximum and minimum values being 23.65 and 8.85 μg at l^{-1}. At St. I, NH_4^+-N concentration decreased gradually from 0600 hrs to 1800 hrs whereas at St. II, NH_4^+-N concentration showed a steady decrease from 0900 hrs to 2100 hrs. At St. III, NH_4^+-N showed no definite pattern in diel changes. At St. I and II, unlike PO_4^{3-}-P, NO_2^--N and NO_3^--N, NH_4^+-N did not show any relationship with tides even during night hours which may indicate that the regeneration and utilization of NH_4^+-N is independent in each biotope.

Primary production

Gross production in the marine (St. I) and mangrove water (St. II) was 82.4 mg C m^{-3} h^{-1} and 262.3 mg C m^{-3} h^{-1} respectively. Net production at Sts. I and II was 31.2 mg C m^{-3} h^{-1} and 136.5 mg C m^{-3} h^{-1} respectively. High primary production at St. II may attributed to the optimum salinity condition (12‰ to 20‰) prevailing in the mangrove waters.

Discussion

Mangrove ecosystems are tidally affected to such an extent that tides are the main ecological factor serving as major energy input (Odum, 1971). It is generally believed that coastal wetlands including marshes and mangroves export large quantities of dissolved inorganic nutrients to the adjacent estuarine or coastal waters through tidal movement (Woodwell & Whitney, 1977; Bacon, 1981). In contrast Boto & Wellington (1988) showed that only dissolved organic materials are exported from the mangroves. They estimated that nearly 24% of the phosphate requirement of the mangrove environment is supplied from coastal waters. Guerrero *et al.* (1988) also confirmed such a phenomenon and reported that not only dissolved inorganic phosphate but also dissolved nitrate was imported to the mangrove environment from coastal waters. The present study shows that concentration of PO_4^{3-}-P, NO_2^--N, and NO_3^--N were more or less similar in the mangrove and bay water at first low tide.

At second low tide concentrations of all these nutrients were slightly higher in the bay water than that of the mangrove. This indicates that though there is a little leaching of dissolved nutrients into the bay from mangrove at low tide this phenomenon seems to be uncommon at all low tides. Secondly, the increased concentration of P O_4^{3-}-P, NO_3^--N and NO_2^--N during the high tide period in the mangrove water indicates that some amount of nutrients are imported to this mangrove environment through the tidal movement. However, this phenomenon is only apparent during the night hours and it is no identifiable during the day time. This might be due to continuous utilization of dissolved inorganic nutrients by phytoplankton which is indicated by high primary production values. In addition, the increased concentration of PO_4^{3-}-P, NO_2^--N, NO_3^--N at night in the bay water suggests that the freshwater flow from Commercial Canal acts as a source of these nutrients. The low concentration of these nutrients in the mangrove water clearly indicates that mangrove acts as a sink for land-driven nutrients.

Heald (1971) and Boto & Wellington (1988) showed that mangrove waters are high in ammonia concentration resulting from continuous decomposition of mangrove litter. Heald (1971) showed that mineralization of mangrove organic matter is rapid in low to moderate salinity. In the present study high concentration of ammonia was observed in the mangrove water when the salinity was moderate. Such an increase in ammonia concentration and the prevalence of optimum salinity, 12‰ to 20‰ (Qasim & Devassy, 1971) might be the reasons for the observed higher primary production in the mangrove waters than in the Kakinada Bay.

Acknowledgement

The first author (VS) thanks the Council of Scientific and Industrial Research, Govt. of India, New Delhi for the financial support.

References

Bacon, P. R., 1981. Nutrient export from a Carribean lagoonal ecosystem. Proc. Intl. Symp. Coastal Lagoons, Brazil: 17–21.

Boto, K. G. & J. T. Wellington, 1988. Seasonal variations in concentrations and fluxes of dissolved organic and inorganic materials in a tropical, tidaly dominated mangrove waterway. Mar. Ecol. Prog. Series. 50: 151–160.

Guerrero, G. R., D. R. Cervantes & I. A. Jimenez, 1988. Nutrient variation during a tidal cycle at the mouth of a coastal lagoon in the Northwest of Mexico. Indian J. mar. Sci. 17: 235–237.

Heald, E., 1971. The production of organic detritus in a South Florida estuary. Univ. Miami Sea Grant Tech. Bull. 6: 110.

Odum, E. P., 1971. Fundamental of Ecology. W.B. Saunders Co, Philadelphia, 574 pp.

Odum, E. P., C. C. McIvor & T. J. Smith, 1982. The ecology of the mangroves of South Florida, a community profile. U.S. Fish and Wild Life Service, Washington D.C., 144 pp.

Parsons, T. R., Y. Maita & C. M. Lalli, 1984. A manual of chemical and biological methods of seawater analysis. Pergamon Press, Oxford, 173 pp.

Qasim, S. Z. & V. P. Devassy, 1971. The influence of salinity on the rate of photosynthesis and abundance of some tropical phytoplankton. Mar. Biol. 12: 200–206.

Saraya, A., 1984. The physico-chemical properties of a mussel farm at Samaekho, Thailand. Proc. Asi. Symp. Mangr. Envir. – Research Manag. 405–428.

Strickland, J. D. H. & T. R. Parsons, 1968. A practical hand book of seawater analysis. Fish. Res. Bd Canada, Ottawa, Bull. 167, 311 pp.

Woodwell, G. M. & D. E. Whitney, 1977. Flux pond ecosystem study: exchange of phosphorus between a slat marsh and the coastal waters of Long Island Sound. Mar. Biol. 41: 1–6.

Hydrobiologia **247**: 187–193, 1992.
V. Jaccarini & E. Martens (eds), The Ecology of Mangrove and Related Ecosystems.

Methanogenic bacteria in mangrove sediments

R. Mohanraju [1] & R. Natarajan [2]
[1] *Centre of Advanced Study in Marine Biology, Annamalai University, Parangipettai-608 502, India;*
[2] *Institute for Ocean Management, Centre for Water Resources, Anna University, Madras-600 025, India*

Key words: methanogenic bacteria, mangrove sediments

Abstract

The occurrence of methanogenic bacteria in the Kodiakkarai (10° 18′ N; 79° 52′ E) mangrove sediments, where *Avicennia* spp are predominant, was studied. Trimethylamine under N_2:CO_2 (80:20% v/v) was used as the substrate. Most Probable Number (MPN) of methanogenic bacteria was determined for a period of one year from July 1987 to June 1988 with monthly sampling. The methanogenic bacterial populations were found to be at the maximum of 1.1×10^5 MPN gm^{-1} of wet sediment during August 1987 and from February to June 1988. The bacterial numbers were found to decrease during October to December 1987 with a minimal value of 3.6×10^2 MPN gm^{-1} during December 1987. Environmental factors were correlated with the methanogenic bacterial population.

Introduction

Mangroves are characteristic of tropical coasts and are the counterparts of the tidal salt marshes of the temperate regions (Walsh, 1974; Chapman, 1976). They are bordered by shallow waters and protected from wave action. This environment is typically dominated by large amounts of organic matter brought in during the monsoon season and forms a complex ecosystem containing plant, animal and microbial life. The conversion of this organic matter buried in sediments to natural gas by bacterial decay under anaerobic conditions yields 'biogenic methane'. This gas is formed by the activities of methanogenic bacteria.

Methanogenic bacteria are of much interest due to their role in the terminal dissimilation of organic carbon and also in nitrogen metabolism by utilization of methylamines (Mathrani & Boone, 1985) and produce methane as their end product (Balch *et al.*, 1979; Daniels *et al.*, 1984). The methanogenic bacteria commonly occur in anaerobic environments, where organic matter undergoes decomposition. Mah *et al.* (1977) observed that environments with an Eh lower than −200 mV have the largest number of methanogens. The methanogens are characterized by their ability to produce methane from $H_2 + CO_2$, formate, acetate, methanol, methylamines, dimethyl sulphide, etc. In marine sediments trimethylamine is the major substrate for methane production and metabolism of methylated amines is of special significance because high concentrations are produced by marine organisms (Yancey *et al.*, 1982) and as a result of bacterial decomposition of choline and glycine betaine (King, 1988). The aim of the present study was to study the methanogenic bacterial population in these mangrove sediments and the possible influence of environmental parameters on them.

Description of study area

The Kodiakkarai mangrove is located along the south east coast of India (10° 18′ N; 79° 52′ E) (Fig. 1). It is bordered by the Bay of Bengal on the northeast and the Palk strait on the southwest with the presence of some mangrove of *Avicennia* spp. When sampled, the sediment was black in colour, saline and smelled strongly of hydrogen sulphide. The mangrove was flooded twice daily.

Material and methods

Monthly sampling of sediment and water was carried out for a period of one year (July 1987 to June 1988) during low tide. Sediment samples were collected using a polycarbonate corer from a depth of 70 cm following the method of Pedersen & Sayler (1981) and transferred to a bottle filling it completely so as to avoid any pockets of air. Surface (overlying) water samples were collected in sterile plastic containers. Both the samples were returned to the laboratory within two hours under iced conditions (temperature < 5 °C).

Sediment and water temperatures were measured *in situ* with a mercury thermometer (accuracy of ± 0.1 °C). The pH and Eh of the sediment were measured using a portable pH/Eh meter (Industrial Electronics Corporation, India). Rainfall data were obtained from the meterological unit of the Avifauna Project (BNHS) at Kodiakkarai. Salinity was determined by the method of Strickland & Parsons (1972). Nitrite and nitrate were also determined by the method of Strickland & Parsons (1972). For total phosphorus, sediment samples were ground and digested by the method of Rochford (1951) and the concentration determined by the single solution method of Murphy & Riley (1962). Organic carbon was estimated by the method of el Wakeel & Riley (1956).

One gram wet sediment was transferred into anaerobic serum bottles containing a mixture of 99 ml of sterile sea water and distilled water (in the ratio of 1:1). All manipulations were done under a constant flow of O_2-free N_2. The bottles were then sealed with butyl rubber stoppers and crimped with aluminium seals.

The culture medium used by Mathrani & Boone (1985) was used. The anaerobic technique of Hungate (1950) for media preparation as modified by Bryant (1972) & Balch *et al.* (1979) using serum bottles (Miller & Wolin, 1974) was followed. The medium contained (in grams per litre

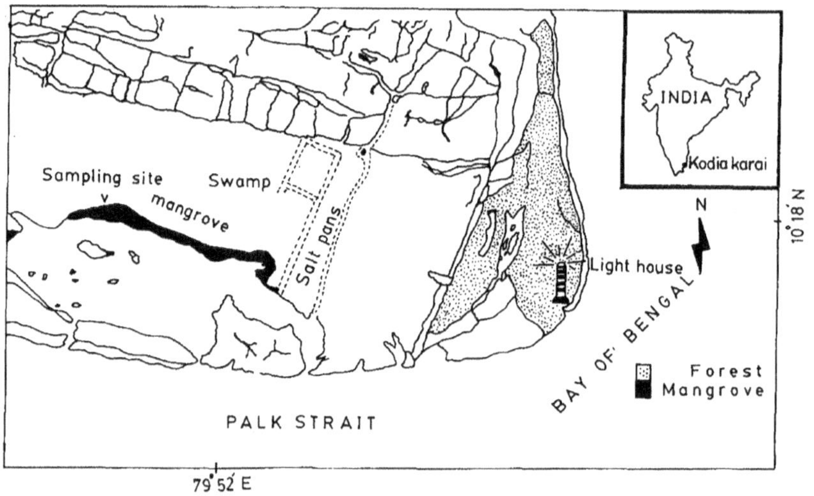

Fig. 1. Location of sampling site.

of 50% estuarine water): NH_4Cl, 1.0; K_2HPO_4, 0.4; $MgCl_2$, 0.1; yeast extract, 0.5; diatomaceous earth, 10.0; Na_2S, 0.2; $NaHCO_3$, 2.0; Trimethylamine hydrochloride, 1.91 (50 mM), trace mineral solution 10 ml (Wolin *et al.*, 1963), vitamin solution 10 ml (Wolin *et al.*, 1963) and resazurin, 0.0001. The final pH was adjusted to 7.2. The gas phase was N_2:CO_2 (80:20 v/v) 15 ml of medium was dispensed into 65 ml capacity serum bottles which were pregassed with N_2:CO_2. The gassing was continued for 3 to 5 minutes to avoid oxygen contamination. The bottles were then stoppered with black butyl stoppers, crimped with aluminium seals and sterilized. Syringes flushed with N_2 were used for all transfers (Macy *et al.*, 1972). Filter-sterilized penicillin solution was added to a final concentration of 3.3 $\mu g\ ml^{-1}$, as antibiotic offers a selective advantage for methanogens (Bock & Kandler, 1985).

The Most Probable Number (MPN) technique was used to enumerate methanogens (Heukelikian & Heinemann, 1939; Siebert & Hattingh, 1967; Belyaev & Finkelshtein, 1973; Zeikus & Winfrey, 1976; Jones & Paynter, 1980; Phelps & Zeikus, 1985; Franklin *et al.*, 1988). A three dilutions three tube MPN was prepared by transferring serial dilution (10^{-2}) of 10, 1.0 and 0.1 ml into the serum bottles using sterile syringes. The bottles were then incubated at 37 °C for 20 days. The MPN bottles were computed using McCrady tables (1918). Methane was measured in a Chemito Gas Chromatograph (Model 3800) with a Porapak Q column equipped with a flame ionization detector.

Results

Rainfall along the southeast coast of India is largely influenced by the northeast monsoon with a heavy rainfall at Kodiakkarai (Fig. 2). The annual rainfall amounted to 530.5 mm with a monthly average of 44.21 mm. The minimum rainfall of 1.8 mm was observed during June 1988 and a maximum of 201 mm in October 1987.

Monthly variations in temperature are shown in Fig. 3. The minimum surface water temperature observed was 25 °C during October–December 1987 and January 1988. The sediment temperature showed a minimum value of 26 °C in December 1987, January and February 1988. The maximum value of 30 °C in July 1987, March and May 1988.

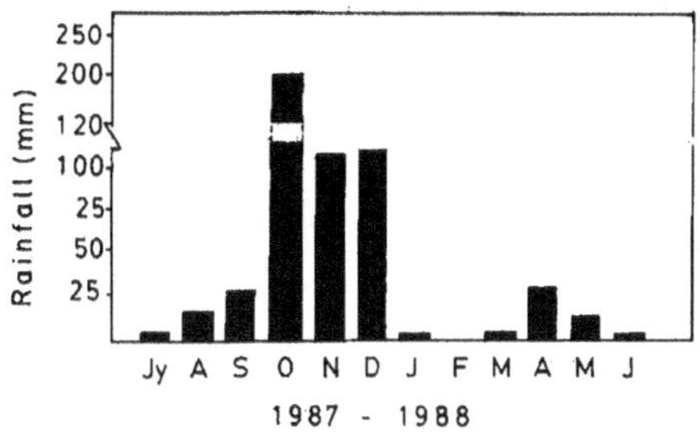

Fig. 2. Monthly variation in rainfall.

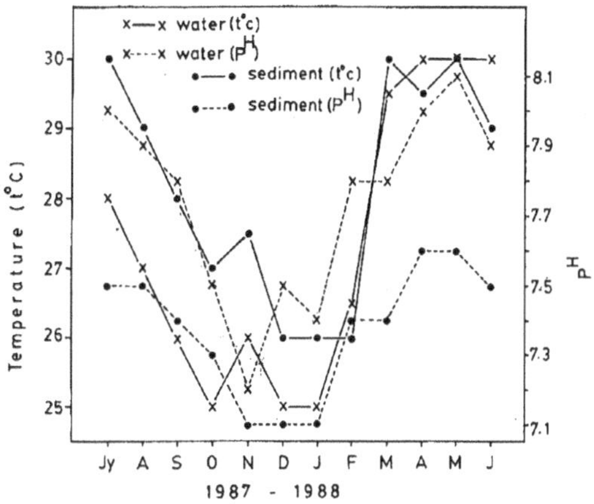

Fig. 3. Monthly variation in temperature and pH.

The pH values of the surface water ranged from a minimum of 7.2 in November, 1987 to a maximum of 8.1 in May, 1988; whereas in sediments it ranged from a minimum of 7.1 in November, December 1987 and January 1988 and a maximum value of 7.6 in April and May, 1988. (Fig. 3).

Redox potential (Eh) values showed a minimum value of −75 mV in November 1987 and a maximum of −129 mV in June 1988 (Fig. 4).

Salinity was found to be lowest (5.5‰) during November, 1987 and highest (61.7‰) in September, 1987 (Fig. 4).

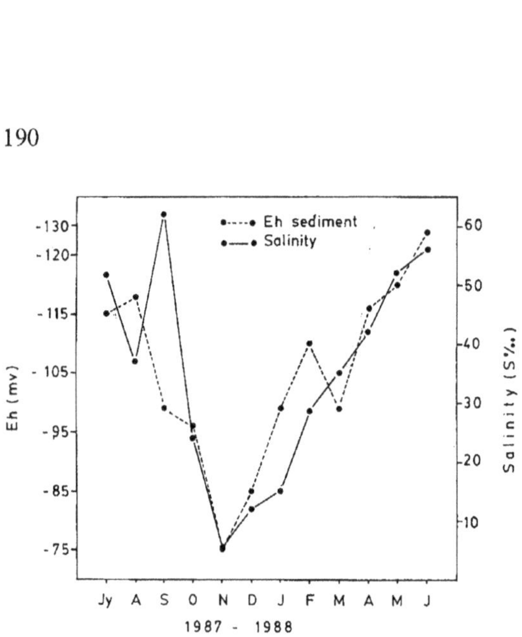

Fig. 4. Monthly variation in salinity and Eh.

The nitrite concentration of the water showed a minimum level of 0.5 μg l^{-1} in June 1988 and a maximum of 6.3 μg l^{-1} in November 1987. The nitrate concentration of water showed a minimum level of 0.2 μg l^{-1} during May and June 1988, and the maximum concentration of 5.5 μg l^{-1} in October and November, 1987. The total phosphorus concentration in sediment showed a minimum of 0.8 μg l^{-1} during May and June 1988, while the maximum value of 9.2 μg l^{-1} was recorded in September and October 1987 (Fig. 5).

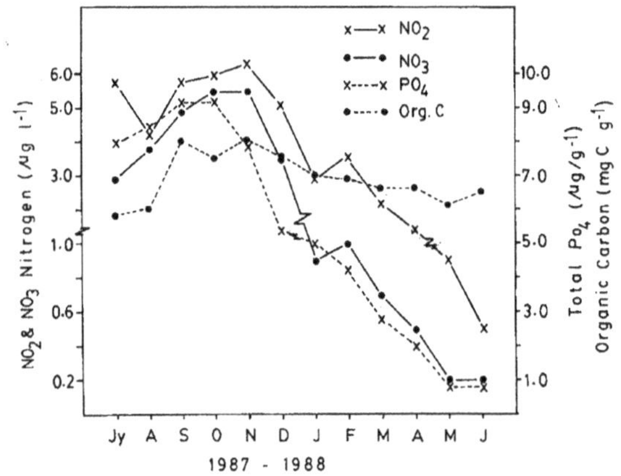

Fig. 5. Monthly variation in Nitrite, Nitrate, Total Phosphorus and Organic Carbon.

The organic carbon content of the sediment ranged from a minimum value of 1.8 mg l^{-1} in July 1987 to a maximum of 4.2 mg g^{-1} in September and November 1987 (Fig. 5).

The Most Probable Number for trimethylamine utilizing methanogens are shown in Table 1.

The methanogenic bacterial population showed a maximum value of $> 1.1 \times 10^5$ MPN g^{-1} of wet sediment during August, 1987. A decreasing trend was observed from then on to reach the minimum value of 3.6×10^2 MPN g^{-1} during December, 1987. Thereafter an increase in the population was noticed, with the maximum value during February to June, 1988. The average methanogenic bacterial population was found to be 5.9×10^4 MPN gm^{-1}.

The correlation co-efficient (r) values calculated between the methanogenic bacteria (MB) and various environmental parameters showed positive correlation with temperature, pH, Eh and

Table 1. Methanogenic bacterial population.

Month	Bacterial numbers (MPN per gm. wet sediment)
July 1987	4.6×10^4
August	$> 1.1 \times 10^5$
September	4.3×10^3
October	3.9×10^3
November	9.1×10^2
December	3.6×10^2
January 1988	2.7×10^3
February	$> 1.1 \times 10^5$
March	$> 1.1 \times 10^5$
April	$> 1.1 \times 10^5$
May	$> 1.1 \times 10^5$
June	$> 1.1 \times 10^5$

Table 2. Correlation co-efficient between methanogenic bacterial population and environmental parameters.

Temperature	pH	Eh	Salinity
0.6746**	0.8842****	0.8554****	0.6343**
Nitrite	Nitrate	Organic Carbon	Total phosphorus
− 0.6530**	− 0.6489**	− 0.7813***	− 0.5120^{NS}

P value: ** < 0.02; *** < 0.01 and **** < 0.001; ^{NS} > 0.05.

salinity; while the nutrients nitrite, nitrate, total phosphorus and organic carbon showed negative correlation (Table 2).

Discussion

Generally, maximum temperature was recorded during summer and minimum during the monsoon as observed earlier by Anbhazhagan (1988), Karthikeyan (1988) and Sampath (1989) for the Kodiakkarai region. No distinct seasonal pattern was observed for pH. The values ranged from 7.0 to 7.5 which are comparable to previous reports by Anbhazhagan (1988) & Karthikeyan (1988). Williams (1981) attributed the variation in pH to photosynthesis by submerged vegetation in the area.

During summer and early premonsoon maximum salinity values were observed. The reasons for high salinity values are due to the addition of harvested salt heaped from salt pans close by and also due to the brine from the salt pan being drained in the vicinity of the study area. The fresh water discharge was also low during this period. With the advent of premonsoon showers the salinity declined slightly because of freshwater run-off from sources like irrigation channels. The northeast monsoon starting in October and continuing up to December decreased the salinity values to the minimum. The same trend has been previously observed by Anbhazhagan (1988), Karthikeyan (1988) & Sampath (1989) in the Kodiakkarai mangroves and by Vasantha (1989) in the Pichavaram mangroves.

The maximum nitrite and nitrate concentrations during the monsoon were apparently due to the influence of seasonal freshwater discharges (Carlucci *et al.*, 1970). Higher concentrations of nitrite than nitrate could be attributed to bacterial reduction of nitrate and oxidation of ammonia by nitrifying bacteria. During summer, the total phosphorus concentration increased gradually and reached a maximum during the monsoon owing to enrichment by monsoonal floods as pointed out by Chandran & Ramamoorthi (1984) for the Pichavaram mangroves. Another reason could be the regeneration of phosphorus in the bottom muds and the subsequent release of the same into the water column due to turbulence and mixing caused by tides as well as the heavy winds prevailing during this season.

No definite trend was observed in the sediment organic carbon. Higher concentrations were observed during monsoon and postmonsoon seasons probably due to the high rate of freshwater inflow enriched with organic debris from land run-off.

The methanogenic bacteria are a major component of the microbial population in the mangrove sediment (Ramamurthy *et al.*, 1990). They are restricted with respect to the nature of the substrate utilized for methane formation. Franklin *et al.* (1988) observed a two to three times higher number of methanogens in a short *Spartina* zone than in a tall *Spartina* zone, and they attributed this to the abundance of more compact *Spartina* roots in the short *Spartina* zone. Bartlett *et al.* (1985) on measuring methane flux from a coastal salt marsh found that temperature increase was accompanied by an increase in the methane flux. Jones & Paynter (1980) observed a decrease in the methanogenic population which utilized hydrogen as a substrate on the mud flats away from the *Spartina* plants, showing a population level of 3.7×10^4 to 1.2×10^6 cells gm^{-1} of dry sediment. King (1988) found that methane production in sediments of a hypersaline pond was associated with algal mats which were apparently sources of methylated amines.

Observations by Phelps & Zeikus (1985) in the sediments of Lake Mendota showed a bacterial population range of 10^3–10^5 cells gm^{-1} of dry sediment. Higher population levels were observed during summer and lower ones during winter. The change in population level was due to the fall in temperature and acetate availability. The present study of methanogenic bacterial population from the mangrove sediments showed higher bacterial numbers during the summer and premonsoon and lower values during the monsoonal months. The difference in population levels is mainly due to the decline of salinity due to freshwater influx, temperature changes and changes in pH and Eh.

Trimethylamine was an important precursor (31.5%–61.1%) in the estuarine intertidal sediments of Lowes Cove, Maine (King *et al.*, 1983). The importance of this compound as a methane precursor was due to the abundance of decomposed plant material in the sediment system. TMA has been confirmed as a non-competitive substrate for methane production (King, 1984) by sulphate reducers. In sediments choline and betaine undergo initial attack by fermentative bacteria rather than sulphate reducers. This results in the production of trimethylamine and acetate which are then respectively channelled to methanogenesis and sulphate reducers.

In conclusion the presence of methanogenic bacteria in mangrove sediment appears to be influenced by environmental parameters such as temperature, pH, Eh, salinity and other nutrients.

Acknowledgement

We wish to thank the Department of Non-conventional Energy Sources, Govt. of India, new Delhi for funding this study.

References

Anbhazhagan, P., 1988. Hydrobiology and benthic ecology of Kodiakkarai coastal sanctuary (Southeast coast of India). Ph.D. Thesis, Annamalai Univ., India, pp. 208.

Balch, W. E., C. E. Fox, L. J. Magrum, C. R. Woese & R. S. Wolfe, 1979. Methanogens: Re-evaluation of a unique biological group. Microbiol. Rev. 43: 260–296.

Bartlett, K. B., R. C. Hariss & D. I. Sebacher, 1985. Methane flux from coastal salt marshes. J. Geo-phy. Res. 90: 5710–5720.

Belyaev, S. S. & Z. I. Finkelshtein, 1973. Method of counting methane producing bacteria in media containing organic substrates. Mikrobiologiya 42: 1102–1107.

Bock, A. & O. Kandler, 1985. Antiobiotic sensitivity of archaebacteria. In: C. R. Woese & R. S. Wolfe (eds), The Bacteria, Vol. 8. Academic Press, New York: 525–544.

Bryant, M. P., 1972. Commentary on the Hungate technique for culture of anaerobic bacteria. Am. J. Clin. Nutr. 25: 1324–1328.

Carlucci, A. F., E. O. Hertwig & P. M. Bowes, 1970. Biological production of nitrite in seawater. Mar. Biol. 7: 161–166.

Chapman, V. J., 1976. Mangrove vegetation. Cramer Vaduz.

Chandran, R. & K. Ramamoorthi, 1984. Hydrobiological studies in the gradient zone of the Vellar estuary. I. Physico-chemical parameters. Mahasagar-Bull. Nat. Inst. Oceanogr. 17: 69–77.

Daniels, L., R. Sparling & G. D. Sprott, 1984. The bioenergetic of methanogenesis. Biochim. Biophys. Acta 768: 113–163.

el Wakeel, S. K. & J. P. Riley, 1956. The determination of organic carbon in marine muds. J. du. Coun. 22: 180–183.

Franklin, M. J., W. J. Wiebe & W. B. Whiman, 1988. Populations of methanogenic bacteria in a Georgia salt marsh. Appl. Envir. Microbiol. 54: 1151–1157.

Heukelikian, H. & B. Heinemann, 1939. Studies on the methane bacteria. I. Development of a method for enumeration. Sew. Wks. J. 11: 426–435.

Hungate, R. E., 1950. The anaerobic mesophilic celluloytic bacteria. Bact. Rev. 14: 1–49.

Jones, W. J. & M. J. B. Paynter, 1980. Populations of methane producing bacteria and *in vitro* methanogenesis in salt marsh and estuarine sediments. Appl. Envir. Microbiol. 39: 864–871.

Karthikeyan, E., 1988. Organochlorine pesticides in the tropical estuary (The Vellar) mangrove (Pichavaram) and swamp (Kodiakkarai) in the east coast of India. Ph.D. Thesis, Annamalai Univ., 151 pp.

King, G. M., 1984. On the metabolism of trimethylamine, choline and glycinebetaine by sulfate reducing and methanogenic bacteria in marine sediments. Appl. Envir. Microbiol. 48: 719–725.

King, G. M., 1988. Methanogenesis from methylated amines in a hypersaline algal mat. Appl. Envir. Microbiol. 54: 130–136.

King, G. M., M. J. Klug & D. R. Lovely, 1983. Metabolism of acetate, methanol and methylated amines in intertidal sediments of Lowes Cove, Maine. Appl. Envir. Microbiol. 45: 1848–1853.

Macy, J. M., J. E. Schnellen & R. E. Hungate, 1972. Use of syringe method for anaerobiosis. J. Clin. Nutr. 25: 1318–1323.

Mah, R. A., D. M. Ward, L. Baresi & T. L. Glass, 1977. Biogenesis of methane. Ann. Rev. Microbiol. 31: 309–341.

Mathrani, I. M. & D. R. Boone, 1985. Isolation and characterization of a moderately halophilic methanogen from a solar saltern. Appl. Envir. Microbiol. 50: 140–143.

McCrady, A. H., 1918. Tables for rapid interpretation of fermentative tube results. Can. J. Publ. Health. 9: 1153–1158.

Miller, T. L. & M. J. Wolin, 1974. A serum bottle modification of the Hungate technique for cultivating obligate anaerobes. Appl. Microbiol. 27: 985–987.

Murphy, J. & J. P. Riley, 1962. A modified single solution method for the determination of phosphate in natural waters. Analyt. Chem. Acta. 12: 162–176.

Pedersen, D. & G. S. Sayler, 1981. Methanogenesis in freshwater sediments, inherent variability and effects of environmental contaminants. Can. J. Microbiol. 27: 198–205.

Phelps, J. & J. G. Zeikus, 1985. Effect of fall turnover an terminal carbon metabolism in Lake Mendota sediments. Appl. Envir. Microbiol. 50: 1285–1291.

Ramamurthy, T., R. Mohanraju & R. Natarajan, 1990. Distribution and ecology of methanogenic bacteria in mangrove sediments of Pichavaram, east coast of India. Ind. J. Mar. Sci. 19: 269–273.

Rochford, D. J., 1951. Studies in Australian estuarine hydrography I. Introductory and comparative features. Aust. J. mar. Freshwat. Res. 2: 1–116.

Sampath, K., 1989. Studies on the ecology of shore birds (Aves: Charadriformes). Ph.D. Thesis, Annamalai Univ., 202 pp.

Siebert, M. L. & W. H. J. Hattingh, 1967. Estimation of methane producing bacterial numbers by the most probable number (MPN) technique. Wat. Res. 1: 13–19.

Strickland, J. D. H. & T. R. Parsons, 1972. A practical handbook of sea water analysis. Bull. Fish. Res. Bd Can. 167: 310.

Vasantha, K., 1989. Studies on hydrobiology and decomposition of macrophytes in Portonovo marine environments (Southeast coast of India). Ph.D. Thesis. Annamalai Univ., 132 pp.

Walsh, G. E., 1974. Mangroves: a review In: R. Reinhold & W. Queen (eds), Ecology of Halophytes. Academic Press, New York, 174 pp.

William, W. D., 1981. The limnology of saline lakes in Western Victoria. A Review of some recent studies. Hydrobiologia 82: 233–259.

Wolin, E. A., M. J. Wolin & R. S. Wolfe, 1963. Formation of methane by bacterial extracts. J. Biol. Chem. 238: 2882–2886.

Yancey, P. H., M. E. Clark, S. C. Hand, B. D. Bowlus & G. N. Somero, 1982. Living with water stress: evolution of osmolyte system. Sci. 217: 1214–1222.

Zeikus, J. G. & M. Winfrey, 1976. Temperature limitation of methanogenesis in aquatic sediments. Appl. Envir. Microbiol. 31: 99–107.

Hydrobiologia **247**: 195–207, 1992.
V. Jaccarini & E. Martens (eds), The Ecology of Mangrove and Related Ecosystems.

Mangroves as a habitat for fish and prawns

A. Sasekumar, V. C. Chong, M. U. Leh & R. D'Cruz
Department of Zoology, University of Malaya, 59100 Kuala Lumpur, Malaysia

Abstract

Mangrove inlets and creeks in Selangor, Malaysia are the habitat for 119 species of fish and 9 species of prawns. The majority of fish and all prawns sampled in the inlets were juveniles. The common fish species in the inlets in terms of weight were *Arius sagor*, *Ambassis gymnocephalus*, *Liza subviridis*, *Toxotes jaculator*, *Sphyraena barracuda* and *Lates calcarifer*. Prawns were represented by juvenile *Penaeus penicillatus*, *P. merguiensis*, *P. indicus*, *Metapenaeus brevicornis* and *M. affinis*. Samples from enclosure traps set on mudflats during ebbing water captured 37 species of fish and 11 species of prawns. The role of mangroves as nursery and feeding grounds for fish and prawns is reviewed in the light of recent work in Selangor. It is apparent that mangroves support fisheries by providing habitat and food.

Introduction

The importance of mangroves as habitat for fish and prawns continues to be a topic of debate among marine ecologists (Macnae, 1974; Robertson & Duke, 1987; Hatcher *et al.*, 1989). The prevailing view of mangrove-offshore interactions is that mangroves export large quantities of detritus to estuarine and near shore waters (Boto & Bunt, 1981; Robertson, 1986). This detritus is either refractory and sinks to the bottom, presumably never enters the food chain, or is dispersed over a large area without any local impact (e.g. magnification of benthic secondary production). However, it should be noted that marine ecologists generally agree that inshore areas constitute a crucial part of the life-support systems of offshore populations by providing nursery and feeding areas (Jansson *et al.*, 1988).

The correlation between offshore prawn catches and areal extent of mangroves (Macnae, 1974; Turner, 1977) has been suggested to be indicative of the dependence of juvenile prawns on mangroves rather than of a food chain link (Hatcher *et al.*, 1989). Though prawns offshore did not carry the mangrove carbon signal, it was shown that several species of juveniles prawns collected from a mangrove inlet consumed mangrove-derived carbon (Rodelli *et al.*, 1984). Comparative analysis of the tissue carbon of the common consumers captured in the inshore and offshore waters in Selangor indicated that prawns in mangrove inlets derived an average of 65% of their carbon from mangrove trees (Rodelli *et al.*, 1984). Eleven species of fish were also found to have derived an average of 60% of their carbon from mangroves.

The coast of central Selangor experiences strong semi-diurnal tides, with ranges of 5 m and above during extreme spring tides. The coast may be classified as a high mesotidal environment. The waters are well mixed vertically, and salinities range from 20 to 32‰ (Chong, 1980). Major dilution occurs from the discharge of freshwater by the Klang and Langat Rivers.

The objective of this paper is to review the role of mangroves as a habitat for fish and prawns based on our previous studies on the mangrove-fringed Selangor coast, and on new evidence from the present study.

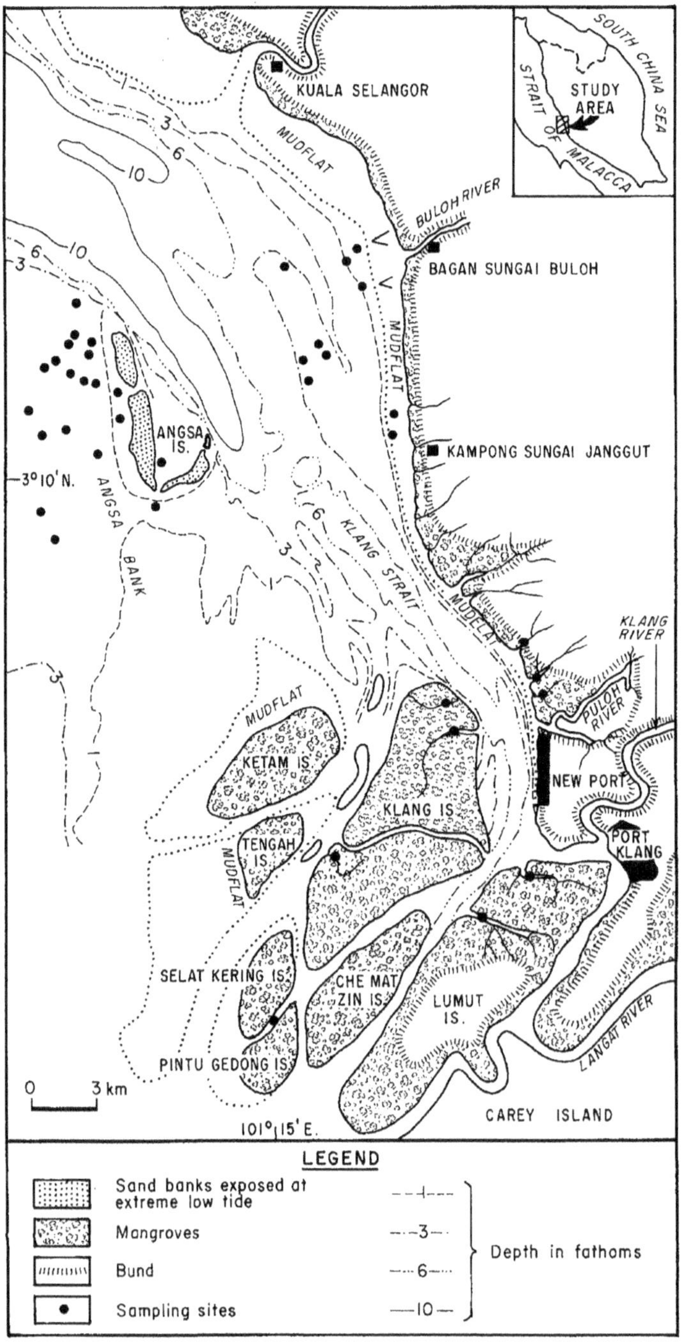

Fig. 1. The study area indicating sampling sites in Angsa Bank, Klang Strait and mangrove inlets on the Selangor coast. The enclosure trap sampling sites are situated north and south of the Buloh River.

Materials and methods

The study site consisted of several habitats on the coast of central Selangor (Fig. 1). Sites sampled included (i) mangrove creeks and inlets, (ii) mudflats, and (iii) the adjacent shallow waters, which consists of the near inshore Klang Straits and the far inshore Angsa Bank. The Angsa Bank and the Klang Straits are less than 10 metres in depth.

Eight mangrove inlets were sampled during daylight, two on mainland Selangor and six on the mangrove islands (Fig. 1). All inlets are on the average 2 to 5 km long, 50 to 150 m wide at the mouth and about 5 m deep in the lower reaches at low tide. The creeks are narrow (5 to 10 m wide), shallow (3 m deep at high tide) and are completely drained of water during low spring tides.

Due to site differences, different sampling techniques were employed. Gill nets of several mesh sizes (1.3 cm, 2.5 cm, 5.1 cm, 7.6 cm and 15.2 cm) and a 2 m wide bean trawl were used in sampling inlets while creeks were samples with a bag net. The fish community of the mudflats at two sites north and south of the mouth of the Buloh River were sampled with a commercially operated enclosure trap. The Angsa Bank (far inshore) and Klang Strait (near inshore) waters were sampled with prawn and fish otter trawls. Details of the sampling techniques and analysis of catches are given in Chong *et al.* (1990) macroscopic examination of the gonads (Kesteven, 1960). For prawns, the ovaries or petasma and spermatophores were examined (Chong, 1980).

Results

Creeks and inlets

A total of 119 species of fish and 9 species of prawns were found in the creeks and inlets of these mangrove swamps (Table 1). The creeks were dominated by schooling fish species belonging to the families Ambassidae, Engraulidae, Clupeidae, Leiognathidae, Eleotridae and Mugilidae (Table 2). The prawns were mainly penaeids, namely *Penaeus penicillatus*, *P. merguiensis*, *P. indicus* as well as *Metapenaeus brevicornis* and *M. affinis*.

The inlets were also dominated by the same schooling fish and 70% of the total catches in gill nets consisted of *Ambassis gymnocephalus*, *Sardinella melanura*, *Thryssa kammalensis*, *T. hamiltoni* and *Stolephorus indicus* (Table 3). However, in terms of biomass, the fish which constituted as much as 62.5% of the catch were *Arius sagor* (15.5%), *Ambassis gmnocephalus* (14%), *Liza subviridis* (8%), *Toxotes jaculator* (5%) and 6 others, including *Lates calcarifer* which made up 3% of the total catch. The beam trawl was effective in capturing the prawns as well some of the demersal fish (Table 4).

Mudflats

Thirty-seven species of fish and 11 species of prawns (Table 5) were captured in the intertidal mudflats. Approximately 80% of the total number of these fish species were also caught in the inlets and creeks. For prawns the species overlap was 55%.

The composition of catches by weight and numbers was different on each sampling occasion (Table 6). The magnitude of the catch appeared to be dependent on the tidal amplitude. A higher catch biomass was obtained during extreme high tides (> 5 m). The percentage composition of fish and prawns were also markedly different on each sampling occasion. Carnivores (catfish and catfish-eel) and detritivores (mullets) dominated the catch on one occasion while the prawn *Palaemon styliferus* was dominant on another occasion (Table 6).

Angsa bank and Klang straits

A total of 102 species of fish and 15 species of prawns were found in the area (Table 1). The fish community was dominated by members of the families Sciaenidae, Cynoglossidae, Engraulidae, Clupeidae and Leiognathidae. The prawn com-

Table 1. Fish and prawn species found in the inshore waters of the Selangor coast, Malaysia. c = creeks/inlets; m = mudflats; n = near inshore waters; f = far inshore waters.

Fish species					
Abalistes stellaris	f	*Gobiidae species A*	c m n f	*Protonibea diacanthus*	m
Ablennes hians	c	*Gobiidae species B*	m f	*Psettodes erumei*	n
Acentrogobius caninus	c	*Gobiidae species C*	m n f	*Pseudorhombus arsius*	c
Acentrogobius sp. A	c	*Harpadon nehereus*	n f	*Pseudorhombus javanicus*	c n f
Alepes djeddaba	f	*Ilisha filigera*	c n f	*Rastrelliger kanagurta*	f
Allaneta forskali	c	*Ilisha megaloptera*	c m n f	*Sardinella gibbosa*	n
Ambassis commersoni	c	*Ilisha melastoma*	c f	*Sardinella melanura*	c m f
Ambassis gymnocephalus	c m n f	*Johnius aneus*	c	*Sardinella fimbriata*	n
Ambassis kopsii	c f	*Johnius belangerii*	c m n f	*Saurida tumbil*	c m n f
Amphotistius imbricatus	c	*Johnius carutta*	c n f	*Saurida undosquamis*	f
Anguilla nebulosa	c	*Johnius coitor*	c m n f	*Scatophagus argus*	c n
Anodontostoma chacunda	c m f	*Johnius dussumieri*	c n	*Scomberoides commersonianus*	c f
Apogon hyalosoma	c	*Johnius soldado*	c m n	*Scyllium marmoratum*	m n
Apogon quadrifasciatus	m f	*Johnius sp.*	m n f	*Secutor insidiator*	c m n f
Arius caelatus	c m	*Laputa cingalensis*	f	*Selaroides leptolepis*	f
Arius maculatus	c n m	*Lates calcarifer*	c	*Secutor ruconius*	f
Arius sagor	c m	*Leiognathus bindus*	c f	*Setipinna taty*	c m n f
Arius tonggol	c m	*Leiognathus brevirostris*	c m n f	*Sideria thyrsoidea*	c m
Arius sp. A	c f	*Leiognathus daura*	c	*Siganus canaliculatus*	c f
Arius venosus	c m f	*Leiognathus fasciatus*	c	*Siganus javus*	c f
Arothron leopardus	c	*Leiognathus lioneolatus*	c	*Siganus vermiculatus*	c
Atropus atropus	f	*Leiognathus splendens*	c	*Sillago sihama*	c m n f
Bagridae sp. I	c	*Liza argentea*	m	*Sphyraena barracuda*	c
Bagridae sp. II	c	*Liza malinoptera*	c m f	*Sphyraena jello*	c
Batrachus grunniens	c n f	*Liza subviridis*	c m n f	*Stigmatogobius sadanundio*	c m n f
Batrachocephalus mino	c	*Lutjanus johni*	c f	*Stolephorus indicus*	c f
Boleopthalmus boddaerti	c	*Lutjanus sanguineus*	c m f	*Stolephorus macroleptus*	m
Butis butis	c	*Macrones gulio*	c m f	*Stolephorus macrops*	c n f
Callionymus schaapi	n f	*Monodactylus argenteus*	c	*Stolephorus tri*	c m n f
Carangoides malabaricus	c m n f	*Muraenesox cinereus*	m n f	*Synaptura commersoniana*	f
Caranx melampygus	c	*Ophichthys rhytidodermatoides*	m f	*Tenualosa sinensis*	c m n f
Caranx sansun	c	*Ophiocara porocephala*	c	*Therapon jarbua*	c m n f
Carangidae sp.	c	*Opisthopterus tardoore*	f	*Thryssa mystax*	c
Chelonodon fluviatilis	c m	*Osteogeneiosus militaris*	c m	*Thryssa dussumieri*	n f
Chiloscyllium indicum	c n f	*Oesteogeneiosus sthenocephalus*	n	*Thryssa hamiltonii*	c m n f
Chirocentrus dorab	c f	*Otolithus argenteus*	n f	*Thryssa kammalensis*	c m n f
Coilia dussumieri	f	*Otolithes ruber*	c m n f	*Thryssa setirostris*	c
Cynoglossus macrolepidotusc	m n f	*Otolithoides brunneus*	f	*Thunnidae*	f
Cynoglossus lingua	c f	*Pampus argenteus*	m n f	*Thysanophrys indicus*	c
Dasyatis uarnak	c f	*Pampus chinensis*	m n f	*Tilapia mossambica*	c
Dasyatis zugei	c m n f	*Pennahia argentata*	m	*Toxotes jaculator*	c
Dendrophyssa russelli	c m n	*Periophthalmodon scholsseri*	c	*Trichiurus savala*	c m n f
Dendroscorpaena sp.	f	*Platax teira*	f	*Trypauchen sp.*	n f
Drepane longimana	c m f	*Platycephalus asper*	f	*Trypauchen vagina*	c m n f
Drepane punctata	c m f	*Platycephalus indicus*	c	*Tylosurus annulatus*	c m
Eleutheronema tetradactyluc	m n f	*Platycephalus scaber*	c m n f	*Tylosurus leiurus*	c
Elops machnata	c	*Platycephalus tuberculatus*	c m f	*Tylosurus strongylurus*	c m
Ephinephelus fario	c	*Plotosus canius*	c m	*Upeneus sulphureus*	m n f
Ephippus orbis	f	*Polynemus plebeius*	c n	*Uranoscopus sp.*	f
Formio niger	n f	*Polynemus sextarius*	c m n f	*Uropterygius concolor*	c
Gastrophysus lunaris	c m n f	*Pomadasys hastga*	c m f	*Zebrias quaga*	m f
Gerres abbreviatus	c m n f	*Pomadasys maculatus*	c m f	*Zenarchopterus buffoni*	c
Gerres filamentosus	c	*Potamolosa sp. A*	c	*Zenachopterus candovittatus*	c
Glossogobius giuris	c m	*Prinobutis koilomatodon*	c		

Table 1. (Continued)

Prawn species					
Macrobrachium equidens	c m	*Parapenaeopsis coromandelica*	m f	*Penaeus indicus*	c m f
Metapenaeus affinis	c m n f	*Parapenaeopsis gracillima*	m f	*Penaeus merguiensis*	c m n f
Metapenaeus brevicornis	c m n f	*Parapenaeopsis hardwickii*	m n f	*Penaeus monodon*	c
Metapenaeus ensis	c	*Parapenaeopsis hungerfordi*	m n f	*Penaeus penicillatus*	c m f
Metapenaeus lysianassa	m	*Parapenaeopsis maxillipedo*	m n f	*Solenocera subnuda*	m n f
Metapenaeus stridulans	f	*Parapenaeopsis sculptilis*	c m n f	*Trachypenaeus fulvus*	m f
Palaemon styliferus	m				

Habitat group	Number of species			
	Creeks/inlets	Mudflat	Near in shore	Far in shore
Fish	119	37	58	92
Prawns	9	11	8	15

Table 2. Percentage abundance by number of major fish species caught by the bag net in two mangrove creeks in S. Sementa Kecil and S. Perepat Kecil from November 1987 to May 1988, (+) means present in small numbers.

Species	Site							
	S. Sementa Kecil				S. Perapat			Pooled mean
	Nov	Dec	Jan	Feb	Jan	Mar	May	
Ambassis gymnocephalus	28.78	+	13.27	3.9		5.61		11.87
Butis butis	+	1.29	1.69	12.99	0.46		1.9	2.18
Chelonodon flaviatilis	0.51	1.13	14.56	+	4.67	4.67	3.18	7.56
Eleutheronema tetradactylum	+		+	+	0.92	0.93	4.44	0.41
Ilisha megaloptera	14.29	3.72	+					3.24
Ilisha melastoma				20.56				1.68
Leiognathus brevirostris	23.37	+						4.09
Liza malinoptera	1.33	1.13		8.44	13.99	27.1	12.06	3.63
Liza subviridis					33.49	18.69	53.65	5.93
Macrones gulio		+			15.6	+	2.22	1.4
Periophthalmodon schlosseri						0.93	+	0.04
Plotosus canius			+		0.92		+	0.19
Pomadasys hasta		+	5.17	1.95	1.38	4.67	2.22	3.05
Scatophagus argus	0.71	+	3.74	1.52			1.59	2.21
Scomberoides commersonianus	+	+	4.47	12.99				3.29
Secutor insidiator	10.41	2.26	16.1					9.83
Stigmatogobius sadanundio		+	+	+	0.46		+	0.19
Stolephorus indicus	4.18	6.3	+	0.87				1.98
Thryssa hamiltonii	5.82	73.02	2.86	11.26		0.93	+	11.35
Toxotes jaculator		0.81	+	+				0.44
Tylosurus strongylurus			+	+	11.47	23.36	+	1.34
Zenarchopterus buffoni	10	2.91	3.05	22.08	+	12.15	13.33	6.32
Zenarchopterus candovittatus		2.1	28.27		20.87			15.5
Others	0.6	5.33	6.82	3.44	0.44	0.96	3.19	2.08

munity was dominated by two penaeid species, *Parapenaeopsis sculptilis* and *P. hungerfordi*.

Sexual maturity

Ninety-seven percent of the fish sampled in the inlets (including creeks) were immature. This was in contrast to 80% over the mudflats and 60% in the nearshore waters. Of the total fish population in the far inshore area, only 29% were immature (Table 7). All the prawns sampled in the inlets were immature. Their carapace lengths ranged from 7 to 28 mm CL. The percentage of immature prawns in the shallow coastal waters of the Klang Straits and Angsa Bank were 8 and 38% respectively.

Discussion

The community of fish found in mangrove inlets in Selangor resembles that of similar habitats

Table 3. Percentage abundance by numbers of the major fish species caught by gill nets in the mangrove inlet S. Sementa Kecil from November 1987 to May 1988, (+) means present in small numbers.

Species	Month							Pooled mean
	Nov	Dec	Jan	May	Aug	Sep	Oct	
Ambassis commersoni		+	0.65					0.2
Ambassis gymnocephalus	74.44	17.32	14.25	64.42	24.87	4.11	21.82	53.36
Anodontostoma chacunda	+	+		+	2.07	+	3.64	0.43
Arius sagor			+	+	+	4.11		0.22
Arius tonggol	+	5.03	0.28					0.31
Arothron leopardus			0.19					0.04
Carangoides malabaricus					1.55			0.06
Chelonodon fluviatilis	0.59							0.23
Dasyatis uarnak							3.64	0.04
Dasyatis zugei						+	1.82	0.06
Dendrophyssa russelli				+	+		5.45	0.37
Eleutheronema tetradactylum					1.04	+	+	0.16
Ilisha megaloptera		+		+		10.96		0.27
Gerres abbreviatus					1.55			0.06
Johnius aneus				1.01				0.29
Johnius coiter			+	+		6.85	5.45	0.22
Johnius soldado	+	5.03	0.19					0.25
Leiognathus brevirostris	+			0.94	29.53	5.48	7.27	1.62
Leiognathus fasciatus	0.29	+	0.65					0.33
Liza malinoptera		+		1.89		+		0.59
Liza subviridis	1.08	3.35	1.02	2.56		4.11	+	1.58
Osteogeniosus militaris					2.07		3.64	0.12
Pomadasys hasta	0.73	2.79	+	+		+		0.49
Sardinella melanura	2.64	17.88	80.76	15.30	28.50			24.28
Secutor insidiator	+			1.95				0.59
Setipinna taty	+			1.15		10.96		0.51
Stolephorus indicus	3.67	5.03	+	+				1.58
Stolephorus macrops		4.47						0.16
Stolephorus tri					+	8.22	38.18	0.55
Thryssa hamiltonii	2.69	20.67	0.19	2.16	1.55	9.59	3.64	2.7
Thryssa kammalensis	10.90	5.03		3.64	4.15	19.18		5.99
Toxotes jaculator	1.12	+	0.37					0.7
Others	1.85	13.4	1.64	4.98	3.12	16.43	5.45	1.66

Table 4. Percentage abundance by numbers of the major and prawn species caught by beam trawl in the mangrove inlet S. Sementa Kecil from Januray 1988 to August 1989. (+) present in small numbers.

Species	Month						Pooled mean
	Jan	Apr	May	Jun	Jul	Aug	
Fish							
Ambassis gymnocephalus	18.46	4.02	89.90	56.18	8.33	3.31	34.94
Thryssa hamiltoni		22.81	1.21	16.38	12.14	1.24	14.53
Ilisha megaloptera		0.43	0.22	5.17	39.52	3.72	2.48
Thryssa kammalensis		43.32	1.31	3.66	+		23.01
Leiognathus brevirostris		1.25		3.66	6.19	26.86	1.90
Glossogobius giuris	10.07	0.76	0.10	0.93	5.24	13.22	1.28
Pomadasys hasta	17.79	0.31	0.06	0.93	3.57	5.37	1.01
Scatophagus argus	9.73	0.52	2.55	4.24	4.76	2.07	1.98
Chelonodon fluviatilis	1.68	3.68	0.38	2.95	2.62	10.74	2.73
Leiognathus daura		18.31	2.45	0.14			10.08
Others	42.27	4.59	1.82	5.76	17.63	33.47	6.06
Prawn							
Penaeus penicillatus	21.21	23.89	20.74	36.15	43.14	22.81	28.65
Metapenaeus brevicornis	20.61	46.02	61.17	17.79	1.96	15.79	30.01
Penaeus merguiensis	9.70	17.70	6.91	31.40	31.37	43.86	21.93
Penaeus indicus	48.48	2.65	2.66	8.97	11.76	1.75	13.53
Metapenaeus mutatus		7.08	5.32	1.85	11.76	10.53	3.88
Macrobrachium equidens		2.65	2.66	1.58		3.51	1.68
Penaeus monodon				0.26		1.75	0.21
Parapenaeopsis sculptilis			0.53				0.10

elsewhere (Little *et al.*, 1988; Pinto, 1988; Robertson & Duke, 1987). Six species of small fish of low and medium economic value comprised 70% of a total of 21670 specimens (119 species) examined. The most abundant, *Ambassis gymnocephalus*, is not eaten by man because of its small size. However, others like the *Sardinella* sp., *Thryssa* spp. and *Stolephorus* sp. are consumed. Although the catfish-eel, *Plotosus canius*, was rarely captured in the gill nets or the beam trawl, it constitutes an important fishery and is normally landed by eel trap or hook and line (Leh, unpublished).

Fish occurring in creeks feed on benthic food resources including crabs, penaeids, insects, mangrove detritus, diatoms, macroalgae, and on pelagic fauna like *Acetes* and copepods (Leh & Sasekumar, 1991). The mullets, *Liza malinoptera* and *L. subviridis* consume significant quantities of mangrove detritus (Chong, 1977; Sasekumar *et al.*, 1984). Predation of benthic mangrove invertebrates including sesarmid crabs (*Chiromanthes* spp.), *Metaplax* spp., bivalves, gastropods and sipunculids by catfish eel and catfish is common (Sasekumar *et al.*, 1984). In addition, mangrove invertebrates liberate larvae into coastal waters that become food for planktivores. Macintosh (1979) estimated that two species of fiddler crabs (*Uca* spp.) in Kuala Selangor produced 200000 megalopae $m^{-2} yr^{-1}$. Crab megalopae are important food for several species of anchovies (Engraulidae) and *Ambassis* spp. (pers. obs.).

The prawns in the mangrove inlets are represented largely by juvenile penaeids as has been observed in Queensland (Robertson & Duke, 1987). Several species of prawns which include the predominant species of *Penaeus merguiensis*, *P. penicillatus* and *P. indicus* have been shown to use these inlets as nursery areas (Chong *et al.*, 1990). *Metapenaeus brevicornis*, *M. affinis* and

Table 5. Percentage abundance by numbers of the major fish and prawns species caught by the enclosure net in intertidal mudflats, north and south of the river the River Buloh, Selangor, Peninsular Malaysia, from January 1988 to August 1989.

Species	Month			Total	Pooled mean
	Jan	Aug	Oct		
Fish					
Macrones gulio	16.82	46.40		63.22	26.50
Cynoglossus macrolepidotus	10.28	14.70	15.24	40.22	13.65
Arius tonggol	20.56	8.21		28.81	9.75
Liza subviridis	14.02	2.45	7.22	23.69	6.99
Pomadasys hasta	3.27	6.34	11.23	20.84	6.72
Arius caelatus	20.56			20.56	5.92
Ambassis gymnocephalus		1.73	15.24	16.97	4.64
Johnius coiter	1.64	11.38	3.74	16.76	6.72
Arius sagor			14.71	14.71	3.70
Thryssa kammalensis		0.29	9.89	10.18	2.62
Others	12.85	8.50	22.73	44.09	12.78
Prawn					
Palaemon styliferus		45.18	75.82	121	70.34
Metapenaeus brevicornis	41.03	3.07	3.62	47.72	4.36
Metapenaeus mutatus	7.69	16.67	12.86	37.22	13.23
Penaeus indicus	25.64	7.02		32.66	1.43
Parapenaeopsis coromandelica	12.82	17.54	0.65	31.01	3.03
Penaeus merguiensis	7.69	4.39	3.68	15.76	3.86
Macrobrachium equidens	5.13	2.19	2.91	10.23	2.87
Penaeus penicillatus		3.51		3.51	0.44
Parapenaeopsis gracillima		0.44		0.44	0.06
Metapenaeus lysianassa			0.26	0.26	0.22
Parapenaeopsis hardwickii			0.19	0.19	0.17

Penaeus merguiensis occurred together both in the inlets and the adjacent mudflats. However, there is evidence of spatial distribution of species within the inlet. The three *Penaeus* species are restricted mainly to the upper and middle reaches of the inlet, whereas the two *Metapenaeus* species are found in the lower reaches and mudflats (Figs 2a & 2b). Spatial partitioning of juvenile prawns in mangrove estuaries has also been observed in East Africa (de Freitas, 1986).

The presence of juvenile prawns in mangrove inlets may be attributed to three main factors: shelter from predators, high food abundance and turbidity. Although the catfish eel and ariid catfish are common in the inlets, prawns could take refuge from these predators amongst the prop roots at high tide. However, prawns are vulnerable at low tide as the waterways do not have sea grass cover or other refugia. Therefore, high food abundance is more likely the main benefit to juvenile prawns. Turbidity (including food availability and temperature) is known to influence the movement and distribution of juvenile fish in estuarine waters (Blaber & Blaber, 1980; Cyrus & Blaber, 1987). Similar studies have not been done with prawns though high turbidity could be important in reducing predation.

Penaeid prawns are opportunistic generalists in their food habits, being able to consume a variety of animal and plant matter and bacteria (Chong & Sasekumar, 1981; Leh & Sasekumar, 1984; Moriarty, 1976).

The abundance of food material on the surface of mud or sand-mud substrates associated with

Table 6. Biomass and numbers of fish and prawns captured in the enclosure trap subsamples (10% of the total catch).

Date	Location	Ht. of high water above C.D. (metres)	Sample		Composition of catch by groups				
			Weight (g)	Numbers	Catfish	Mullets	Prawns	Other fish	Crabs
23 Jan '88	Mudflat, south of Sg. Buloh	4.6	4495.6	467	937.4 (20.8%)	1914.8 (42.5%)	133.7 (3.9%)	1510.3 (33.6%)	0.0 (0.0%)
1 Aug '88	Mudflat, north of Sg. Buloh	5.2	7056.1	934	268.2 (3.8%)	481.0 (6.8%)	588.6 (8.3%)	4936.3 (69.9%)	782.0 (11.1%)
25 Oct '88	Mudflat, north of Sg. Buloh	5.5	6349.9	1945	499.1 (7.8%)	825.4 (13.0%)	3069.9 (48.3%)	1456.5 (2.9%)	499.0 (7.9%)

Table 7. Percentage composition of the maturity stages of fish and prawns sampled in the study area (Selangor, Peninsular Malaysia).

Habitat		Total numbers	Immature	Maturing	Mature
Inlets	fish	15894	97.4	1.4	1.2
	prawn	935	100	0	0
Mudflats	fish	2909	55.7	18.2	26.1
	prawn	2319	49.2	13.4	37.4
Near inshore	fish	1619	60.7	15.6	23.7
	prawn	253	8.3	68.9	22.8
Far inshore	fish	8760	29.1	41.5	29.4
	prawn	3886	37.8	31.4	30.8

the mangrove environment provides a conducive environment for the growth of prawns. On average there are 10^6 meiofauna m^{-2} (1–2 g m^{-2} dry mass) on the surface layers of subtidal mud substrates (Coull & Bell, 1979). The mud substrate in mangroves contains $0.5–1.0 \times 10^6$ meiofauna m^{-2} (Sasekumar, 1981). Microfauna such as ciliates and flagellates are also abundant in mangrove mud in densities of $1.2\ (\pm 2.3) \times 10^6\ m^{-2}$ and $1.5\ (\pm 2.6) \times 10^6\ m^{-2}$ respectively (Alongi, 1986). Mean bacterial abundance and productivity of mangrove sediments is estimated at 1.1×10^{11} cells g^{-1} dry weight and $1.6\ g\ C\ m^{-2}\ d^{-1}$ respectively (Alongi, 1988). In contrast, the surfaces of sandy substrates are an impoverished environment, where the meiofauna are distributed within a broad depth of 10 to 15 cm below the surface (Coull & Bell, 1979).

Hatcher *et al.* (1989) have suggested that factors related to freshwater outflow may be important in the correlation between offshore catches of penaeid prawns and mangrove swamp area. Our studies in Selangor indicate that the occurrence of juvenile prawns cannot be related to freshwater outflow as they also occur in abundance in inlets located on mangrove islands (off Port Klang) where there are no rivers or land runoff (except following heavy rainfall).

Given the present knowledge of the distribution of prawns, it appears that some species of prawns are more linked to mangroves than oth-

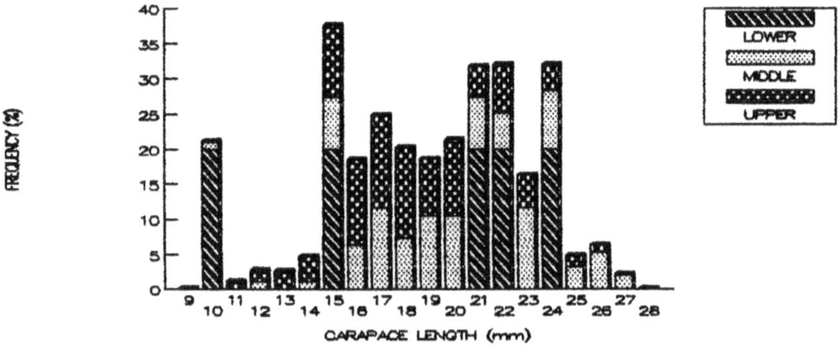

Penaeus penicillatus

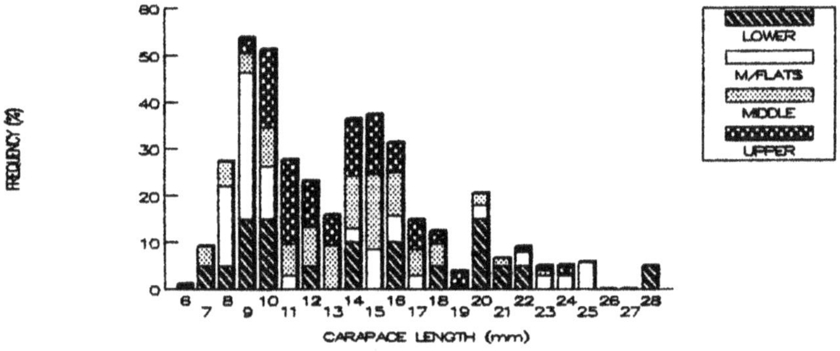

Penaeus merguiensis

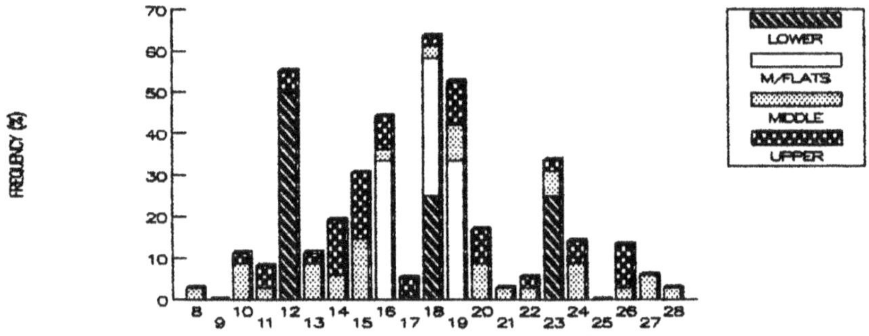

Penaeus indicus

Fig. 2a & 2b. Size frequency distribution of six prawn species by location in the upper, middle and lower reaches of the mangrove inlet (Sungai Sementa Kecil) and the adjacent mudflat.

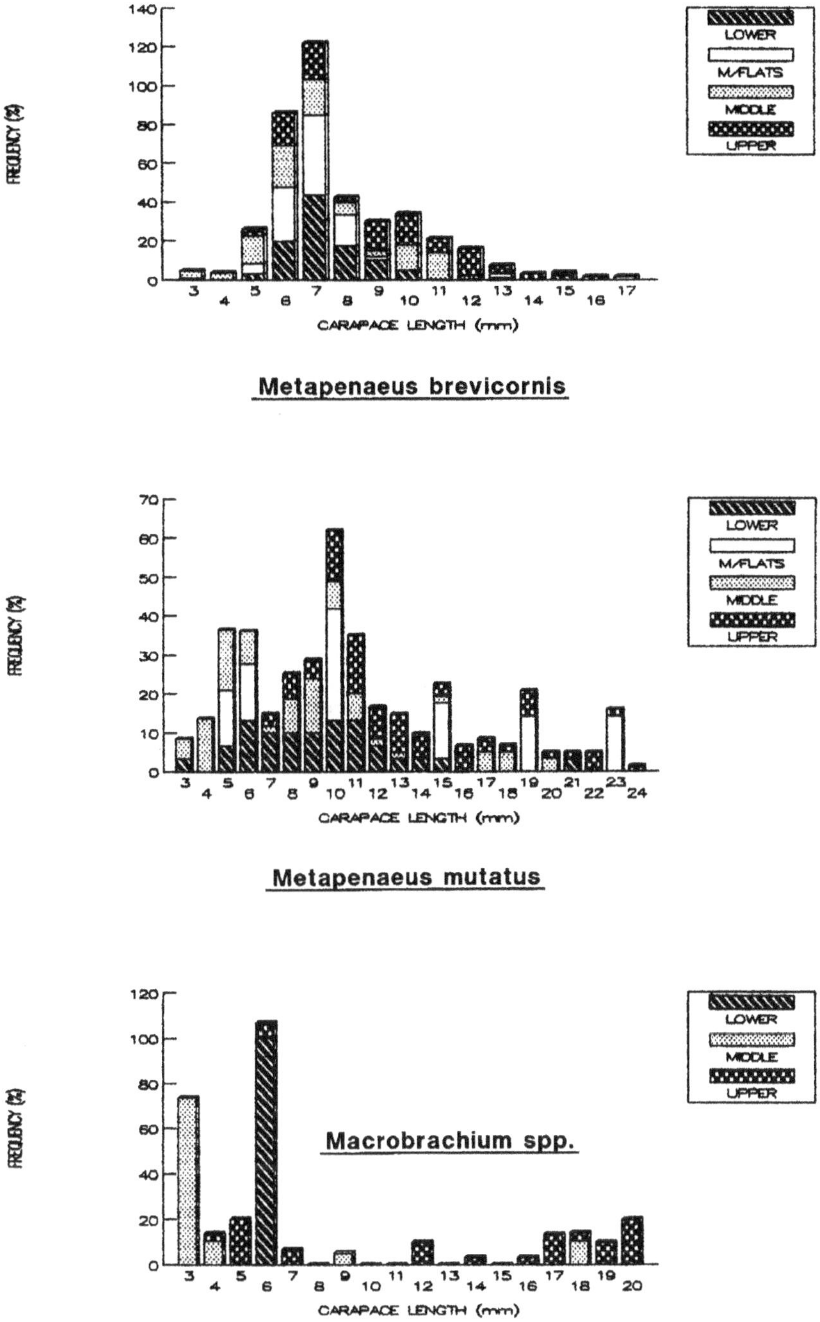
FREQUENCY (%)
CARAPACE LENGTH (mm)
LOWER
M/FLATS
MIDDLE
UPPER
Metapenaeus brevicornis
FREQUENCY (%)
CARAPACE LENGTH (mm)
LOWER
M/FLATS
MIDDLE
UPPER
Metapenaeus mutatus
FREQUENCY (%)
CARAPACE LENGTH (mm)
LOWER
MIDDLE
UPPER
Macrobrachium spp.

Fig. 2b.

ers. *Penaeus merguiensis*, *P. penicillatus*, *P. indicus*, *Metapenaeus brevicornis* and *M. affinis* may be more dependent on mangrove estuaries than other species. Thus it is possible that the species composition of offshore catches of prawns in Malaysia may change with the reclamation or destruction of large tracts of mangroves.

Fish and prawns generally occur in greater numbers and biomass in mangrove habitats than in adjacent near shore habitats (Thayer *et al.*, 1987; Robertson & Duke, 1987; Blaber *et al.*, 1989; Chong *et al.*, 1990). The abundance of a variety of food items and the turbid waters appear to be the main contributing factors. Although juvenile prawns may not derive nutrition directly from mangrove detritus (Stoner & Zimmerman, 1988), it is apparent that significant quantities of mangrove carbon are being assimilated through secondary sources (Rodelli *et al.*, 1984). We therefore conclude that mangroves provide significant support for fish and prawn communities.

Acknowledgements

We thank Jalal Mokhtar and Jamri Tohid for field assistance, and Cheong Nge, G. H. Wee, P. K. Geetha and Siti Nuraini for assistance in the laboratory. We also thank Dr. Susan Lim for her critical comments on the manuscript. Our studies were supported by research grants from AIDAB (Australia) and R & D funds from the University of Malaya.

References

Alongi, D. M., 1986. Quantitative estimates of benthic protozoa in tropical marine systems using silica gel: a comparison of methods. Estuar. coast. Shelf Sci. 23: 443–450.

Alongi, D. M., 1988. Bacterial productivity and microbial biomass in tropical mangrove sediments. Microb. Ecol. 15: 59–79.

Blaber, S. J. M. & T. J. Blaber, 1980. Factors affecting the distribution of juvenile estuarine and inshore fish. J. Fish. Biol. 17: 143–162.

Blaber, S. J. M., D. T. Brewer & J. P. Salini, 1989. Species composition and biomass of fishes in different habitats of a tropical Northern Australian estuary: their occurrence in the adjoining sea and estuarine dependence. Estuar. coast. Shelf Sci. 29: 509–531.

Boto, K. G. & J. S. Bunt, 1981. Tidal export of particulate organic matter from a Northern Australian mangrove system. Estuar. coast. Shelf Sci. 13: 247–255.

Chong, V. C., 1977. Studies on the small grey mullet *Liza malinoptera*. J. Fish. Biol. 11: 293–308.

Chong, V. C., 1980. The biology of the white prawn *Penaeus merguiensis* de Man (Crustacea: Penaeidae) in the Pulau Angsa Klang Strait waters (Straits of Malacca). M. Sc. thesis. Department of Zoology, University of Malaya, Kuala Lumpur. 141 pp.

Chong, V. C. & A. Sasekumar, 1981. Food and feeding habits of the white prawn *Penaeus merguiensis*. Mar. Ecol. Prog. Series 5: 185–191.

Chong, V. C., A. Sasekumar, M. U. Leh & R. D'Cruz, 1990. The fish and prawn communities of a Malaysian coastal mangrove system, with comparisons to adjacent mudflats and inshore waters. Estuar. coast. Shelf Sci. 31: 703–722.

Coull, B. C. & S. S. Bell, 1979. Perspectives in marine meiofaunal ecology. In R. J. Livingston (ed.), Ecological Processes in Coastal and Marine Systems. Plenum Press, New York: 189–216.

Cyrus, D. P. & S. J. M. Blaber, 1987. The influence of turbidity on juvenile marine fishes in estuaries. Part I. Field studies at Lake St. Lucia on the southeastern coast of Africa. J. exp. mar. Biol. Ecol. 109: 53–70.

de Freitas, A. J., 1986. Selection of nursery areas by six Southeast African Penaeidae. Estuar. coast. Shelf Sci. 23: 901–908.

Hatcher, B. G., R. E. Johannes & A. I. Robertson, 1989. Review of research relevant to the conservation of shallow tropical marine ecosystems. Oceanogr. Mar. Biol. annu. rev. 27: 337–414.

Jansson, B. O., A. D. McIntyre, S. W. Nixon, M. M. Pamatmat, B. Zeitzschel & J. J. Zijlstra, 1988. Coastal-offshore interactions – an evaluation of presented evidence. In B. O. Jansson (ed.), Coastal-Offshore Ecosystem Interactions. Springer-Verlag: 357–363.

Kesteven, G. L., 1960. Manual of field methods in fisheries biology. FAO Manuals in Fisheries Science. No. 1, FAO, Rome. 152 pp.

Leh, M. U. & A. Sasekumar, 1984. Feeding ecology of prawns in shallow waters adjoining mangrove shores. In E. Soepadmo *et al.* (eds), Proc. Asian Symp. Mangrove Environment – Res. & Management: 331–353.

Leh, M. U. & A. Sasekumar, 1991. Ingression of fish into mangrove creeks. In Alcala (ed.), Proc. Asean-Australian 'Living Resources in Coastal Areas' Project. 30 Jan–1 Feb. 1989. Manila: 495–501.

Little, C., P. J. Reay & S. J. Grove, 1988. The fish community of an East African mangrove creek. J. Fish. Biol. 32: 729–747.

Macintosh, D. J., 1979. Predation of fiddler crabs (*Uca* spp.) in estuarine mangroves. In Srivastava, P. B. *et al.* (eds),

Proc. Symp. mangrove and estuarine vegetation in South-east Asia. University of Agriculture, Malaysia: 101–110.

Macnae, W., 1974. Mangrove forests and fisheries. FAO, Rome. 35 pp.

Moriarity, D. J. M., 1976. Quantitative studies on bacteria and algae in the food of the mullet *Mugil cephalus* L. and the prawn *Metapenaeus bennettae* (Racek & Dall). J. exp. mar. Biol. Ecol. 22: 131–143.

Pinto, L., 1988. Population dynamics and community structure of fish in the mangroves of Pagbilao, Phillipines. J. Fish. Biol. 33 (Supplement A): 35–43.

Robertson, A. I., 1986. Leaf-burying crabs: their influence on energy flow and export from mixed mangrove forests (*Rhizophora* spp.) in northeastern Australia. J. exp. mar. Biol. Ecol. 102: 237–248.

Robertson, A. I. & N. C. Duke, 1987. Mangroves as nursery sites: comparisons of the abundance and species composition of fish and crustaceans in mangroves and other nearshore habitats in tropical Australia. Mar. Biol. 96: 193–205.

Rodelli, M. R., J. N. Gearing, P. J. Gearing, N. Marshall & A. Sasekumar, 1984. Stable isotope ratio as a tracer of mangrove carbon in Malaysian ecosystems. Oecologia 61: 326–333.

Sasekumar, A., 1981. The ecology of meiofauna on a Malayan mangrove shore. Ph. D. thesis. Department of Zoology, University of Malaya, Kuala Lumpur. 183 pp.

Sasekumar, A., T. L. Ong & K. L. Thong, 1984. Predation of mangrove fauna by marine fishes. In E. Soepadmo *et al.* (eds), Proc. Asian Symp. Mangrove Environment – Res. and Management: 378–384.

Stoner, A. L. & R. J. Zimmerman, 1988. Food pathways associated with penaeid shrimps in a mangrove-fringed estuary. Fish. Bull. 86: 543–551.

Thayer, G. W., D. R. Colby & W. F. Hettler, Jr., 1987. Utilization of red mangrove prop habitat by fishes in South Florida. Mar. Ecol. Prog. Series 35: 25–38.

Turner, R. E., 1977. Intertidal vegetation and commercial yields of penaeid shrimp. Trans. am. Fish. Soc. 106: 411–416.

Hydrobiologia **247**: 209–214, 1992.
V. Jaccarini & E. Martens (eds), The Ecology of Mangrove and Related Ecosystems.

Mariculture development in Kenya: alternatives to siting ponds in the mangrove ecosystem

Joseph Rasowo
Moi University, Zoology Department, Box 3900, Eldoret, Kenya

Key words: Artemia, salinas, mangrove, pond, shrimp, prawn

Abstract

Pond culture of brackishwater prawns and shrimps has recently generated much interest in Kenya. The mangrove areas are the target zones for the construction of these ponds. With the increasing awareness of the unique ecological role played by the mangroves, there is an urgent need to stop the conversion of mangrove swamps into aquaculture ponds. To develop pond aquaculture without destroying the mangroves, a shift from tide-fed to pump-fed pond systems is recommended in order to divert the farming from the mangroves to higher grounds. 'Mangrove-friendly' mariculture practices like pen, cage and raft culture are discussed. Methods of efficiently utilising the already destroyed mangrove areas are considered.

Introduction

Mangroves and other coastal wetlands are highly complex and productive ecosystems which serve a wide variety of useful functions including prevention of coastal erosion, encouraging soil deposition, providing food, shelter and sanctuary for birds and mammals, providing spawning, nursery and forage ground for a wide variety of aquatic organisms, and an organic food factory through its litter production (Bailey, 1988; Singh, 1987; Saenger *et al.*, 1983; MacNae, 1974; Turner, 1977). Mangroves thus support important coastal and nearshore fisheries and indeed good correlations have been found between prawn and fish yields and the area of mangroves adjacent to the fishing ground (Brusher, 1974; MacNae, 1974). Despite these unique physical, biological and economical functions, mangroves and other coastal wetlands are being destroyed with little regard for the consequences. This has been due to an upsurge of interest in the pond culture of brackishwater prawns and shrimps as a result of the high demand of these products in the markets of Western Europe, Japan and USA. The export potential has increased so much that mariculture has become established as a significant source of foreign exchange earnings in a number of developing countries. The success story of countries like Ecuador has attracted world wide attention and encouraged many developing tropical countries with large stands of mangrove forests to invest in shrimp mariculture development.

Kenya has a total of about 59980 ha of mangrove forests on its 450 km long semi arid coastline (Doute *et al.*, 1981). These mangroves have been cleared mainly for salt farming and for wood (fuel and building purposes). Although aquaculture was introduced in Kenya soon after the first world war, well over 250 ha of the total pond area in the country has been devoted for the culture of freshwater finfish especially *Tilapia* (Allela, 1986).

The mariculture subsector has been lagging behind and it was in 1978 that a pilot project was set up to test the technical and economic viability of brackishwater aquaculture of shrimps on the Kenya coast. The production levels of an average of 525 kg ha^{-1} yr^{-1} has created considerable interest among private sector entrepreneurs and national policymakers resulting in plans being drawn up for large scale commercial operations.

This paper proposes a cautious approach to the envisaged mariculture development which would involve the integration of aquaculture with the management of mangroves and the efficient utilisation of already reclaimed mangrove areas.

Efficient utilisation of already reclaimed mangrove lands

Pond aquaculture is currently recognized worldwide as the major culprit in the large scale clearing of mangrove forests and other coastal wetlands (Bailey, 1988; Singh, 1987). So far in Kenya, it is salt farming which has been mainly responsible for the destruction of large tracts of mangroves with 50% of the total salt farm acreage in the coast being located within former mangrove areas (Jira, pers. comm.). The Ngomeni aquaculture pilot project executed under the FAO/UNDP funding is responsible for the destruction of about 60 ha of mangrove forests in the Ngomeni creek. Utilising the reclaimed mangrove land more efficiently rather than further clearing of more mangroves is an avenue which should be considered. One viable alternative is for the solar salt producing companies to diversify into aquaculture. Another alternative is the intensification of pond aquaculture instead of the extensive method which was used in the FAO/UNDP pilot project.

Integrated solar salt – aquaculture production

Irrespective of locality, the solar salt making process is fairly uniform, with sea water being evaporated in controlled areas to produce brine for crystallising salt. A typical saltworks in Kenya consists of a series of ponds where the brine is transferred as it gets progressively more saline. As many as six transfers are involved from the first reservoir through a series of evaporation ponds to a serving pond and finally to the crystallisers (Fig. 1). The first reservoir ponds which generally occupy 7–15% of the saltworks total area are indeed ready-made extensive ponds suitable for fish and shrimp farming. The sea water in the reservoir is always freshly drawn and has a salinity which does not differ much from that of the ocean since the pumps are constantly in operation. Using the estimates of Sammy & Tyler (1986) that some 65 tonnes of sea water are needed to produce one tonne of salt, a solar salt producer could culture plankton feeding animals

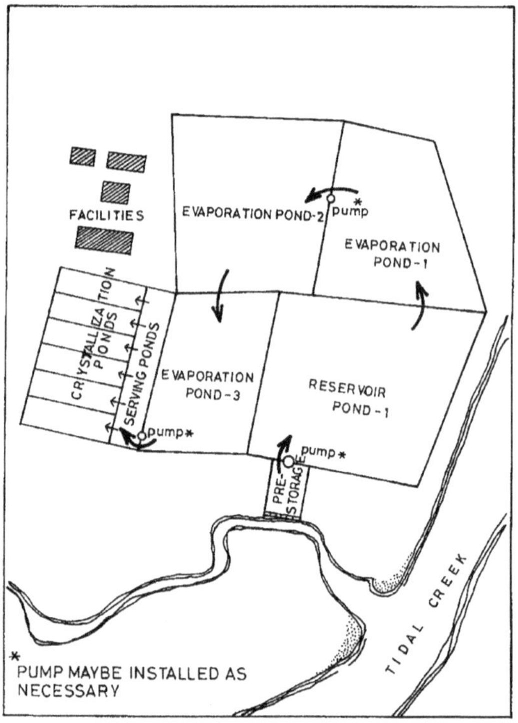

Fig. 1. Generalized layout of a saltwork in Kenya (Source: Yap & Landoy, 1986).

like oysters, mussels and milkfish on the plankton content of this volume of pumped sea water. These reservoir ponds could also be used as extensive shrimp ponds in which the shrimps could subsist and grow on the naturally occurring food in the ponds. Considering that the physical facilities, civil works, and equipment required is common with that for the salt industry and that there is an existing pool of manpower with actual experience in all aspects of pond construction and maintenance, the establishment of an aquaculture facility in the first reservoir makes good economic sense. The necessity and cost of pumping sea water are inherent in the operation of a salt field hence no extra costs would be incurred in obtaining the plankton biomass.

Integrated solar salt – Artemia *culture*

As the sea water gets progressively more saline with passage through a salt field there is a natural mortality of organisms because of the increasing salinity. Those organisms that survive tend to proliferate through lack of competition. One of these organisms is the brine shrimp *Artemia* which thrives in the evaporator ponds where the brine attains a concentration of 60–300‰ (Fig. 2). In the low salinity ponds (100–150 g salt l^{-1}) the *Artemia* reproduces by ovoviviparity (free-swimming larvae) while at higher salinities (>150 g l^{-1}) the animal responds to the sub-optimal conditions by producing cysts (oviparity). Jones *et al.* (1981) have shown that *Artemia*

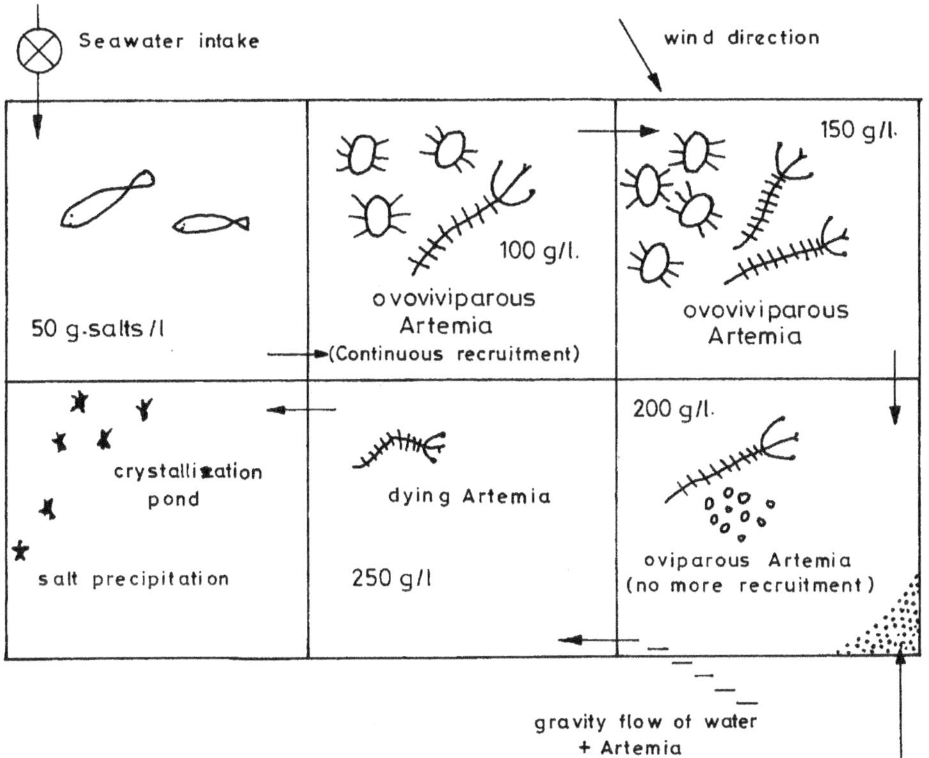

Fig. 2. Schematic diagram of solar salt operation with natural occurrence of *Artemia* (Source: Sorgeloos, 1983).

plays a fundamental role in the salt making process. The continuously non-selective filter feeding *Artemia* keeps under control the algal blooms which usually prevent early precipitation of gypsum and in extreme situations even hamper salt crystallisation. The *Artemia* metabolites or the decaying *Artemia* also provide a suitable substrate for the halophilic microbes of the genus *Halobacterium*, the presence of which assures a red coloration of the cyrstallisers. Dark red crystallisers assures much better heat absorption and thus faster salt crystallisation than in a situation where the white salt at the bottom of the pond reflects most of the light (Sorgeloos, 1983; Jones *et al.*, 1981). In crustacean hatcheries *Artemia* is used as a practical and suitable larval food source either in the form of nauplii that can easily be hatched from the dry cysts or as adults (biomass) that can be harvested from evaporator ponds. *Artemia* biomass would alleviate the feed supply problem in the ponds while the *Artemia* cysts which can command as high as $50 per kilogram in the world market could be a source of foreign exchange earnings. *Artemia* culture in the coastal saltfarms has already been investigated and found viable (Rasowo & Radul, 1986) and only needs management to enhance cyst and biomass production.

Intensification of existing pond aquaculture

Maximising the output in the existing ponds would negate the need for expanding the pond area. Thus intensification in place of extensification of the pond culture seems to be a means of saving the mangroves from further destruction. Since intensive farming involves great production costs because of the use of sophisticated technology, Kaptesky (1987) points out that it can only be a viable management solution if appropriate local technology and manpower trained to utilise it is available. Kenya lacks this technology and hence semi-intensive methods which also include feeding, pond fertilization, predator control and aeration should be adopted.

Aquaculture in the mangroves

Some types of aquaculture activities can be practiced in mangrove areas without adverse effects on the mangrove ecosystem. Pen, cage, raft, long line, and tray culture are good alternatives to pond culture in the mangroves although the methods require biological, technological and infrastructural inputs. Oyster culture is currently being practiced in one creek in Kenya where a simple and inexpensive culture system using a combination of rack and raft method for growth and roofing tiles for spat collection has given very impressive results with the local species *Crassostrea cucullata*. Kenya possesses 22 extensive mangrove creeks covering 52 000 ha in surface area, thus the potential for growing oysters is vast provided that a market can be established. Some species of *Tilapia*, e.g., *T. spilurus* and *T. mossambica*, show remarkable tolerance to salinity. The feasibility of tilapia cage culture within the many sheltered tidal creeks should be studied. The culture of brackishwater fish like rabbit fish (*Siganus* spp.), mullets, and milkfish in floating cages or fish pens should be encouraged. Bwathondi (1986) reports successful attempts at cage culture of *Siganus canaliculatus* in neighbouring Tanzania, though the identity of the species in question is probably *S. sutor* (Fischer & Whitehead, 1974).

Pump-fed ponds

Almost all tropical countries with mangrove forests have enacted legislation regulating the utilisation of the mangrove swamps. But as pointed out by Csavas (1988), conservation measures however strictly enforced cannot completely stop further destruction of mangroves. Hence, alternative solutions have to be found. Diverting coastal aquaculture development out of the mangrove and onto adjacent higher grounds could be an answer. Locating ponds on higher grounds requires pumping to provide water circulation. It has been shown that the pumping costs are more than counterbalanced by the savings on construction costs as there is a marked reduction in the volume of earthwork and structures. Higher production is also achieved with the improved circulation independent of tidal rise and fall (Gedney *et al.*, 1984). Since the ponds are easier to drain a more efficient and complete harvesting facilitating the drying and treatment of the pond bottom is possible. Construction out of the mangrove areas means a substantially decreased danger of hitting acid sulphate soils, especially as the ponds are also only shallowly excavated.

Conclusion

Many tropical countries with extensive areas of inter-tidal land suitable for shrimp production have began to develop the resource for shrimp production with a view to foreign exchange earnings, agricultural diversification and employment generation. As a result, large tracts of mangrove forests are being converted into aquaculture ponds. This large-scale conversion may eventually be detrimental to the aquaculture industry itself because the viability of shrimp mariculture is closely linked to the status of mangroves in the local ecosystem. Indeed, in the majority of these countries the hatchery systems are not well developed hence much of the shrimp culture industry is dependent on the mangrove areas to supply wild fry for stocking the ponds. A number of conservation measures may be employed to minimize the loss of the mangrove habitats: ponds should be sited on the landward side of the mangrove area so as to have the more productive seaward area undisturbed. Where cleared intertidal land is available this should be used in preference to intact mangroves. Mangroves should be retained and planted on the banks of ponds. There is a clear indication that aquaculture activity can exist in harmony with the mangrove ecosystem when 'mangrove-friendly' aquaculture methods like pen, cage, raft and long-line are employed. Ultimately emphasis should be laid on a more efficient use of existing pond areas by encouraging intensification as well as integrated aquaculture practices.

Acknowledgement

This paper was prepared with the support of a grant from Moi University.

References

Allela, S. O., 1986. Development and research of aquaculture in Kenya. In E. A. Huisman (ed.), Aquaculture Research in the Africa Region. Pudoc, Wageningen, 274 pp.

Bailey, C., 1988. The social consequences of tropical shrimp mariculture development. Ocean & Shoreline Management 11: 31–44.

Brusher, H. A., 1974. The magnitude, distribution and availability of prawn (Penaeidae) resources in coastal and estuaries of Kenya. J. mar. Biol. Ass. India 16: 335–348.

Bwathondi, P. O., 1986. A review of aquaculture practices in Tanzania. In E. A. Huisman (ed.), Aquaculture Research in the Africa Region. Pudoc, Wageningen, 274 pp.

Csavas, I., 1988. Shrimp farming developments in Asia. Infofish International 2: 11–16.

Doute, R., N. Ochanda & H. Epp, 1981. A forest inventory of Kenya using remote sensing techniques. KREMU Technical Report Series No. 30. Kenya Ministry of Environment and Natural Resources.

Fischer, W. & P. J. P. Whitehead (eds), 1974. FAO Species Identification Sheets for Fishery Purposes Vol. 4. Rome, FAO, Pag. Var.

Gedney, R. H., Y. C. Shang & H. L. Cook, 1984. Comparative study of tidal and pumped water supply for brackish-water aquaculture ponds in Malaysia. Coastal Aquaculture Demonstration and Training Project, FI:DP/MAL/77/008. Field Document No. 2: 119–60.

Jones, A. G., C. M. Ewing & M. V. Melvin, 1981. Biotechnology of solar salt fields. Hydrobiologia 81: 391–406.

Kapetsky, J. M., 1987. Conversion of mangroves for pond aquaculture: some short-term and long-term remedies. Workshop on the Conversion of Mangrove Areas to Aquaculture. UNDP/UNESCO Regional Project RAS/79/002.

MacNae, W., 1974. Mangrove forests and fisheries. Indian Ocean Programme, Indian Ocean Fishery Commission, Roma, IOFC/DEV/74/34.

Rasowo, J. & J. Radul, 1986. Inoculation of brine shrimp, *Artemia salina*, in Kenya; expected impact on aquaculture development. In E. A. Huisman (ed.), Aquaculture Research in the Africa Region. Pudoc, Wageningen 274 pp.

Saenger, P., E. J. Hegerl & J. D. S. Davie, 1983. Global status of mangrove ecosystems. Commission on Ecology Papers No. 3, International Union for Conservation of Nature and Natural Resources, 88 pp.

Sammy, N. & J. Tyler, 1986. Opportunity knocks for solar salt field aquaculture. Australian Fisheries, November 1986, 2–4.

Singh, T., 1987. The use of mangrove-managed areas for aquaculture. Workshop on the conversion of Mangrove areas to Aquaculture. UNDP/UNESCO Regional Project RAS/79/002.

Sorgeloos, P., 1983. Potential of the mass production of brine shrimp *Artemia*. Journal of the Society for Underwater Technology. Spring 1983, 27–29.

Turner, R. E., 1977. Intertidal vegetation and commercial yields of penaeid shrimp. Trans. am. Fish. Soc., 106: 411–6.

Yap, G. W. & R. J. Landoy, 1986. Development of Coastal Aquaculture. Report on a survey of the coastal areas of Kenya for shrimp farm development. KEN/80/018, Fisheries Department, FAO, Rome.

Hydrobiologia **247**: 215–222, 1992.
V. Jaccarini & E. Martens (eds), The Ecology of Mangrove and Related Ecosystems.

The effect of oocytic atresia on fecundity estimates of the rabbit fish *Siganus sutor* (Pisces: Siganidae) of Kenyan marine inshore waters

M. J. Ntiba & V. Jaccarini
Department of Zoology, University of Nairobi, P.O. Box 30197, Nairobi, Kenya

Key words: oocytic atresia, fecundity, *Siganus sutor*, Kenya coast

Abstract

In the strongly group-synchronized oocyte development of *Siganus sutor* (Valenciennes, 1835) the group of oocytes to be released in the following spawning, is identified. The smallest size of oocyte belonging to this group was identified by the presence of cytoplasmic vacuoles in oocytes in histological sections. These vacuolated oocytes corresponded to oocytes of 150 μm diameter obtained by treatment with Gilson's fixative. The mean number of such oocytes in stage 4 (late developing) ovaries was found to be 638 000. The proportion of these oocytes removed by atresia before spawning was determined on histological sections to be 5%. The corrected estimate of mean fecundity was thus 606 000 oocytes per spawning.

Introduction

A major problem in fecundity studies of fish is how to identify those oocytes which are potentially capable of release in the coming spawning season (Macer, 1974). Some workers count the actual number of oocytes which are released in a single spawning season. The difficulty with this method is that it is often not easy to determine the number of spawning seasons for a female in a year. So most other workers use the presence of yolk in the oocytes as the criterion for determining the oocytes to be spawned. However, Macer (*op. cit.*), working with the carangid *Trachurus trachurus*, argued that the presence of vacuoles in the cytoplasm is an earlier indication of active maturation of the oocytes than the presence of yolk. He pointed out that using the acquisition of yolk as the sole criterion of oocyte development could seriously underestimate fecundity by omitting a later batch of oocytes which are actively developing but have not yet laid any yolk and which may be spawned together with the more advanced yolked oocytes.

Another complicating factor in the estimation of fecundity is the phenomenon of atresia (Yamamoto, 1956; Macer, 1974; Guraya *et al.*, 1975; Monaco *et al.*, 1978; Wallace & Selman, 1981; Cyrus & Blaber, 1984). This has mostly been disregarded in the estimation of fecundity, in spite of the fact that it may affect a considerable proportion of the oocytes at all stages of development (Macer, 1974; Wallace & Selman, 1981; Cyrus & Blaber, 1984). Most studies of atresia in the teleost ovary have failed to investigate its causes, and no work has so far quantified its role in regulating the numbers of oocytes as they mature in tropical fishes. Yet it is clear that atresia can have an important effect on recruitment into the fishery.

The object of this study is to determine fecundity in *S. sutor*, which is by far the most important siganid of the Kenyan inshore water (Ntiba, 1986), taking into account the effect of atresia.

The siganids contribute 50% of the total artisanal catch of the reef fisheries of Kenya (Nzioka, 1984). The mangrove-lined creeks provide nursery grounds for these economically important fish (Grove *et al.*, 1986).

Materials and methods

Fish were caught using traditional bottom traps in shallow water (< 14 m depth) in the vicinity of Mombasa, Kenya. Sampling was done fortnightly around neap tides throughout 1985, with an occasional additional sample taken on spring tides. The total number of fish caught was 904. All fish were sexed and maturity stages assigned according to a modified Nikolsky (1963) scheme, which has been described elsewhere (Ntiba & Jaccarini, 1990). In this scheme six maturity stages are distinguished, *viz.* (1) virgin, (2a) developing virgin, (2b) resting and recovering, (3) early developing, (4) late developing, (5) ripe, sometimes running, (6) spent.

Histological techniques

For histology, portions of ovaries were either fixed for 24 h in Smith's formal dichromate, washed in running tap water for another 24 h, and stored in 10% formalin, or fixed and stored in Bouin's fluid. The stored material was dehydrated in graded alcohols, cleared in xylene and embedded either in paraffin wax or in ester wax (Steedman, 1960). Sections were cut at 4–10 μm and stained in iron haematoxylin and eosin.

Gilson's treatment and counting of oocytes

Pieces of ovary were cut longitudinally and turned inside out and treated with Gilson's fluid in 50 ml specimen bottles for three months. The bottles were vigorously shaken from time to time to aid in the release of oocytes from the ovarian walls. Before counting, the contents of the bottles were poured into a petri dish and those oocytes which had not been liberated from the ovarian tissue were released by teasing. The debris of the ovarian walls was then removed and the remaining fluid containing the oocytes poured into 1 l beakers. The oocytes were washed in tap water which was changed several times by decanting after allowing the oocytes to settle. Each time the water was decanted, the supernatant was checked under a dissecting microscope for any small oocytes.

For counting, the clean oocytes were poured into a 1 l beaker and water made up to a known volume. To ensure the even distribution of oocytes in the water column, a plastic ruler was used to stir the oocytes vigorously with a to and fro motion. A subsample was taken after 10 strokes of the ruler by a Labsystem finelet 1 ml pipette. One aliquot usually gave sufficient numbers of the large and small oocytes to yield satisfactory counts and diameter distributions.

The oocytes were pipetted into a solid watch glass and their diameter measured along a horizontal axis, regardless of their shape, under a standard microscope using a calibrated eye-piece graticule using a total magnification of $\times 40$. The accuracy of the subsampling method was tested by taking seven replicates and calculating the coefficient of variation, which was found to be 6.2%. The total number (N) of oocytes in any size class in the ovary was calculated as follows:-

$$N = \frac{V}{V_1} n \times \frac{W}{W_1}, \quad \text{where}$$

n = number of oocytes of a given size class in the subsample;

V = volume of sample;

V_1 = volume of the subsample;

W = weight of whole ovary;

W_1 = weight of portion of ovary preserved.

Distribution of oocytes within the ovary

To establish whether there were any differences in the distribution of oocytes of different stages within the ovary, counts were made on portions of the material treated with Gilson's taken from the anterior, mid- and posterior regions of the ovary and from the right and left lobes. An analysis of variance showed that there were no significant differences in oocyte diameter distribution among the different regions of the ovary ($p > 0.05$). Therefore portions for oocyte counts could be cut from any region of the ovary. Throughout this work, however, portions were cut from the mid-region of left or right lobes.

The oocytes to be spawned

In order to make accurate fecundity estimates, it is necessary to identify the oocytes that will be spawned in the season following capture. This was done in the first place by examining the oocyte frequency distribution of Gilson's treated material for all the maturity stages of *S. sutor* (Fig. 1). It is seen that stage 4 and 5 ovaries have an oocyte diameter distribution that is strongly bimodal, with small and large oocytes. In some ovaries intermediate-sized oocytes are completely missing. In contrast, stage 1, 2, 3, and 6 ovaries have only one mode, that of small oocytes.

With such a strongly bimodal distribution of oocytes in stage 4 and 5 ovaries, it is reasonable to assume that the mode of larger oocytes will be spawned in the season following capture. The question now arises: what is the smallest size to include in the mode of large oocytes? Following Macer (*op. cit.*), it was decided that the most reasonable answer is to include those oocytes which have started to form cytoplasmic vacuoles.

The size of such oocytes was determined on histological sections. Eight stage 4 ovaries were examined at ×40, and the size frequency distribution of all oocytes in randomly selected sections was determined. The sections were moved along the *y*-axis by means of a mechanical stage and all the oocytes cut through the nucleus and falling within the graticule were measured. Care was taken to avoid counting the same oocyte twice by calculating how many sections had to be skipped to avoid revisiting even the largest oocytes.

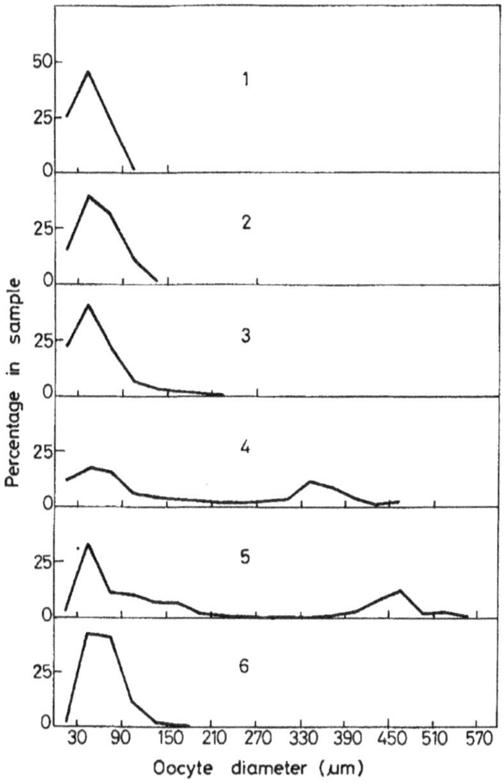

Fig. 1. Oocyte frequency distribution in Gilson's treated ovaries of *Siganus sutor* in various maturity stages, indicated by number in each polygon. Total counts of oocytes from 6 ovaries were made from each maturity stage.

Results

Figure 2 shows the percentage of oocytes containing vacuoles in each oocyte size class. Since the smallest size of oocyte to show vacuoles was, of necessity, determined on sectioned material, account had to be taken of the possible effect of differential shrinkage between material fixed in Bouin's for sectioning and Gilson's treated material used for the isolation of oocytes. To measure this effect the mean diameters of oocytes of the same stage of development were determined using six stage 4 ovaries, portions of each of which were fixed in Bouin's and Gilson's, respectively. It was found that Bouin's fixative shrunk material 9.3% more than Gilson's fluid. All measurements

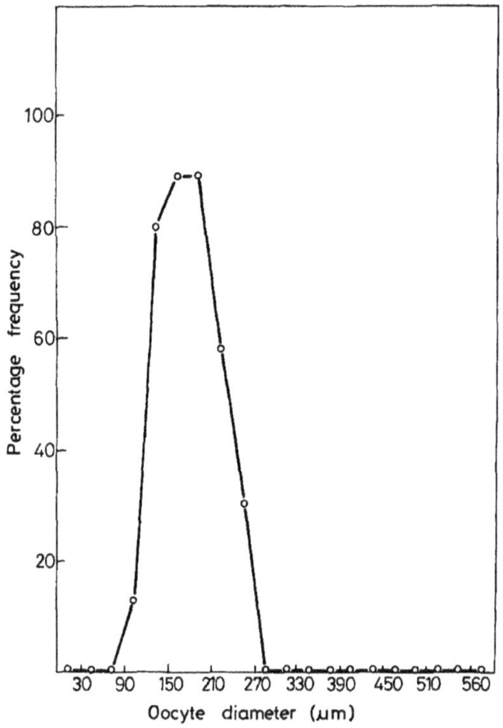

Fig. 2. Percentage of oocytes containing cytoplasmic vacuoles in histological sections of eight stage 4 ovaries of *Siganus sutor*.

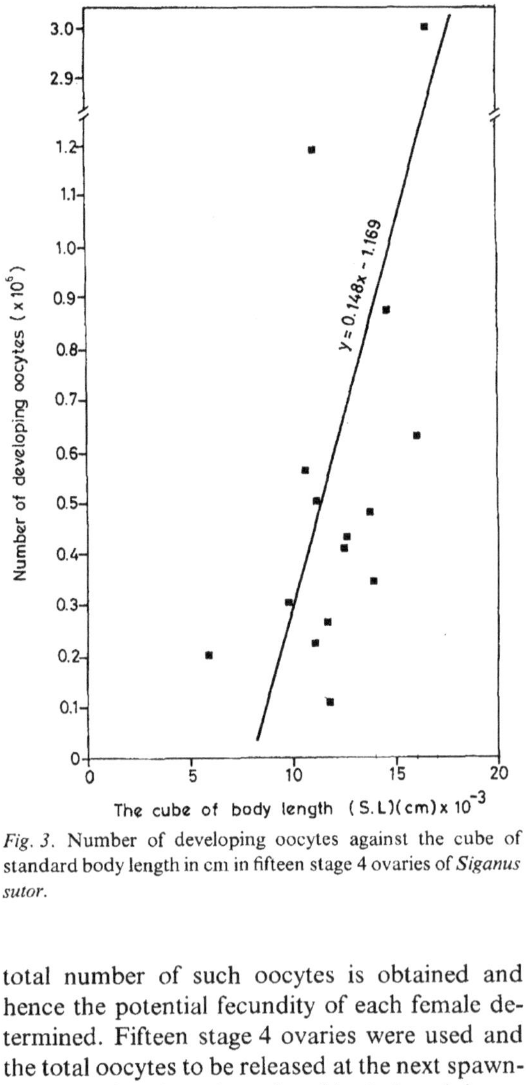

Fig. 3. Number of developing oocytes against the cube of standard body length in cm in fifteen stage 4 ovaries of *Siganus sutor*.

done on histological material were thus increased by this percentage to obtain Gilson's equivalents, which are used for plotting Fig. 2. This plot shows that vacuoles start appearing in the cytoplasm of the oocytes on their reaching a diameter of 90 μm, and that, at 150 μm, about 87% of the oocytes have a vacuolated cytoplasm. From about 200 μm the cytoplasm fills with yolk granules and the vacuoles are obliterated, so that at about 290 μm no vacuolated oocytes remain. Therefore, for fecundity estimates, all oocytes isolated by Gilson's treatment which were ≥ 150 μm were counted.

Counting the oocytes to be spawned

Applying the equation $N = (V/V_1)n \times (W/W_1)$ on the numbers of oocytes ≥ 150 μm in diameter obtained by Gilson's treatment, an estimate of the total number of such oocytes is obtained and hence the potential fecundity of each female determined. Fifteen stage 4 ovaries were used and the total oocytes to be released at the next spawning plotted against the cube of body length in cm (Fig. 3). A correlation was found between fecundity and cube of body length ($r = 0.546$, d.f. 13, $P < 0.05$). A mean of 638 000 oocytes to be released in the next spawning was determined.

Effect of egg atresia on fecundity estimates

Atretic oocytes have a chorion which initially becomes less distinct, then disintegrates and sinks into the cytoplasm of the oocyte. At the same time the yolk is resorbed and the oocytes collapse. Associated with such oocytes is a multitude of squamous follicular cells and many blood vessels which invade the atretic oocytes. These observations are in close agreement with those of Cyrus & Blaber (1984) working with *Gerres* species.

Figure 4 gives the percentage frequency of atretic oocytes for each maturity stage of the ovary. It is apparent from this figure that the degree of oocyte atresia is lowest in stage 1 (< 20%), rises to a minor peak of 45% in stage 3, and a major peak of 58% in stage 6. Stages 4 and 5 have a rate of atresia of 27 and 32%, respectively.

While some stages 4 and 5 ovaries treated with Gilson's fluid had the intermediate-sized oocytes missing altogether, this was never the case with histological sections of stage 4 ovaries, which always showed developing oocytes of all sizes including an appreciable proportion of oocytes of intermediate size. This apparent discrepancy can be explained by the observation that many of the intermediate-sized oocytes (as well as some smaller and larger ones) were atretic and had a disintegrated chorion and were already invaded by a number of squamous follicular cells. One would expect these oocytes to disintegrate under Gilson's treatment. A comparison of oocyte diameter frequency distributions from portions of the same ovaries fixed in Bouin's and in Gilson's fluid separately supports this interpretation. Figure 5 shows the proportion of atretic oocytes in different size classes in 10 ovaries in histological sections of material fixed in Bouin's and the size frequency distribution for the same 10 ovaries fixed in Gilson's fluid. It is clear that the batch of the intermediate-sized oocytes (210–330 μm), present in the histological sections of stage 4 ovaries, is almost completely missing from the oocyte population derived from Gilson's treated material. As seen from histological sections, atresia peaks precisely in the oocyte-size class missing from Gilson's treated material.

It follows that the estimate of 638 000 as the mean fecundity of *S. sutor* in Kenyan waters derived from stage 4 ovaries must be corrected by the proportion of oocytes lost through atresia by

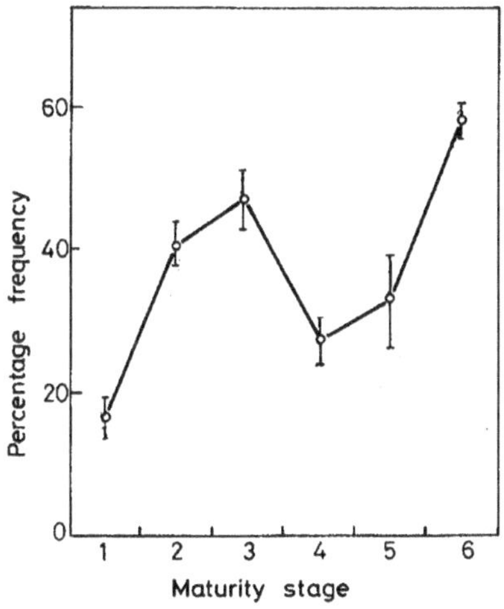

Fig. 4. The mean (± sem) percentage of atretic oocytes in six ovaries of *Siganus sutor* in each maturity stage.

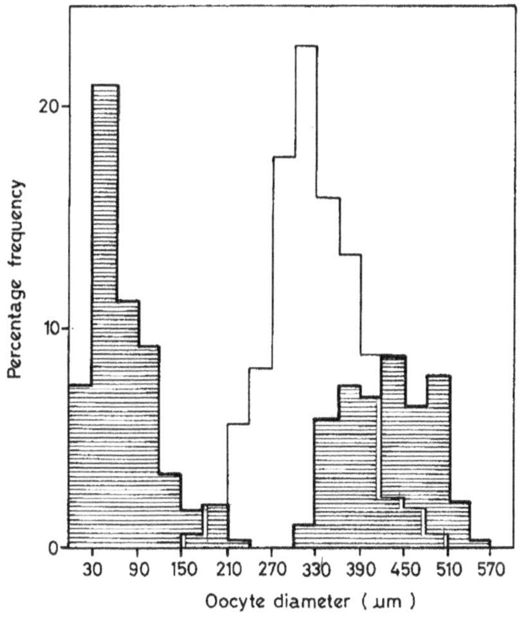

Fig. 5. Cross-hatched histogram gives the percentage frequency distribution of the various oocyte classes (*non-atretic*) from 10 ovaries of *Siganus sutor* isolated with Gilson's fluid. Open histogram gives the percentage frequency distribution of the *atretic oocytes* in histological sections of the same ovaries. One lobe of each ovary was treated with Gilson's fluid, while the other lobe was fixed in Smith's fluid.

the time they reach stage 5, i.e., before they are spawned. From Fig. 4, it is seen that this amounts to 5% over the stage 4 level. Applying that correction factor the estimated resultant fecundity becomes 606000 oocytes per spawning.

Though there are two spawning peaks per year in the population of *S. sutor* (Ntiba & Jaccarini, 1990), it is not clear from the evidence available whether individual fish spawn once or twice a year. On the latter assumption, the mean corrected fecundity estimate would be about 1.2 million.

Discussion

Fecundity

The strongly bimodal size frequency distribution of oocytes in stage 4 and 5 ovaries of *S. sutor* suggests that oocyte development in this species is group-synchronized. In this type of development two populations of oocytes are distinguished: a more advanced, fairly synchronous, population of large oocytes, and a less advanced heterogeneous population of small oocytes from which the larger ones are recruited (Wallace & Selman, 1981). Taking the fecundity for the immediately following spawning season to be the total number of all developing oocytes in the ovary at stage 4 minus what will be lost to atresia by stage 5, the estimated mean for *S. sutor* in Kenyan coastal waters is 606000. De Souza (1988) obtained a total fecundity of 700000 eggs for *S. sutor* at the Kenya coast by counting only yolked eggs from stage 4 and 5 ovaries treated with Gilson's fluid. His counts therefore involved only non-atretic oocytes. De Souza's estimate agrees numerically fairly well with ours, but this is hard to explain since we counted both yolked oocytes and the earlier vacuolated ones to obtain our estimate. Monacop (1937, cited in Lam, 1974) working with *S. canaliculatus*, a species closely related to *S. sutor*, reports 300000–500000 oocytes in its ovary. In the ovaries of artificially spawned *S. canaliculatus* individual variation ranging from 50000 to over 2 million eggs were found (Bryan *et al.*, 1975), while Hasse *et al.* (1977) reported a mean per female of 300000 eggs with a mean diameter of 0.5 mm. This wide range agrees with our findings in which the fecundity ranges from 200000 in a fish of 18.0 cm to well over 3 million in one of 25.5 cm SL. A similar situation was found for *Trachurus trachurus* of the North Sea (Macer, *op. cit.*). Macer found a correlation between fecundity and the cube of body-length ($P < 0.001$). Though we also see such a correlation ($P < 0.05$) in *S. sutor* inspection of residuals shows that the linear regression model was not a good fit. But Fisher's z transform shows no significant difference ($z = 1.31$) between Macer's and our correlation coefficients. So more data are required to test the apparent discrepancy with the model. A possible explanation of our findings is that *S. sutor* is short-lived, e.g. the largest fish in a sample of over 900 was less than two years old (Ntiba & Jaccarini, 1988). One perhaps might not expect a well-defined trend in fecundity to be established in the short time available.

Some workers use already running fish obtained from commercial fishermen (Hasse *et al.*, 1977; De Souza, 1988). But the uncertainties introduced by this procedure are obvious, since such fish may have already shed some eggs.

It is certain that *S. sutor* populations spawn twice a year in East Africa waters (Ntiba & Jaccarini, 1990 and this Symposium). This is indicated by the temporal variation in the weights of the gonads, the seasonal appearance of juveniles, the temporal changes in the relative condition factor, K_n, and especially by the seasonal occurrence of spent males and females in the catches. This is in agreement with results of other workers studying different species of siganids in Palau, Singapore, and the Philippines (Lam, 1974; Kami & Ikehara, 1976; Hasse *et al.*, 1977). However whether the individual *S. sutor* spawn once or twice a year it is impossible to decide on the present evidence.

Functional significance of atresia

In group-synchronous ovaries, a variety of oocyte recruitment strategies exist (Wallace & Selman,

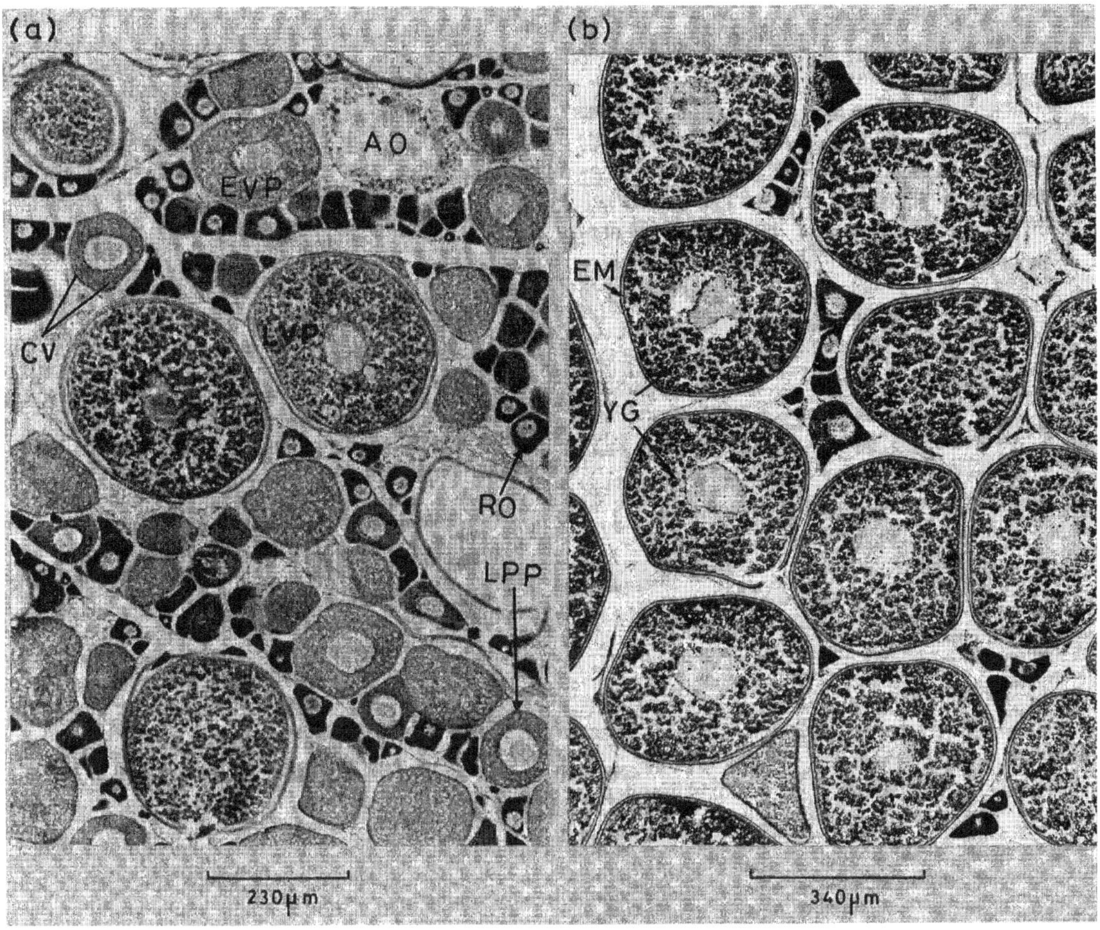

Fig. 6. Sections from stages 4 (a) and 5 (b) ovaries of *Siganus sutor*. Iron haematoxylin and eosin. AO: atretic oocyte; RO: resting oocyte; LPP: late protoplasmic phase; EVP: early vitellogenic phase; LVP: late vitellogenic phase; CV: cytoplasmic vacuoles; EM: egg membrane; YG: yolk granules.

1981). Recruitment can be (a) directly from oogonia, at the end of the gonadotropin-independent stage, (b) from pre-vitellogenic to vitellogenic stages, or (c) from oocytes which have terminated the vitellogenic stage. In *S. sutor* recruitment seems to be at the end of the gonadotropin-independent stage, since oocytes in this stage are always present in the ovary. This type of oocyte recruitment was shown in the herring, *Clupea harengus* (Bowers & Holliday, 1961). Considering *S. sutor* (Fig. 1), it is clear that at stage 4 there is a considerable number of the intermediate-sized oocytes, while at stage 5 there are two distinct modes separated by a gap. These results suggest that during development from stages 4 to 5, the group of largest oocytes is augmented in numbers by the later development of the smaller ones. This is further supported by the histological appearance of stages 4 and 5 ovaries (Figs 6a and b) in which the former present a wide variety of oocyte stages, from early resting ones, through vacuolated oocytes of a wide range of sizes, to advanced vitellogenic oocytes; while stage 5 ovaries present an almost uniform distribution of advanced vitellogenic oocytes, with some scattered groups of much smaller resting oocytes situated in the interstices.

Atresia of oocytes is prevalent in both pre- and post-spawning stages of *S. sutor*. While in the latter atresia aids in the removal of unwanted material from the ovary and helps to re-utilize valuable material, the functional significance of pre-spawning atresia is not so clear. Wallace & Selman (1981) suggest that poor nutrition and hormonal imbalances, especially of the gonadotropins, may cause involution of developing oocytes in the pre-spawning phase. The ovaries may function as storage organs. The occurrence of atresia at all oocyte maturation stages may suggest that this storage function is a finely-tuned process and that this fish may be responding to one or more physiological and environmental cues (Mommsen & Walsh, 1988).

Acknowledgements

We thank Mr. Allela, Director, Kenya Marine and Fisheries Research Institute, Mombasa, for use of facilities. This work was supported by a German DAAD research grant and another grant from the National Council for Science and Technology, Kenya. We thank Mr. Kigondu and Mr. Oswago, Department of Zoology, University of Nairobi, for help with the illustrations. We also thank the Botany Department for help with the micrographs.

References

Bowers, A. B. & F. G. T. Holliday, 1961. Histological changes in the gonad associated with the reproductive cycle of the herring (*Clupea harengus* L.). Mar. Res. 5: 1–16.

Bryan, P. G., B. B. Madraisau & J. P. McVey, 1975. Hormone-induced and natural spawning of captive *Siganus canaliculatus* (Pisces: Siganidae) year round. Micronesica 11: 199–204.

Cyrus, D. P. & S. J. M. Blaber, 1984. The reproductive biology of *Gerres* in Natal estuaries. J. Fish Biol. 24: 491–504.

De Souza, T. F., 1988. Reproduction, length-weight relationship and condition factor in *Siganus sutor* (Valenciennes, 1835) (Pisces: Siganidae) from the Kenyan waters of the western Indian Ocean. Kenya J. Sci. Ser. B. 9: 89–101.

Grove, S. J., M. C. Little & P. J. Reay, 1986. Tudor Creek Mombasa. The early life-history stages of fish and prawns 1985. Overseas Development Administration, Research Project R 3888, UK, 133 pp.

Guraya, S. S., R. Kaur & P. K. Saxena, 1975. Morphology of the ovarian changes during the reproductive cycle of the fish, *Mystus tengara* (Ham.). Acta anat. 91: 222–260.

Hasse, J. J., B. B. Madraisau & J. P. McVey, 1977. Some aspects of the life history of *Siganus canaliculatus* (Park) (Pisces: Siganidae) in Palau. Micronesica 13: 297–312.

Kami, H. T. & I. I. Ikehara, 1976. Notes on the annual juvenile siganid harvest in Guam. Micronesica 12: 323–325.

Lam, T. J., 1974. Siganids: their biology and maricultural potential. Aquaculture 3: 325–354.

Macer, C. T., 1974. The reproductive biology of the horse mackerel, *Trachurus trachurus* (L.) in the North Sea and English Channel. J. Fish Biol. 6: 415–438.

Mommsen, T. P. & P. J. Walsh, 1988. Vitellogenesis and oocyte assembly. In W. S. Hoar & D. J. Randall (eds), Fish Physiology 11A. Academic Press, New York: 374–406.

Monaco, P. J., E. M. Rasch & J. S. Balsano, 1978. Cytological evidence of temporal differences during asynchronous ovarian maturation of bisexual and unisexual fishes of the genus *Poecilia*. J. Fish Biol. 13: 33–44.

Monacop, P., 1937. The artificial fertilization of dangit, *Amphacanthus oramin* (Bloch & Schneider) (Probably = *Siganus canaliculatus*). Philip. J. Sci. 62: 229–237.

Nikolsky, J. R., 1963. Ekologiya Ryb. Moscow: Vysshaya Schkola Press, 368 pp.

Ntiba, M. J., 1986. The biology of the Kenya reef fish of the genus *Siganus*. M.Sc. Thesis, University of Nairobi, xviii + 143 pp.

Ntiba, M. J. & V. Jaccarini, 1988. Age and growth parameters of *Siganus sutor* of Kenya marine inshore waters derived from numbers of otolith microbands and fish lengths. J. Fish Biol. 33: 465–470.

Ntiba, M. J. & V. Jaccarini, 1990. Gonad maturation and spawning times of *Siganus sutor* off the Kenya coast: evidence for definite spawning seasons in a tropical fish. J. Fish Biol. 37: 315–325.

Nzioka, R. M., 1984. The evaluation of marine fisheries resources in Kenya. In S. A. Iversen & S. Myklevoll (eds), Proceedings of the NORAD-Kenya seminar to review the marine fish stocks and fisheries in Kenya. Mombasa, Kenya, 13–15 March 1984. Institute of Marine Research, Bergen, Norway: 27–32.

Steedman, H. F., 1960. Section cutting in microscopy. Oxford: Blackwell, 172 pp.

Wallace, R. A. & K. Selman, 1981. Cellular and dynamic aspects of oocyte growth in teleosts. Am. Zool. 21: 325–342.

Yamamoto, K., 1956. Studies on the formation of fish eggs. 1. Annual cycle in the development of the ovarian eggs in the flounder, *Liopsetta obscura*. J. Fac. Sci. Hokkaido University Ser. 6, Zoology 12: 352–373.

Hydrobiologia **247**: 223–227, 1992.
V. Jaccarini & E. Martens (eds), The Ecology of Mangrove and Related Ecosystems.

Differences in biomass production and carrageenan yields among four strains of farmed carrageenophytes in Northern Bohol, Philippines

Gavino C. Trono, Jr. & Arturo O. Lluisma
Marine Science Institute, College of Science, University of the Philippines, Diliman, Quezon City, Philippines

Key words: carrageenan, carrageenophytes, biomass production, farming, Philippines

Abstract

Comparative studies on the biomass and carrageenan production of two strains of *Eucheuma denticulatum* and two strains of *Kappaphycus alvarezii* were made to assess the seasonality in their production capacities.

The high and similar refined carrageenan (RC) yields (43–53% of dry wt.) of the four strains in the first cropping season (June–October) coincided with their high biomass production with plants averaging from 1.1 to 1.8 kg each at harvest. The poor RC yields (21–33%) recorded in the second cropping (October–February) coincided with their season of low biomass (av. wt: 0.34 to 1.0 kg). The four strains, however, recorded contrasting performance in the third cropping season (February–July) with the two *E. denticulatum* strains recording high RC yields (43 and 42.5%) together with high biomass (av. wt: 1.5 and 1.6 kg) in contrast to the low RC yields (30 and 39%) and low biomass (av. wt. 0.21 and 0.28 kg) of the two *K. alvarezii* strains. Records for semi-refined carrageenan (SRC) yields in the second and third cropping seasons were quite consistent and similar for the four strains (42–55%), except in the second cropping where the two *K. alvarezii* strains recorded low SRC. These differences in production potentials highlight the need for cropping management of the four strains to improve their cropping performance.

Introduction

The great demand for carrageenans on the world market has encouraged the commercial production of carrageenophytes through mariculture which is presently one of the few important livelihoods among the coastal populations in the Philippines. The industry, however, is faced with problems like the apparent decrease in vigor of farmed stocks resulting in low farm production outputs, poor quality of the produce caused by poor post harvest handling practices, e.g., high moisture content, presence of extraneous materials such as other seaweeds, sand, and plastic materials, and low yields of refined carrageenans from farm grown crops.

The different strains of carrageenophytes presently farmed in Danajon Reef, Northern Bohol show distinct differences in their growth rates and production (biomass) during the different seasons of the year (Trono & Ohno, 1989). The current cropping practice of farmers is to mix-plant the different strains and select seedstocks for the next cropping season from the present harvest. Because the different strains differ in their production capacities during the different parts of the year, the seedstocks selected for the succeeding cropping season are not at all acclimated to this season resulting in low production yields of pure carrageenan of crops.

The green (GSP1) and the brown (BSP1) strains of *Eucheuma denticulatum* (Burman) Collins et Hervey and the green (GST1) and the brown (BST1) strains of *Kappaphycus alvarezii* (Doty) Doty are commonly farmed throughout the year. *E. denticulatum* produces *iota* carrageenan and is called the 'spinosum' of commerce while *K. alvarezii* produces *kappa* carrageenan and is the 'cottonii' of commerce. We report here the seasonal differences in biomass production and carrageenan yields (semi-refined and refined) of the four strains from three different cropping seasons with the view to illucidating the seasonality in their production potentials as basis for their cropping management.

Materials and methods

This research was conducted at the Antonio-Agro Seaweed Farm on Danajon Reef, Northern Bohol, Philippines (10° 16′ N. Lat; 124° 3′ E. Long.) from March 1983 to September 1984.

The fixed bottom monoline method (Trono & Fortes, 1988) was used in the culture of the four strains. The experimental plots for each strain consisted of 20 monolines, each monoline planted to 20 plants/cuttings, the individual cuttings weighing 100 g.

Four croppings were made, the first was from 13 June to 5 October 1983; the second from 12 October 1983 to February 1984; the third from 7 April to 20 July 1984. The fourth cropping from 5 August to September 1984 was destroyed by a typhoon. The growing period lasted from 12 to 18 weeks.

The sampling for biomass was done one week after planting and weekly thereafter. Sampling was done by randomly harvesting the whole monoline from experimental plots of each strain. The average biomass (fresh) of the plants in the monoline was recorded. The harvests (biomass) were sun dried. The dried materials starting with the fourth week harvest were analyzed for semi-refined (SRC) and refined carrageenan (RC) yields. The extraction of SRC was done by Marine Colloids Philippines, Inc. in Cebu City. RC was processed by the alcohol precipitation method (Towle, 1973).

The data on standing crop and carrageenan yields of the four strains from the different cropping periods were analyzed and compared.

Results

No significant differences in the amount of RC between strains were observed in the first cropping (June–October) (Table 1a). The RC yields of the four strains were within the narrow range of 43–53% of dry weight and were the highest among values recorded in the three cropping seasons. The biomass of the plants was also the highest recorded as all strains showed good growth.

The second cropping (Oct–Feb) was a period of poor growth for all strains as the strains suffered the 'ice-ice' disease, with the exception of the brown strain of *K. alvarezii* (BST1) which recorded relatively good growth (attaining more than 1000 g maximum average biomass). The maximum average biomass attained by the other strains was barely 500 g.

Differences between strains in amount of carrageenan yields in the second cropping were observed. The *E. denticulatum* strains (BSP1 and GSP1) yielded significantly higher amount of SRC than did the *K. alvarezii* strains (Table 1b). The amount of RC appeared to be significantly different among the four strains although the green strains (GSP1 and GST1) seemed to yield more RC than did the brown strains (BSP1 and BST1) (Table 1a). The yields of RC relative to SRC from the four strains during this cropping season differed, i.e., the RC yields of the *K. alvarezii* strains were almost as much as the SRC in contrast to those of *E. denticulatum* strains which were considerably less than that of the SRC.

In the third cropping (Feb–Jul), the RC yields of the *E. denticulatum* strains were significantly

Table 1. Comparison of the RC (Table 1a) and SRC (Table 1b) yields (± std. deviation), in per cent of dry weight, of the two strains of *Eucheuma* (BSP1 and GSP1) and *Kappaphycus* (BST1 and GST1). Results of the ANOVA-Duncan's Multiple Range Test are shown on the left (for comparisons between cropping seasons within a strain) and on the right (for comparisons between strains within a cropping season) of the values; means with the same letter are not significantly different from each other ($p < 0.05$). For Table 1b, the t-Test was used to compare the yields of a strain from the two cropping seasons; s = significant, ns = not significant ($p < 0.05$).

Table 1a

Strain	Cropping season		
	Jun–Oct	Oct–Feb	Feb–Jul
GST1	a 51.30 (± 4.65) a	b 33.18 (± 5.27) a	b 37.05 (± 3.88) a
BST1	a 49.74 (± 2.22) a	b 26.51 (± 7.55) b	b 29.98 (± 2.82) b
GSP1	a 49.68 (± 3.66) a	b 30.82 (± 6.16) ab	c 42.47 (± 3.87) c
BSP1	a 49.88 (± 3.65) a	b 20.86 (± 6.18) c	c 43.00 (± 5.46) c

Table 1b

Strain	Cropping season	
	Oct–Feb	Feb–Jul
GST1	s 33.68 (± 6.04) a	54.86 (± 4.71) a
BST1	s 31.40 (± 7.55) a	41.48 (± 5.00) b
GSP1	ns 52.84 (± 3.35) b	49.16 (± 6.40) c
BSP1	ns 52.28 (± 7.25) b	47.63 (± 9.41) c

higher than those of the *K. alvarezii* strains (Table 1a). All strains, however, yielded fairly the same amount of SRC but the amount of RC from the *K. alvarezii* strains was significantly less than that of the SRC while the amount of RC from the *E. denticulatum* strains was almost as much as their SRC.

The four strains appeared to follow a common pattern of intercropping differences in the yields of RC, i.e., highest in the first cropping, lowest in the second, and intermediate in the third (Table 1a). These differences appeared significant among the *E. denticulatum* strains while they did not appear substantial in the *K. alvarezii* strains in the third cropping.

Differences in biomass production of the different strains during the three cropping seasons were also observed. The *E. denticulatum* strains showed a trend similar to that of their RC yields: high in the first and third croppings but low in the second cropping although a longer period (16 weeks) was required to attain maximum biomass in the third cropping compared to the first cropping (7 weeks). In contrast, high biomass production among the *K. alvarezii* strains was observed only in the first cropping periods. Their RC yields appeared to follow the trend in their biomass production, e.g., high during the first and low during the second and the third croppings.

Discussion

The low quality of farm produce (low RC yields) has always been attributed to poor post-harvest processes. Despite efforts to induce farmers to improve the quality of their produce such as through graded pricing based on the physical quality, processors still receive batches of seaweeds with low RC yields. The results of this study show that the state of the crop at the time of harvest, i.e., whether healthy or not, appeared to be an important factor affecting the SRC and RC yields. The amount of yields appeared to vary with the growth performance of the strains, as indicated by biomass production. High RC yields of the four strains were recorded in the June–October cropping period when growth was good (high standing crop) and significantly low yields in the October–February cropping period when growth was poor. Thus, the observed intercropping differences in the yields of RC appear to take exception to the general observation that rapid growth of carrageenophytes, as a result of elevated nutrient levels is accompanied by less carrageenan production (Dawes *et al.*, 1974a, b; McCandless & Craigie, 1979; Chopin *et al.*, 1990). The lack of agreement between the results of the present studies and those of the other workers might have been caused by differences in ambient environmental conditions. Dawes *et al.* (1974a, b) reported that maximum growth and minimum carrageenan yields occurred in period of low temperature. Trono & Ohno (1989) reported high growth and biomass during periods of relatively high temperature regimes.

Differences in the RC yields and biomass production observed among the four strains appeared to be species-specific. The two *E. denticulatum* strains showed similarly seasonality in biomass and RC yields as did the two strains of *K. alvarezii*. Trono & Ohno (*loc. cit.*) reported the same species-specificity in the seasonality of biomass production of the different strains. But differences in RC yields between the brown and green strains of the two species were also noted during the second cropping periods with the green strains of the two species recording higher RC yields than their brown counterparts.

The distinctly higher ratio of SRC to RC yields of crops during cropping seasons of low biomass production and lower ratio during cropping seasons of high biomass production are very apparent. The incidence of what is known as the 'ice-ice' disease is preceded by the paling of the color, the loss of the glossy surface and thinning of the branches, and occurrence of epiphytes and other weeds. The SRC yields during this period remain approximately similar to that recorded in crops from good cropping period but the RC yields were significantly lower. A large portion of the carrageenan in diseased or sickly crops apparently is degraded to other products and cannot be recovered by alcohol precipitation during processing of RC. But during SRC processing, this portion is apparently retained. The presence of these degraded products may reduce the gel strength and viscosity of the carrageenan. This important problem needs clarification.

The results of this study show that the production potentials of the different farmed strains varied with species and with cropping seasons. The production in the first cropping period was not at all affected by choice of species or strains as all showed good performance (high biomass and carrageenan yields). In the second cropping period, however, the green strains of both species appeared to record better crop performance than did the brown strains. In the last cropping, both strains of *E. denticulatum* recorded good growth compared to those of *K. alvarezii*. This information can be used as basis for cropping management of different strains to improve the productivity of the farms.

Acknowledgements

This study was supported by the International Development and Research Center (IDRC) – Canada and the Philippine Council for Aquatic and Marine Research and Development (PCAMRD).

References

Chopin, T., M. D. Hanisak, F. E. Koehn, J. Mollion & S. Moreau, 1990. Studies on carrageenans and effects of seawater phosphorus concentration on carrageenan content and growth of *Agardhiella subulata* (C. Agardh) Kraft and Wynne (Rhodophyceae, Solieriaceae). J. appl. Phycol. 2: 3–16.

Dawes, C. J., A. C. Mathieson & D. P. Cheney, 1974a. Ecological studies of Floridean *Eucheuma* (Rhodophyta, Gigartinales) I. Seasonal growth and reproduction. Bull. mar. Sci. 24: 235–273.

Dawes, C. J., J. M. Lawrence, D. P. Cheney & A. C. Mathieson, 1974b. Ecological studies of Floridean *Eucheuma* (Rhodophyta, Gigartinales) III. Seasonal variation of carrageenan, total carbohydrate, protein and lipid. Bull. mar. Sci. 24: 286–299.

McCandless, E. L. & J. S. Craigie, 1979. Sulfated polysaccharides in red and brown algae. Ann. Rev. Pl. Physiol. 30: 41–53.

Towle, Gordon, A., 1973. Carrageenan. In R. L. Whistler & J. N. BeMiller (eds), Industrial Gums: Polysaccharides and their Derivatives. Academic Press, New York/London: 83–113.

Trono, G. C. Jr. & E. G. Fortes, 1988. Philippine Seaweeds. National Bookstore, Metro Manila. 330 pp.

Trono, G. C. Jr. & M. Ohno, 1989. Seasonality in the biomass production of the *Eucheuma* strains in Northern Bohol, Philippines. In I. Umezaki (ed.), Scientific Survey of Marine Algae and their Resources in the Philippine Islands. Monbushio International Scientific Research Program, Japan.

Hydrobiologia **247**: 229–234, 1992.
V. Jaccarini & E. Martens (eds), The Ecology of Mangrove and Related Ecosystems.

Status and exploitation of the mangrove and associated fishery resources in Zanzibar

M. A. K. Ngoile & J. P. Shunula
Institute of Marine Sciences, P.O. Box 668, Zanzibar

Key words: exploitation, mangroves, fishery resources, Zanzibar

Abstract

The distribution and growth forms of the mangroves on Zanzibar island (Unguja and Pemba) were investigated. Pemba island, with approximately 12000 ha of mangroves, had nine species; and Unguja island, with 6000 ha, had eight. Average cropping intensities in Pemba island varied between 350–1937 cut plants per ha while in Unguja these were between 777–3567 cut plants per ha. In Pemba, mangrove forests were relatively less heavily exploited compared to those in Unguja, where clear felling was common. Charcoal, lime and salt production are the major activities which involve heavy utilization of mangrove wood.

The fishery resources (shellfish and finfish) associated with the mangrove vegetation were also investigated by examining the composition of fish catches at landing stations in mangrove as well as non-mangrove zones of fishing districts. A survey on the fauna associated with mangrove forests was also conducted. A number of the local commercial fish species were found to be associated with the mangrove vegetation. The fishes of the genus *Lethrinus* and the rabbit fish *Siganus* were found to constitute the bulk of the fish caught from waters adjacent to mangroves using movable traps. The mollusc *Pyrazus* sp., which is commonly used by the local people for food and bait, was found only in the mangroves.

Introduction

Studies on the mangroves of Zanzibar islands (4° 50′–6° 30′ S and 38° 10′–39° 55′ E) are few. Griffith (1949) enumerated the mangrove species found in several stands in both Unguja and Pemba islands, and attempted a visual estimation of the percentage abundance of the various species in addition to estimating the areas under mangrove cover using a planimeter. Since his time, there were virtually no other studies on the mangroves of Zanzibar. Most studies on the mangroves of Tanzania have concentrated on the mainland coastal stands.

The information on mangroves and fisheries of Zanzibar islands reported in this paper have been extracted from completed and ongoing research work by the authors. These studies were initiated in view of recent developments towards environmental conservation, marine coastal resources exploitation and management, hence the need for inventory and assessment.

Materials and methods

The sites in which the studies were made included Unguja-Ukuu Kisakasaka, Pete, Chwaka and Muwanda in Unguja island (See Fig. 1), while in Pemba islands the sites were Micheweni, Chakechake and Wete (See Fig. 2).

A minimum of two visits were made to each

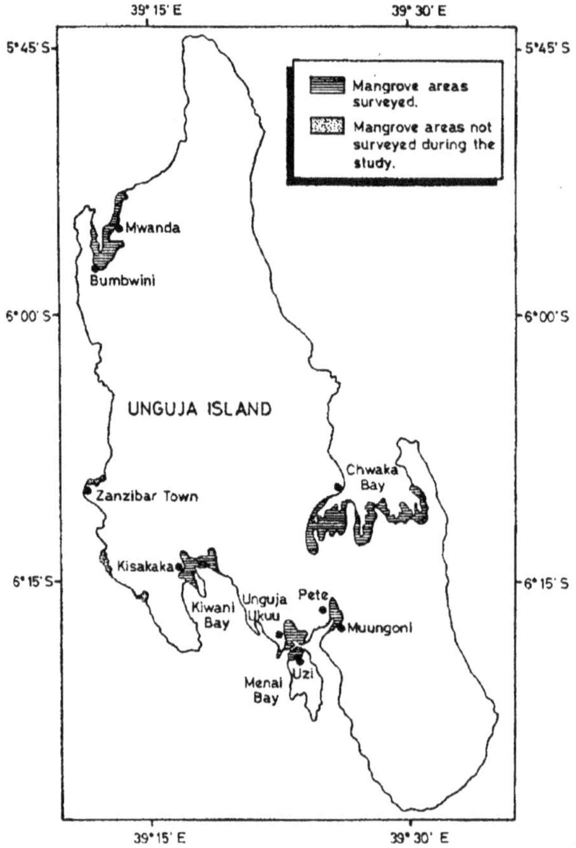

Fig. 1. Map of Unguja island showing the distribution of mangroves.

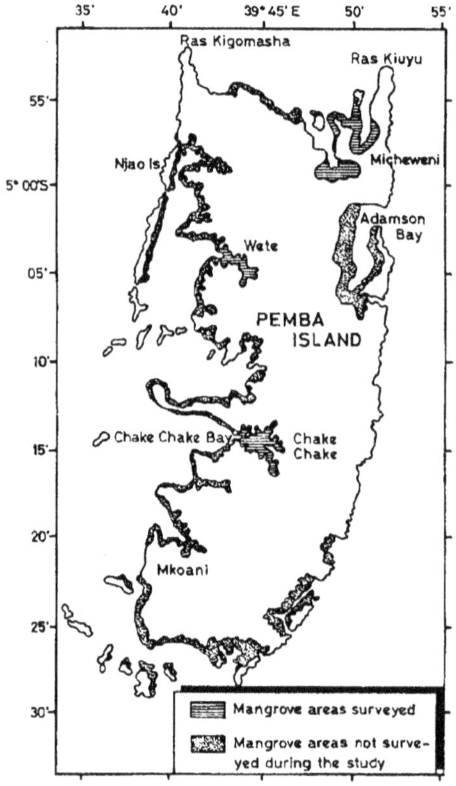

Fig. 2. Map of Pemba island showing the distribution of mangroves.

study site. The information collected at each site included a list of mangrove species, an estimate of cut mangrove trees in random sample plots of 5 m × 5 m, percent coverage by species, diameters and average lengths of randomly sampled mangrove poles, and the fauna associated with the mangroves. Information regarding the local uses of the various mangrove species was gathered from local residents through interviews. The species composition of the fish caught by artisanal fishermen was obtained by identification of fish species at landing sites and calculation of the percentage composition of the species for each fishery type e.g. *dema* trap, *towe* trap, hand lines etc.

Results and discussion

Mangroves

The distribution of the species of mangroves together with estimates of their percentage abundance and cutting intensities at the study sites is given in Table 1.

Prior to these studies, eight species of mangroves were recorded for Tanzania (Semesi, 1986) and these were found both on the mainland and the islands. *Xylocarpus moluccensis* (Lam) Roem was identified during these studies and has so far been recorded only in Pemba islands (J.P.S.) and Tanga area on the mainland (Semesi, 1988). The other species include *Bruguiera gymnorrhiza* (L) Lam, *Rhizophora mucronata* Lam., *Ceriops tagal*

Table 1. Information on mangroves on Unguja and Pemba, Zanzibar.

Item	Use[a]	Unguja sites[b]					Pemba sites[b]		
		Unguja-Ukuu	Kisaka-saka	Pete	Chwaka	Muwanda	Miche-weni	Chake-chake	Wete
Total mangrove areas (ha)[c]	–	344	460	320	3240	520	1288	1324	1128
Cutting (Nr/ha)[d]	–	2317	1463	777	n.a.	3567	1937	350	335
Coverage by species (in %):									
A. marina	F, C	15	15	10	10	25	10	5	20
B. gymnorrhiza	F, C, P	30	20	25	30	4	30	0	2
C. tagal	P, F	10	15	15	15	20	10	3	3
H. littoralis	F, M	0	2	0	2	3	1	3	2
L. racemosa	F, C, P	5	5	5	0	5	15	2	3
R. mucronata	P, C, D, L, T	20	30	35	25	7	25	15	8
S. alba	P, B, F	20	8	5	15	35	6	70	60
X. granatum	F, M	0	5	5	3	1	2	2	2
X. moluccensis	M(?)	0	0	0	0	0	1	0	0
Size distribution of poles									
Average diameters (cm)	–	5.7	n.a	n.a	6.7	n.a	11.5	6.7	n.a
Average length (m)	–	3.50	n.a	n.a	4.00	n.a	3.45	3.57	n.a
Sample size	–	418			326		374	345	

[a] F = firewood, C = charcoal, P = poles, B = boat building, D = dye, L = leather tanning, T = fish traps, M = medicine.
[b] Underlined % value refer to species most affected by cutting at the location indicated.
[c] From Griffith (1949).
[d] Average of three plots in each area.

(Perr.) C.B. Robinson, *Avicennia marina* (Forsk) Vierh, *Lumnitzera racemosa* Willd, *Xylocarpus granatum* Koen, *Sonneratia alba* J. Smith, and *Heritiera littoralis* Dryand in Ait. The most common species found on the sites are *R. mucronata*, *B. gymnorrhiza*, and *S. alba*.

About two thirds of the mangrove forest stands on Zanzibar are located in Pemba island while one third is found in Unguja island. The areas with mangroves in Unguja (See Fig. 1) are located mostly to the west and include Menai bay (Unguja Ukuu, Pete and Muungoni, Kiwani bay (Kisakasaka) and Makoba bay (Muwanda and Bumbwini). On the eastern side of Unguja island there is only one stand of mangroves, found around Chwaka bay. There are also small stands at Maruhubi and Chukwani to the north and south of Zanzibar town respectively. The Unguja mangrove stands cover about 6000 ha. However, more accurate data about the area and cover of mangroves would be available after the interpretation of more recent aerial photographs.

On Pemba island (See Fig. 2) the mangrove forests are similarly found on the western side. They stretch southward from Njao island north of Wete town through Chakechake to Mkoani. More than three quarters of the mangroves lie in this region. Other comparatively less extensive but important stands are found at Adamson bay on the east coast, and in the bay between Ras Kiuyu and Ras Kigomasha on the northern part of the island. The total area covered by mangrove vegetation here has been estimated (Griffith, 1949) at 12000 ha. This figure will also be revised after interpretation of more recent data.

The various local use of the mangrove species are enumerated in Table 1. The most destructive of these are firewood and charcoal production because the trees are cut indiscriminately resulting in wide gaps in the area. Several affected areas were seen at Unguja Ukuu, Kisakasaka, Muwanda, around Chwaka bay (Unguja) and to some extent at Micheweni (Pemba) where cutting was done for lime production. Salt production by boiling seawater, although less practised now has caused damage to some areas, e.g. the Muwanda area along Makoba bay (Unguja). This area was known for salt production and collection of holothurians (*bêche de mer*) for export in the early seventies, both activities involve boiling treatment, thus requiring firewood.

The use of mangrove timber is also important especially in Unguja Ukuu, Chwaka, Kisakasaka, and to some extent at Micheweni and Chakechake (as shown in Table 1). Pole production was qualitatively estimated to be more intense in Unguja than in Pemba, as judged by the amount of poles seen staked at various points on the islands. This is probably due to a higher demand for construction materials because of the relatively higher population density on Unguja in contrast to Pemba. The larger area of mangroves in Pemba and the low cutting pressure from the smaller human population probably explains the superior growth form in terms of tree stem thickness, observable in most stands (as shown in Table 1).

Leather tanning is another use of the mangrove. The bark of *R. mucronata* is stripped off the trees, pounded and mixed with lime, and soaked in metal or concrete troughs into which animal skins are immersed and left to soften for one to two weeks. The skins are then ready for use in the manufacture of shoes (*makubadhi*), in small local shoe factories. Leather tanning was common at Uzi island in Unguja; and at Micheweni, Wete and Chakechake in Pemba. Debarked trees die on drying. The problem is serious because the trees affected are the mature seed producing plants.

The making of fixed stake traps utilizing the roots of *R. mucronata* is another destructive practice. It removes the supporting roots, sometimes of an entire plant leading to the collapse of the trees. In the making of the trap the stilt roots of the mangrove are cut and split into thin trips which are then woven together in a long line. These long lines are then stuck erect in the mud or sand in mid-shore at low tide to trap fish moving out with the water on the next low tide. Although some other mangrove species are also used e.g. *L. racemosa*, *C. tagal* and the mangrove associate, *Pemphis acidua*, the roots of *R. mucronata* are the most commonly used.

For medicinal purpose, only *X. granatum* and *H. littoralis* are used. A powder obtained from the seed coat from both species is dissolved and orally administered for the treatment of abdominal ailments, however, overdose of this concoction can be dangerous.

The utilization of mangroves in Zanzibar is thus carried out at as shown in various intensities (Table 1). Some of the stands are under more serious threat than others. From interviews with villagers it was apparent that in some places, indiscriminate mangrove cutting had already resulted in serious exposure of the coastline to erosion. The Muwanda village, for example, was reported to have been shifting landward away from the approaching ocean. Ruins of the older settlements that stand surrounded by water could be seen as evidence of loss of land to the sea. This phenomenon has been partly due to the cleared mangrove vegetation.

Associated fishery resource

During the survey of the fauna associated with mangroves, the most common fish encountered were *Lethrinus harak* (Forskal) parrot fishes, *Chanos chanos* (Forskal) and fishes of the genus *Siganus*. Crustaceans included *Scylla serrata* (Forskal), hermit crabs, and prawns; and molluscs included *Pyrazus* sp., *Pinctada* sp., *Modiolus* sp., *Crassostrea* sp.

The association of fish with mangrove vegetation has been pointed elsewhere (Ahmad, 1966), and the potential of several tropical fisheries have been known to correspond to the respective area of mangrove vegetation (MacNae, 1974). It was evident also from the interviews conducted during the study that a considerable amount of conventional as well as non-conventional fishing takes place either within or in the vicinity of the mangrove stands. Notable among the non-conventional fishery is the collection of molluscs. Evidence for the magnitude of edible mollusc collected was provided by the large heaps of empty

Table 2. Species composition of fish landings at three stations on Unguja island, Zanzibar (Tarbit, 1976).

Fish species/group	% Species composition					
	Malindi (non-mangrove)		Chwaka (mangrove)		Mkokotoni (mangrove)	
	Towe	Dema	Towe	Dema	Towe	Dema
Lutjanidae	0.4	0.4	–	1.0	4.2	9.6
Lethrinidae	–	10.5	35.5	6.7	39.0	31.7
Siganus	13.7	20.1	–	45.7	26.3	37.5
Scardae	26.2	13.0	46.1	24.1	10.1	5.8
Scolopsia	36.1	39.3	–	0.6	3.0	–
Gaterin	0.7	0.3	0.3	3.0	0.8	0.2
Acanthurus	–	–	–	4.7	0.4	–
Epinephelus	0.2	1.2	1.2	6.0	3.4	8.0
Nemipterus	–	–	–	0.1	0.3	0.5
Rajidae	–	–	–	0.1	2.8	–
Caesionidae	–	–	15.1	0.4	–	–
Parupeneus	22.6	10.2	–	7.1	7.6	4.2
Others	0.1	5.2	1.8	0.5	2.1	2.5

shells left at various points on the beach after the removal of the meat. Villagers reported that most of the molluscs are used to supplement protein especially for those families that cannot afford finfish. A small proportion is used as bait in hook and line fishing. Molluscs are also collected for their opercula especially those from the genus *Murex*. These opercula cost as much as US$ 40 per kilogram. At Chakechake for instance the trade in this article was reported to have been quite attractive and that the products were collected and exported to Mombasa, Kenya.

Three fishery types (based on the type of fishing gear) were found to be associated with the mangroves and adjacent habitats. These included fixed stake traps (*uzio*), movable traps (*madema* and *towe*) and hand line fisheries (Hoekstra *et al.*, 1990). A comparison of trap fishery landings at three stations (Malindi, Chwaka and Mkokotoni) indicates that more fish are caught from mangrove-lined areas than elsewhere (Tarbit, 1976). Important fish species belonging to the families Lethrinidae, Siganidae and Scaridae are caught in greater numbers in mangrove-lined area (as shown in Table 2). However, fish landed at Malindi (Zanzibar Town), which are caught mostly from non-mangrove lined areas, were found to contain less of the mangrove associated groups mentioned above, instead, the coralline fishes of the genera *Scolopsis* and *Parupeneus* dominated the catches.

Cintron & Schaeffer-Novelli (1983) observed that due to the presence of juvenile organisms and their migration in mangroves as well as seagrass beds, these regions make excellent fishing grounds even for larger predatory fishes such as red snappers, yellow fin tuna, sharks and other highly carnivorous species that forage these regions. On Zanzibar islands, 62% and 82% of all the fixed stake trap fishery units operate on the west coast of Unguja and Pemba, respectively. Of the movable traps, 77% and 85% operate on the respective west coasts of Unguja and Pemba. Likewise, more troll line fishery units (89.5% in Unguja and 76.6% in Pemba), a gear which is used to catch large pelagic fish, operate on the west coasts where extensive mangrove stands are found (Hoekstra *et al.*, 1990).

Conclusion

Mangrove swamps support several types of fisheries on Zanzibar islands and indeed worldwide. Mangroves, also provide direct resources to the coastal communities and play an important role in protecting the coastline from erosion. However, it is not realized that over-exploitation of the mangrove forests leads to a decline in the fisheries resources and severe environmental damage. There is therefore an urgent need for the introduction of effective management measures that would guarantee sustained utilization of the mangroves.

Acknowledgement

We wish to express our deep appreciation to the Commission of Lands and Environment (CoLE), Zanzibar, for suggesting and providing funding to support the mangrove survey on Unguja and Pemba island through FINNIDA'S financial assistance. We would also like to thank all those who have contributed in one way or another in the preparation of this paper.

References

Ahmad, N., 1966. Fish and fisheries of the Sundarbans. In Scientific problems of humid tropical zone deltas and their implications. Proceedings of the Daccar Symposium February/March 1964, UNESCO Paris: 271–276.

Cintron, G. & Y. Schaeffer-Novelli, 1983. Mangrove forests: Ecology and response to natural and man induced stressors. UNESCO Report in Marine Science # 23 ROST-LAC. 133 pp.

Griffith, L. A., 1949. Reconnaissance report on the forest problems of Zanzibar protectorate. Govt. Printer, Zanzibar. 21 pp.

Hoekstra, M., M. A. K. Ngoile, N. S. Jiddawi & C. Rotteglia, 1990. Census of the marine fishing units of Zanzibar in 1989 FAO/UNDP: RAF/87/008/DR/60/99/E. 119 pp.

MacNae, W., 1974. Mangrove forest and fisheries. Indian Ocean Program Publication Number 34, Rome, Indian Ocean Fishery Commission. 35 pp.

Semesi, A. K., 1988. Status and utilization of mangrove along the coast of Tanga, Tanzania. In Mainoya, J. R. (ed.), Proceedings of workshop on Ecology and Bioproductivity of marine coastal waters of Eastern African, Dar es Salaam, Tanzania, 18–20 January 1988. Faculty of Science, University of Dar es Salaam: 174–181.

Semesi, A. K., 1986. Zonation and vegetation structure of mangrove communities in Tanzania. In Mainoya, J. R. and P. R. Siegel (eds), Proceedings of Workshop on 'Save the mangrove ecosystems in Tanzania'. Dar es Salaam, Tanzania, 21–22 February 1986. Faculty of Science, University of Dar es Salaam: 15–35.

Tarbit, J., 1976. Demersal Fisheries Research Terminal Report 1974–76. EAMFRO Annual Report 1976, EAMFRO, Zanzibar.

Hydrobiologia **247**: 235–241, 1992.
V. Jaccarini & E. Martens (eds), The Ecology of Mangrove and Related Ecosystems.

The management of Insular Caribbean mangroves in relation to site location and community type

Peter R. Bacon & Gerard P. Alleng
Zoology Department, University of the West Indies, Mona, Kingston 7, Jamaica

Key words: assessment, Caribbean, ecosystem interactions, management, mangroves, site variability

Abstract

Mangrove ecosystems occupy different locations on Caribbean island coasts, ranging from open bays (fringe mangals) to totally enclosed salt ponds and salinas. On geomorphologically active coastlines, such as south Jamaica, systems are at varying degrees of maturity and productivity. Furthermore, because of system variability, the interactions between mangroves and associated marine systems, such as coral reefs and seagrass beds, are developed to different degrees.

Community structure and productivity of a range of mangals on different islands of the Greater and Lesser Antilles are discussed. Forcing functions and levels of interaction with the marine environment are identified.

The rational choice of management options must be based on the range of goods and services provided by the different systems; and a good understanding of their ecology is essential when choosing sites for protection, waste disposal, landfill, marina development and fisheries enhancement. Examples are given from current studies in Jamaica, St. Lucia, the British Virgin Islands and Trinidad, of a flexible management response to mangrove ecosystem diversity.

Introduction

The importance of wetland conservation is widely recognised and Caribbean island governments are working to produce inventories and management plans. Mangrove dominated wetlands, or mangals, are the most common type in the Insular Caribbean, particularly in the Lesser Antilles. Mangroves are threatened and impacted by development activity, particularly because of their coastal location and the importance of coastal economies in the Caribbean (Hudson, 1983; Bacon, 1987). In most coastal areas the mangals contain only three species of mangrove trees, which in order of importance are *Rhizophora mangle* L., *Avicennia germinans* (L.) Stearn and *Laguncularia racemosa* (L.) Gaertn.; with *R. harrisonii* Leechman, *R. racemosa* G. F. W. Meyer and *A. schaueriana* Stapf & Leechman at scattered locations in the Southeastern Caribbean. The buttonwood, *Conocarpus erectus* L., is common throughout the region on the drier wetland margins. The invertebrate, fish and waterfowl faunas show a high degree of commonality and Caribbean mangals carry out the classic roles of coastal protection, organic matter production, provision of fish nursery area and wildlife habitat. This suggests that a uniform mangrove management strategy can be applied and that this might follow the globally applicable principles laid down by Hamilton & Snedaker (1984).

However, as Lugo & Snedaker (1974) demonstrated, a range of mangrove types occurs in the Neotropics; the riverine, fringe, basin and scrub types, which have different ecological attributes. Furthermore, Bacon (1986) pointed out that these mangals were associated with a variety of coastal ecosystems whose geographic characteristics determined their ecology and productivity. This paper examines the influence of site location on the ecological and economic attributes of mangals in selected Caribbean islands and discusses how these features can be incorporated into the management decision-making process.

Description of sites studied

Mangrove systems were surveyed on a 50 km stretch of coastline near Kingston, south Jamaica; these ranged from open bay fringe mangal to enclosed salina with scrub mangroves (see Fig. 1). On this geomorphologically active coast, barrier and lagoon development has influenced mangrove distribution and successional processes; so that there are pioneering, mature and senile mangal communities (Bacon, 1986). Site aspect also influences the degree of hurricane impact, which is a further factor controlling mangrove forest structure and productivity (Alleng, 1990).

Mangrove areas studied in St. Lucia showed a similar site variability on both east and west coasts of this geomorphologically different island (see Fig. 2). The generally small mangrove stands occupied open bay fringes or were isolated from the sea behind beach barriers which breach only in times of heavy flooding. Site characteristics influence plant distribution, tidal exchanges, salinity range, groundwater and riverine inputs, sediment grain size and composition and nutrient status.

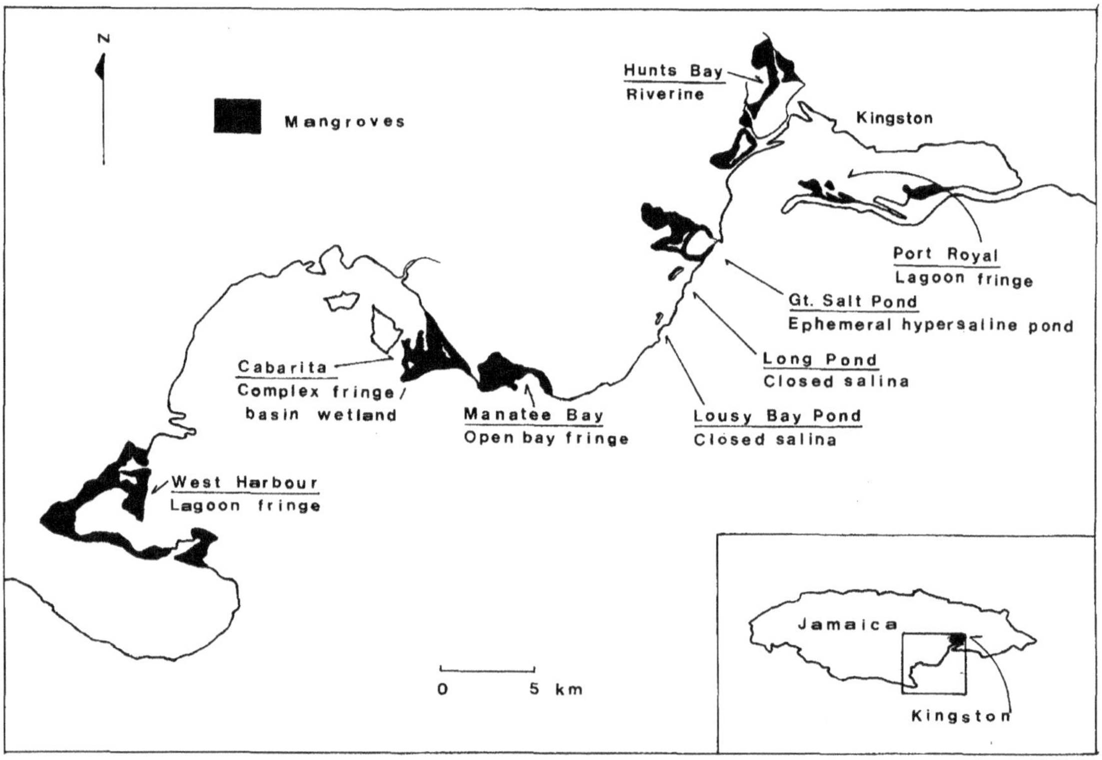

Fig. 1. Selected mangrove sites in south Jamaica.

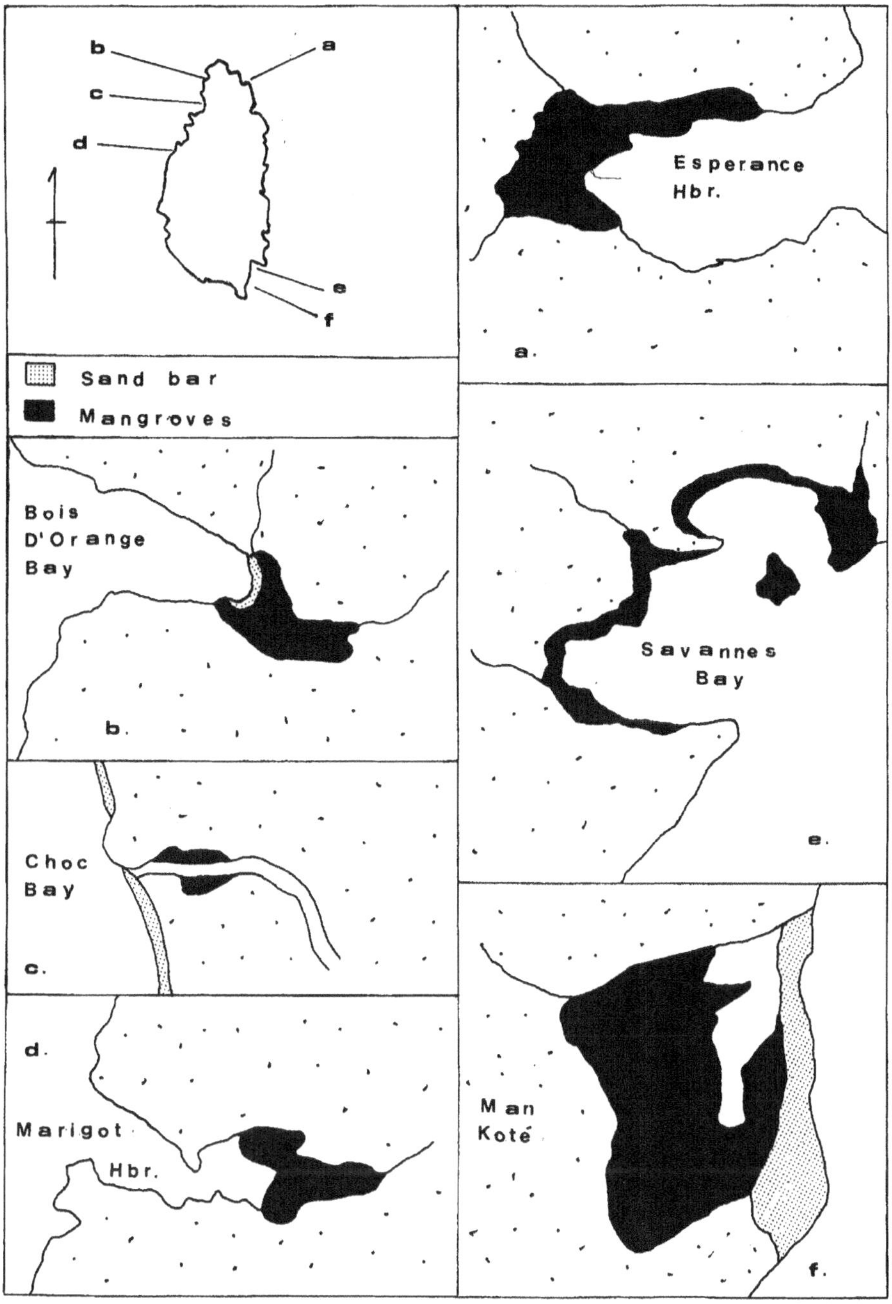

Fig. 2. Selected mangrove sites in St. Lucia. Scale for a–c, & f = 1:10 000; Scale for d & e = 1:25 000.

Methodology for mangrove area assessment

The ecological and socioeconomic values of these two groups of mangals were assessed using a methodology adapted from Adamus (1983). The method assigns a rating to attributes and functions of these wetlands in order to obtain a ranking that can be used for management decision-making (see Table 1). The rating system used was simple, scoring High (3), Medium (2), Low (1) and None (0), because of the relative paucity of data for some categories. Field observation and mapping by the senior author and reports on the Jamaican sites (Chow, 1989; McCain & Bacon, 1989; Alleng, 1990) and St. Lucian sites (Portecop & Benito-Espinal, 1985) provided detailed infirmation on areal extent, plant distribution and use of the habitats by wildlife, but assignments under the socioeconomic categories (such as, importance to fisheries, education value) were based largely on informed opinion.

Results and discussion

The danger of generalizing about mangrove management is apparent when this assessment methodology permits the ranking of systems according to their respective values. All the sites were positive for some values, but the ranking identified areas of priority for conservation while singling out sites with lower value that might be available for use as marinas, dump sites or landfill. All isolated and semi-isolated sites scored low values for wildlife habitat and fisheries, and the potential for mariculture, educational and recreational use (see the WH, EC & RE categories in Table 1). Embayment and Lagoon-type sites also received the highest rankings when this evaluation system was used in the British Virgin Islands (Blok-Meeuwig, 1990); in Tortola, for example, top priority sites for conservation were Paraquita (lagoon fringe), Witches Brew (lagoon fringe) and Hodges Creek (embayment fringe).

Table 1. Assessment of ecological & socioeconomic values of selected mangrove sites in Jamaica & St. Lucia.

Site	SZ	CD	VD	OW	SD	SL	WH	EC	RE	Value index
A. Jamaica										
Kingston Hbr.	3	3	2	3	3	3	2	2	2	2.3
West Hbr.	3	3	2	3	3	3	2	2	2	2.3
GSP/Flashes	3	3	2	2	3	3	2	2	2	2.2
Cabarita	3	2	2	3	2	2	2	2	2	2.0
Manatee Bay	3	1	2	1	3	2	1	1	1	1.5
Long Pond	2	2	2	1	3	1	1	0	1	1.3
Lousy Bay	1	1	1	1	1	1	1	0	0	0.7
B. St. Lucia										
Savannes Bay	3	1	2	1	3	2	2	2	2	1.8
Esperance Hbr.	3	2	2	1	3	2	2	1	1	1.7
Marigot Hbr.	2	1	2	1	2	2	2	1	1	1.4
Man Kote	3	2	3	1	0	1	1	1	1	1.3
Bois d'Orange	2	2	2	1	0	1	1	0	1	1.0
Choc	1	1	1	1	3	1	1	0	0	0.9

Key: SZ = size of mangal
VD = vegetation dispersion
SD = species dominance
WH = wildlife habitat
RE = recreational/educational value
CD = community diversity
OW = open water dispersion
SL = sytem linkages
EC = economic values

Score: High = 3, Medium = 2, Low = 1, None = 0.

Open mangals permit movement of marine and estuarine organisms in and out of the system and export of organic matter. Ogden & Gladfelter (1983) noted that interaction between mangals, seagrass beds and coral reefs was a common feature of Caribbean coasts, but did not comment on the degree of variability which exists in such linkages. The nature of the association between three of the south coast Jamaica mangals and neighbouring marine ecosystems is shown in Fig. 3, to illustrate this point. The communities at these locations are fringe mangroves dominated by *Rhizophora*. In Port Royal they occupy approximately 6 ha, with mean canopy height 4.6 m (Alleng, 1990), while in the Great Salt Pond/Flashes area they occupy approximately 25 ha, with mean canopy height 5.4 m. Fringe mangroves in West Harbour occupy approximately 260 ha, with mean canopy height at 6.0 m. Apart from variations in areal extent, other structural parameters are within the same order of magnitude and accord well with the description of fringe mangroves given by Lugo & Snedaker (1974).

However, the geographic setting and hydrography influence the fate of the organic matter and nutrients generated within the three systems, with implications for the nature of their linkages with contiguous ecosystems. Site geography indicates that the impacts of exported materials are probably in inverse relation to the area of mangroves at each site. Port Royal mangroves are functionally part of the Kingston Harbour system, which has strong fluviatile influences, wind-driven and tidal circulation patterns. Kingston Harbour is a major exporter of nutrients and organic matter to the inshore environment and its effluent strongly influences the neighbouring reefs and seagrass beds (Head & Hendry, 1986; Bacon *et al.*, 1989). Furthermore, this system is an open, continuously exporting system. In contrast, the Great Salt Pond/Flashes mangroves developed around an hypersaline pond closed by an ephemeral sand bar. A lagoon formed briefly during the wet season when flood water breached the barrier permitting entry of sea water. Now that there is a permanent artificial channel stable, more marine conditions have been established in the pond. However, it still acts as a source of nitrite and phosphorus to the inshore waters when there is positive net discharge only during the wet season and particularly after storms (McCain & Bacon, 1989). This situation is similar to the Nariva mangrove lagoon, Trinidad, where discharge to the Atlantic Ocean occurs only during two to three months of the wet season, when the effluent stream can carry up to 100 t org.C day^{-1} from the 90 km^2 wetland (Bacon, 1981). Such pulses of nutrient-rich, brackish water have a seasonal, or intermittent impact on neighbouring marine habitats.

As West Harbour lacks riverine inputs, has prevailing onshore winds, a weak circulation pattern, and experiences a high evaporation rate because it is located in one of the driest areas of Jamaica, a net outflow of water containing dissolved and particulate organic matter has not been detected. This mangal probably has little direct influence on the ecology of the adjacent Portland Bight, but functions as a net accumulator of organic matter. Such lagoonal systems which act as sinks for organic matter occur in several parts of the Jamaican coast, but are poorly described. Mangrove production appears to be available to consumers only if they enter the system, hence West Harbour's importance as fish habitat and nursery environment and its potential for mariculture.

The relative roles of the three fringe mangals shown in Fig. 3 have been assessed as indicated in Table 1 using differences in community structure and ecological forcing functions (Lugo & Snedaker, 1974) and degree of site variability (Bacon, 1986). This allows these ecologically and economically important systems to be rated for management purposes.

What has not been done is to include in the assessment methodology criteria which express the value of the mangals *to* associated marine habitats. Although the value of the mangal itself has been assessed, the relationship between particular mangrove sites and the continued health and productivity of an island's fisheries, seagrass beds or coral reefs is a critical factor in coastal resources management. If this dimension can be

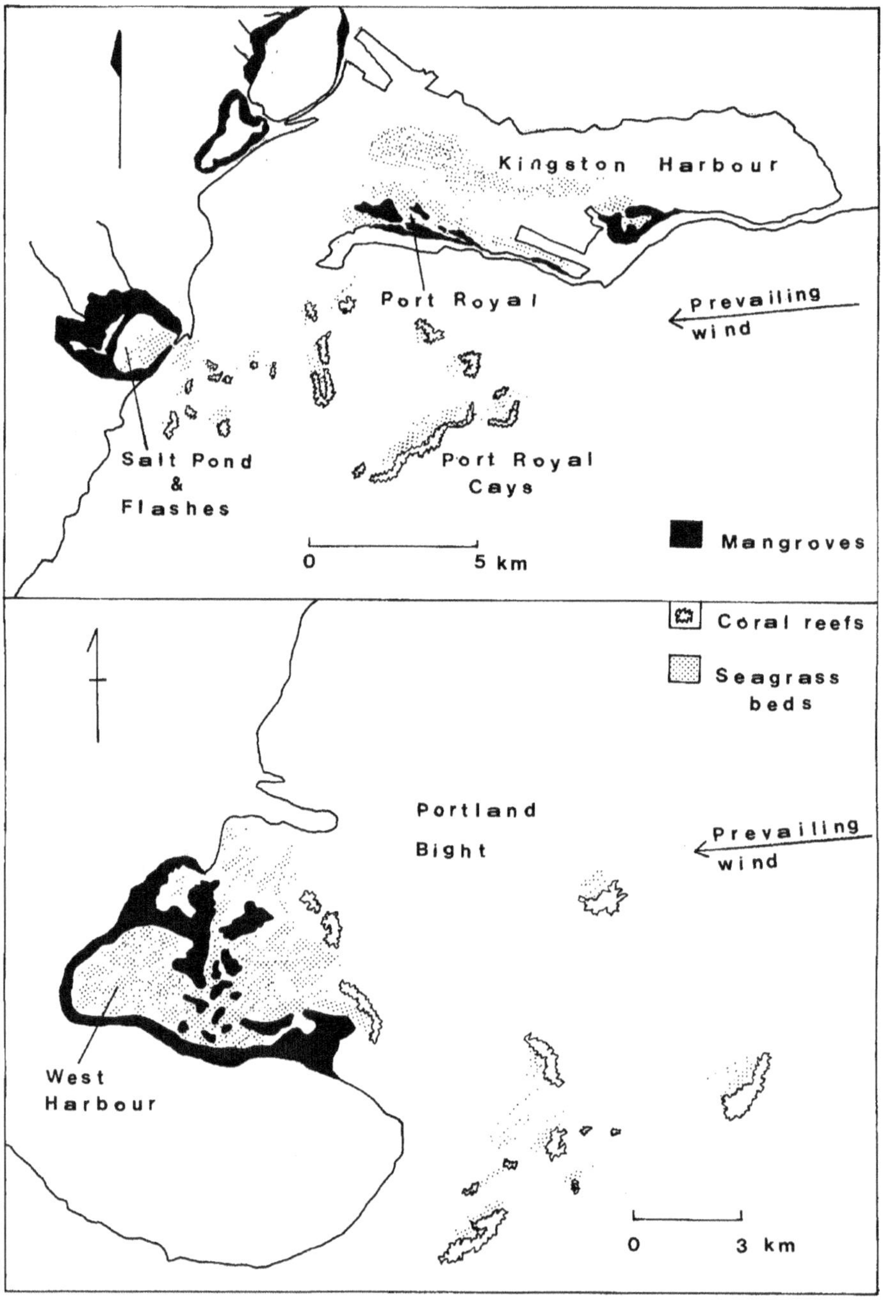

Fig. 3. Associations of three Jamaican mangals with neighbouring coastal ecosystems.

Table 2. Role of mangals in interactions with neighbouring coastal and marine ecosystems.

1. Physical interactions	
1.1. Filtering terrestrially derived sediment	
	Reduction in sediment load reaching the sea
1.2. Buffering salinity changes	
	Reduction in volume of freshwater reaching the sea
2. Nutrient interactions	
2.1. Export of dissolved & particulate organic matter	
2.2. Sinks for particulate organic matter	
3. Biotic interactions	
3.1. Export of meroplanktonic larvae to marine food chains	
3.2. provision of feeding habitat	
	– marine invertebrates & fin fish (diurnal & seasonal migrations)
	– seabirds, shorebirds, marine mammals
3.3. Provision of juvenile (nursery) habitat	
	– crustacea (shrimp, lobster) & molluscs (conch)
	– fin fish (reef fish, seagrass bed fish, pelagics)
	– seabirds & shorebirds nesting (also roosting)
	– marine mammals (dolphins)

added, a more flexible system of mangrove site assessment would be produced which could respond to conservation requirements in other sectors of the economy. Table 2 lists some of the identified interactions with associated ecosystems which influence the value that should be attached to mangrove sites.

The nature of interactions between these systems is understood (Ogden & Gladfelter, 1983), but insufficient data is available on which to base evaluations. Further research is needed to quantify (a) the interactive physical forcing functions, (b) the magnitude of nutrient and biotic exchanges and (c) the degree of nursery dependence between key Caribbean ecosystems, in order to improve the mangrove site assessment methodologies currently in use.

Acknowledgements

Thanks are due to the Government, British Virgin Islands, and the Caribbean Environmental Health Institute, St. Lucia, for opportunities to test the assessment methodology in the field.

References

Adamus, P. R., 1983. A method for wetland functional assessment. U.S. Department of Transport, Washington. 1: 176 pp.

Alleng, G. P., 1990. Historical development, present status and management guidelines for Port Royal mangal, Kingston, Jamaica. M. Phil. Thesis, Zoology Department, University of the West Indies, Mona: 171 pp.

Bacon, P. R., 1981. Nutrient export from a Caribbean lagoonal ecosystem. Int. Symp. Coastal Lagoons, UNESCO, Bordeaux, Sept. 1981. Abstracts p. 17.

Bacon, P. R., 1986. Mangrove ecosystem responses to gradients in factors limiting primary production in the Caribbean. UNESCO Rep. Mar. Sci. 41: 19–24.

Bacon, P. R., 1987. Use of wetlands for tourism in the Insular Caribbean. Ann. Tourism Res. 14: 104–117.

Bacon, P., I. Goodbody, A. Greenaway, M. Hendry, P. Devi-Prasad & J. Woodley, 1989. The Caribbean Coastal Management Study: Investigations on the Hellshire coastline. Proc. 1st. Nat. Conf. Sci. & Tech., Kingston, April 1987. Scientific Research Council, Jamaica: 276–295.

Blok-Meeuwig, J., 1990. Mangrove systems of the British Virgin Islands: Resource mapping and assignment to protection categories. Conservation Office, Government of the British Virgin Island, Tech. Rep. 5: 45 pp.

Chow, B., 1989. Evaluation of the importance of mangroves to the ecology of West Harbour, Jamaica. M. Phil. Thesis, Zoology Department, University of the West Indies, Mona, Jamaica: 261 pp.

Hamilton, L. S. & S. C. Snedaker, 1984. Handbook for mangrove area management. UNEP & East West Centre: 123 pp.

Head, S. M. & M. D. Hendry, 1986. Development of the Hellshire Hills, Jamaica: implications for management strategy of reefs and related ecosystems. Proc. 5th Coral Reef Congress, Tahiti, 1985. 2: 519–524.

Hudson, B. J., 1983. Wetland reclamation in Jamaica. Carib. Geog. 1: 75–88.

Lugo, A. E., S. C. Snedaker, 1974. The ecology of mangroves. Ann. Rev. Ecol. Syst. 5: 39–64.

McCain, C. & P. Bacon, 1989. The Great Salt Pond. Caribbean Coastal Management Study: The Hellshire coast, St. Catherine Jamaica. Marine Science Unit Res. Rep. 1: 100–128.

Ogden, J. C. & E. H. Gladfelter, 1983. Coral reefs, seagrass beds and mangroves: their interactions in the coastal zones of the Caribbean. UNESCO Rep. Mar. Sci. 23: 133 pp.

Portecop, J. & E. Benito-Espinal, 1985. The mangroves of St. Lucia: A preliminary survey. ECNAMP, St. Lucia: 56 pp.

Hydrobiologia **247**: 243–251, 1992.
V. Jaccarini & E. Martens (eds), The Ecology of Mangrove and Related Ecosystems.

Regeneration strategy of mangroves along the Kenya coast: a first approach

D. Van Speybroeck
University of Nairobi, Botany Department, P.O. Box 30197, Nairobi, Kenya

Key words: mangroves, zonation, regeneration, seedlings

Abstract

In June 1990 a research project sponsored by DANIDA and AWF was carried out by botany students of Nairobi University to investigate the regeneration strategy of mangroves at Gazi bay and Mida creek.

Statistical analysis of 449 quadrats (5×5 m) sampled along 35 line transects in 4 mangrove forests showed that mangrove seedlings follow the same distribution pattern in the intertidal zone as their parent trees. In other words, mangrove seedlings mainly develop within a well-defined species specific zone. These distribution zones for the various mangrove trees and their seedlings are defined in terms of elevation above the mean low water level of spring tides.

Through assigning mangrove seedlings in the intertidal zone to the categories (i) 'fixed' or 'not-fixed', and (ii) 'covered' or 'not-covered', evidence was found that propagule dispersal followed both the 'self-planting' theory and the 'stranding' theory. The 'self-planting' theory appeared to be the major mechanism of propagule dispersal in 'undisturbed' mangrove forests, whilst the 'stranding theory' proved to be predominant in colonizing 'over-exploited' and cleared mangrove forests.

It is concluded that re-afforestation of mangrove seedlings in the intertidal zone will be most successful when the seedlings are planted in their specific distribution zones under fixed conditions.

Introduction

More than 50 years ago the zonation pattern of the East African mangroves was described by Graham (1929) and by Walter & Steiner (1936), and it was only in recent years that more studies concentrated on the zonation pattern of mangroves along the Kenya coast; at Gazi bay (Gallin, 1988; Coppejans & Gallin, 1989; Gallin *et al.*, 1989; Beeckman *et al.*, 1990; Ruwa, 1990a, 1990b; Van Speybroeck, 1990), and at Mida creek (Gang & Agatsiva, 1990; Van Speybroeck, 1990); revealing a topography-determined zonation pattern, with *Sonneratia alba* Sm. growing closest to the low water line, followed by *Rhizophora mucronata* Lam., *Bruguiera gymnorrhiza* (L.) Lam., *Ceriops tagal* (Perr.) C. B. Robinson, *Avicennia marina* (Forrsk.) Vierh., *Lumnitzera racemosa* Willd. and *Xylocarpus granatum* Koen.

The distribution of mangroves along the Kenya coast also had been studied by Isaac, W. & Isaac, F. (1968), Moorjani (1977), Kokwaro (1980, 1985) and Ruwa & Polk (1986).

So far, however, the regeneration strategy of mangroves along the Kenya coast has not been studied in detail. Only Graham (1929) briefly discusses the life cycle of Kenyan mangrove trees. An understanding of regeneration strategies, nevertheless, forms the basis for foresters, ecologists and policy makers to develop a management plan, which emphasizes the sustainable use of mangrove resources along the Kenya coast. This paper is a first contribution to the understanding of mangrove regeneration strategies in Kenya.

Do mangrove propagules follow any species specific distribution pattern for anchorage in the intertidal slope and further development? Do mangrove seedlings mainly develop under the parental trees or are they rather washed away by the tidal action and dispersed by the water to anchorage places far away from their parental trees?

When discussing the dispersal of mangrove propagules, Rabinowitz (1978a, 1978b), Dawes (1981) and Tomlinson (1986) adhered to the 'stranding' theory, stressing that all mangrove propagules are water dispersed before establishment in the substratum takes place. Rabinowitz (1978b: 123) even demonstrated that mangrove seedlings in transplant experiments 'generally show better survival in other mangroves' habitats than below their own adults'.

Much less attention, however, has been paid to the 'self-planting' theory (Dawes, 1981: 518; Tomlinson, 1986: 159), highlighting that mangrove propagules are pointed to promote self-planting as they fall from the tree.

The present study concentrates on mangrove forests under different anthropogenic pressures. In total 449 vegetation quadrats were sampled in undisturbed, used and over-exploited mangrove forest along 37 line transects perpendicular to the coastline. From each quadrat the mangrove species composition (adult trees and seedlings) was recorded as well as their respective positions above MLWS. In the vegetation quadrats the growing conditions of the mangrove seedlings were examined and simultaneously assigned to the categories (i) fixed or not-fixed, and (ii) covered or not-covered (explained below).

Statistical analysis of these data allow us (i) to describe the zonation pattern of mangrove trees and mangrove seedlings, and (ii) to estimate the relative importance of the 'stranding' and the 'self-planting' strategy in the dispersal of mangrove propagules in the intertidal zone. Differences in regeneration strategy of undisturbed and over-exploited mangrove forests will be discussed throughout the text.

Definition of terms used in the text:

- covered/not-covered: whether the mangrove seedling is covered by the parent tree or not.
- fixed/not-fixed: whether the seedling uses a fixed object, such as a pneumatophore or a rock in the substratum for anchorage and protection or not).
- undisturbed mangrove forest: canopy cover $\geq 80\%$, obviously no logging activities.
- used mangrove forest: canopy cover $\pm 50\%$, logging activities apparent.
- over-exploited mangrove forest: canopy cover $\pm 15\%$, heavily logged by man.
- MLWS: mean low water level of spring tides (0 cm = datum level).

Description of sites studied

Field work was carried out in June 1990 at Gazi bay (Kenya: 4° 25′ S, 39° 50′ E, Fig. 1) and Mida creek (Kenya: 3° 20′ S, 40° 00′ E, Fig. 1). At Gazi bay 79 vegetation quadrats were sampled along 9 line transects in an undisturbed mangrove forest (Gazi 1; N of the fishermen's hut) whilst 147 vegetation quadrats were sampled along 9 line transects in an over-exploited mangrove forest (Gazi 2; S of the fishermen's hut). At Mida creek 223 vegetation quadrats were sampled along 17 line transects in used mangrove forests (Mida 1 and 2; at the E and NNW of the creek).

Materials and methods

Since the elevation above the mean low water level is considered to be the predominant environmental factor effecting the distribution of mangrove trees (Walter & Steiner, 1936; Clarke & Hannon, 1970), line transects following this altitude gradient (perpendicular to the coastline) were used to sample the mangrove vegetation. Line transects ranged between 100 m and 360 m in length, covering the width of the mangrove forests at the various sites. At least ten 5×5 m quadrats were pegged out at regular intervals

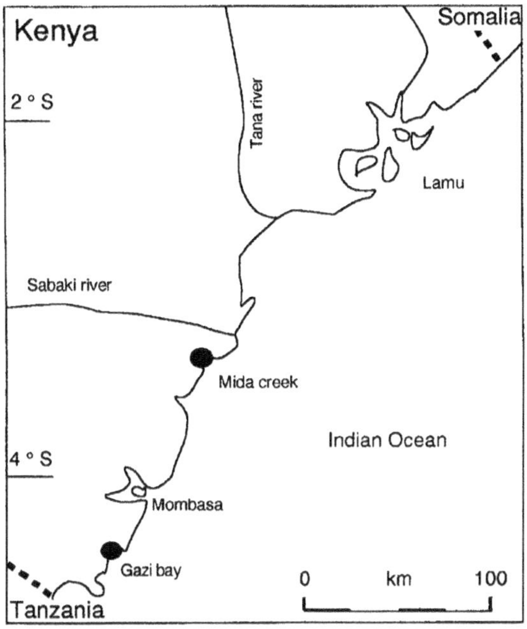

Fig. 1. Map of Kenya situating the study sites at Gazi bay (226 vegetation quadrats) and Mida creek (223 vegetation quadrats).

along each line transect. In each quadrat the mangrove vegetation was described according to the principles of the floristic-sociological approach (Westhoff & van der Maarel, 1978). Species cover and abundance were estimated by the ordinal scale of van der Maarel (1979). All the mangrove seedlings (= young trees having at least 4 leaves, and less than 1 m high) present in the quadrats were counted and assigned to the categories fixed or not-fixed and covered or not-covered.

The elevation in cm above MLWS was measured for each quadrat, according to the line level method (Dawes, 1981). Heights above MLWS were computed using the tide table (June 1990) for Kilindini harbour (Mombasa).

Graphical representation of data distributions in Figs 1 and 2 follows the box plot display (Chambers *et al.*, 1983). A box plot summarizes the distribution of a parameter by means of 5 specific values from the data set: the median (= the 50 percentile), the upper and lower quartile (the 75 and 25 percentiles) and the upper and lower extreme (the 90 and 10 percentiles). Outlier values in the box plot fall above and below the upper and lower extremes, and are represented by isolated points.

Factor analysis (Hotelling, 1933; Orlóci & Kenkel, 1985) is used to visualize the relationship between the mangrove forests studied, in terms of their respective seedling regeneration strategies (Table 1 and Fig. 5).

Results

Zonation of mangrove trees and mangrove seedlings

Figure 2 (Gazi bay) and Fig. 3 (Mida creek) summarize the distribution in the intertidal zone of 7 species of mangrove trees and their respective seedlings in terms of height above MLWS.

Inspection of Fig. 2 shows that mangrove trees and mangrove seedlings clearly grow in species specific distribution zones. Median elevation values (in cm) for adult trees (tr.) and seedlings (sd.) for the respective mangrove species are: *Sonneratia alba* (tr. = 160, sd. = 148), *Rhizophora mucronata* (tr. = 212, sd. = 198), *Bruguiera gymnorrhiza* (tr. = 212, sd. = 209), *Ceriops tagal* (tr. = 245, sd. = 240), *Avicennia marina* (tr. = 258, sd. = 256), *Lumnitzera racemosa* (tr. = 271, sd. = 299) and *Xylocarpus granatum* (tr. = 282, sd. = 270). Figure 2 suggests the existence of monospecific mangrove belts along the intertidal slope, starting with *Sonneratia alba* at the seaward side, but it nevertheless allows mixed mangrove stands as well, since there is a major overlap in distribution pattern between *Rhizophora mucronata* and *Bruguiera gymnorrhiza*, between *Ceriops tagal* and *Avicennia marina* and between *Lumnitzera racemosa* and *Xylocarpus granatum*.

Figure 3 shows the box plots for the distribution of mangrove trees and mangrove seedlings at Mida creek. The sampled mangrove forests at Mida creek are described as used mangrove forests. *Lumnitzera racemosa* and *Xylocarpus granatum*, although present at Mida creek, did not figure in our transects. Only 6 *Sonneratia alba* trees were present in the line transects with a

Table 1. Percentage of seedlings growing under 'fixed' and 'covered' conditions as observed at Gazi bay and Mida creek. Gazi 1 refers to an 'undisturbed' mangrove forest, Mida 1 and Mida 2 to 'used' mangrove forests and Gazi 2 to an 'over-exploited' mangrove forest. The percentage of seedlings growing under 'not-fixed' and 'not-covered' conditions are not listed since they add up to 100% with the 'fixed' and 'covered' seedlings respectively. The table summarizes growing conditions of 9479 seedlings in the four line transects.

		Used or undisturbed			Over-exploited
		Gazi 1	Mida 1	Mida 2	Gazi 2
Sonneratia alba	% fixed :	–	–	–	91
	% covered :	–	–	–	17
Rhizophora mucronata	% fixed :	88	90	67	76
	% covered :	98	93	87	57
Bruguiera gymnorrhiza	% fixed :	60	99	92	–
	% covered :	100	32	23	–
Ceriops tagal	% fixed :	58	36	73	79
	% covered :	67	93	32	77
Avicennia marina	% fixed :	12	37	21	87
	% covered :	75	68	90	19

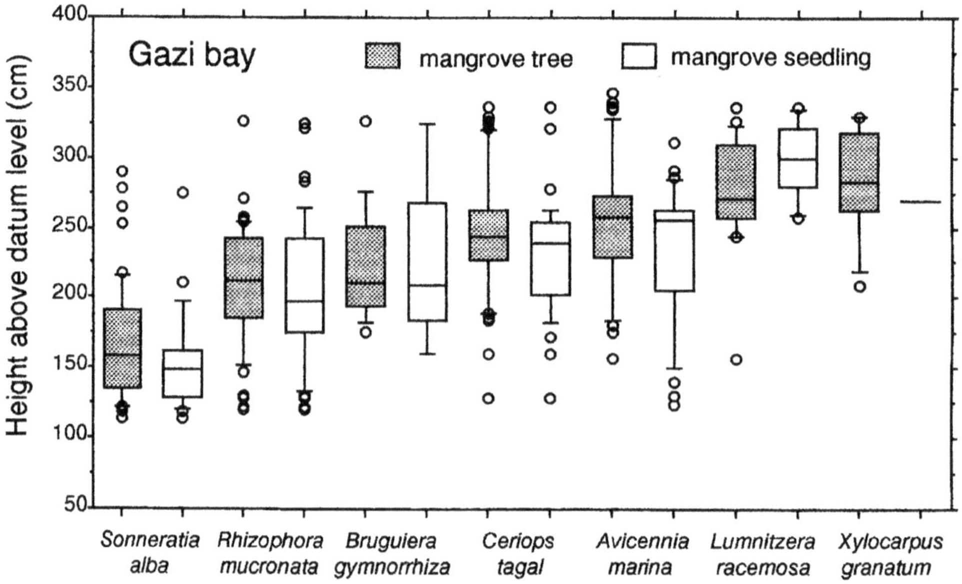

Fig. 2. Box plots showing the position above MLWS of mangrove trees and mangrove seedlings at Gazi bay. The figure includes data from 79 quadrats from an undisturbed mangrove forest (Gazi 1) and 147 quadrats from an over-exploited mangrove forest (Gazi 2).

median elevation above MLWS of 101 cm, but no *Sonneratia alba* seedlings were found. From Fig. 3 it is seen as well that mangrove trees and their seedlings tend to grow in species specific distribution zones.

Although no direct evidence is given in the present paper, no difference was observed in the zonation pattern of mangrove trees and seedlings in undisturbed, used and over-exploited mangrove forests.

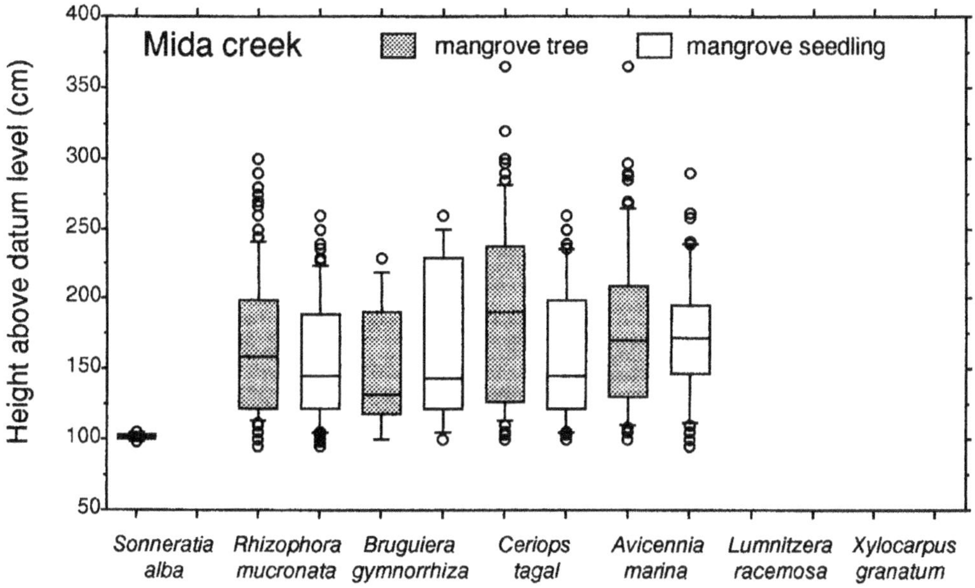

Fig. 3. Box plots showing the position above MLWS of mangrove trees and mangrove seedlings at Mida creek. The figure includes data from respectively 103 quadrats (Mida 1) and 120 quadrats (Mida 2) from used mangrove forests.

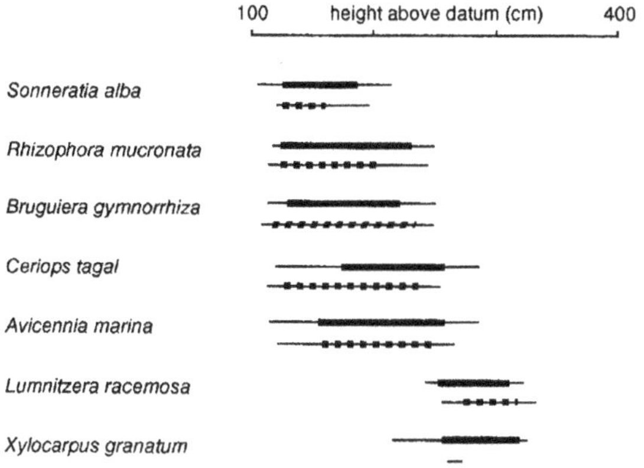

Fig. 4. Profile diagram showing the overall zonation pattern of mangrove trees (full lines) and mangrove seedlings (dashed lines) at Gazi bay and Mida creek. Thick lines are defined by the upper and lower quartiles, whilst thin lines are defined by the upper and lower extremes.

Figure 4 shows the profile diagram of mangrove trees and mangrove seedlings along the intertidal slope at Gazi bay and Mida creek. Mangrove forests occupy the intertidal zone between 100 and 370 cm above MLWS. Although there is a substantial overlap between the distribution patterns in the intertidal zone for the various mangrove species, a clear zonation pattern is apparent and easily recognized in the field. If present, *Sonneratia alba* forms the outermost forest margin towards the open water, followed by pure stands of *Rhizophora mucronata* or mixed stands of *R. mucronata* and *Bruguiera gymnorrhiza*, then followed by pure or mixed stands of *Ceriops tagal* and *Avicennia marina*, and with at the most landward margin of the forest stands of *Lumnitzera racemosa* and a very small margin, if still present, of *Xylocarpus granatum*. Seedling development is restricted to the species specific distribution zones along the intertidal slope.

Albeit not the case in our transects, it should be noticed that *Avicennia marina* may replace *Sonneratia alba* at the seaward side of mangrove stands in small channels and creeks (Graham, 1929; MacNae & Kalk, 1962).

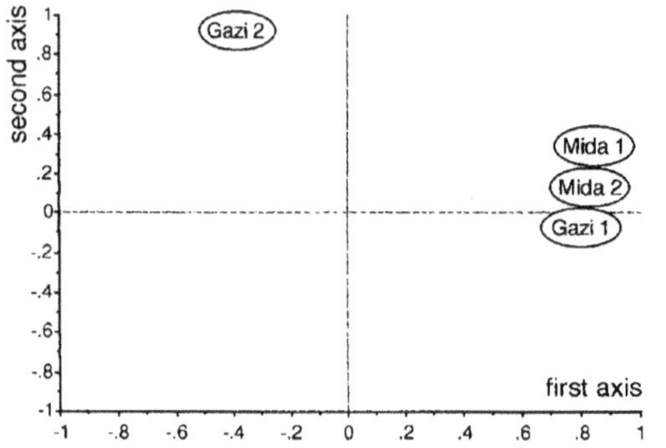

Fig. 5. Factor analysis of the four mangrove forests studied, using the figures of table 1 on seedling growing conditions as input. The first axis explains 60% ($\lambda_1 = 2.5$) of the original variance, whilst the second axis accounts for 20% of the original variance ($\lambda_2 = 1.0$).

Establishment of mangrove seedlings

Table 1 gives the percentages of fixed and covered seedlings for the 5 most dominant mangrove species at Gazi 1 (undisturbed), Gazi 2 (over-exploited), and Mida 1 and 2 (used).

Factor analysis is used to facilitate interpretation of the multivariate data represented in Table 1 (Jongman *et al.*, 1987). The relationships between Gazi 1, Gazi 2, Mida 1 and Mida 2 (in terms of fixed and covered seedlings) are summarized in Fig. 5, showing the position of the four sites in a two-dimensional space determined by the two principal axes extracted by factor analysis. It is clear that Gazi 1 (undisturbed) and Mida 1 and Mida 2 (used) are very similar in terms of percentage fixed and covered seedlings, and that Gazi 2 (over-exploited) is quite different from the first three sites. Indeed, according to Table 1, most of the seedlings at Gazi 2 (over-exploited) are fixed and not-covered, whilst seedlings in the former three sites are predominantly not-fixed and covered.

Discussion

Zonation of mangrove trees and mangrove seedlings

The zonation pattern of mangrove trees described in this study confirms earlier observations made in the East African region (Graham, 1929; Walter & Steiner, 1936; MacNae & Kalk, 1962; Beeckman *et al.*, 1990). Innovative for the Kenya coast however is the observation that mangrove seedlings do not develop at random in the intertidal zone, but are restricted to sites where their parental trees are found. Even when the parental trees have been felled, such as in the over-exploited mangrove forest Gazi 2, mangrove seedling development only takes place in the species specific distribution zones.

This observation supports the idea already promoted by Rabinowitz (1978a) who demonstrated that requirements of propagules for natural establishment correlate with position in the zonation. But we cannot support her idea of tidal sorting being the only mechanism for the control of zonation (Rabinowitz, 1978b). If it were true that propagule establishment is controlled by propagule size (and its movement in shallow water), then we would expect the zonation pattern of a specific mangrove species to be determined by the maximal amplitude of the tidal movements. Knowing that the tidal amplitude at the Kenya coast varies between less than 0.5 m to more than 4 m, tidal sorting only can not explain mangrove distribution zones of less than 4 m. From Figs 2, 3 and 4 it is clear that mangrove seedlings do not develop below 1 m above MLWS, and that the distribution of any given mangrove species in the intertidal zone hardly exceeds a range of 2 m elevation in the intertidal.

Establishment of mangrove seedlings

When considering the ratio of fixed/not-fixed and covered/not-covered seedlings in the studied mangrove forests (Table 1, Figs 6 and 7), arguments for both the 'self-planting' theory and the 'stranding' theory can be formulated.

It is assumed that dispersal according to the 'self-planting' theory will lead to a majority of covered seedlings growing either in a fixed or not-fixed condition, whilst dispersal according to the 'stranding' theory rather will lead to a majority of fixed seedlings growing either in a covered or not-

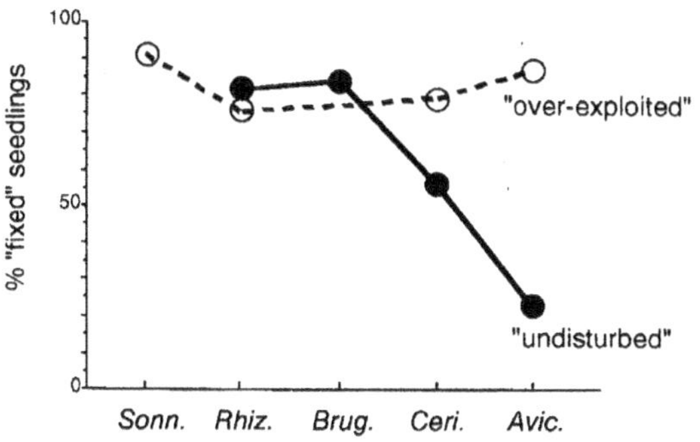

Fig. 6. Percentage seedlings growing under fixed conditions in an over-exploited (Gazi 2; dashed line) and an undisturbed mangrove forest (Gazi 1, Mida 1 and Mida 2; full line). Mangrove species are listed according to their position on the intertidal slope, with *Sonneratia alba* (Sonn.) closest to the low water line, followed by *Rhizophora mucronata* (Rhiz.), *Bruguiera gymnorrhiza* (Brug.), *Ceriops tagal* (Ceri.) and *Avicennia marina* (Avic.).

covered situation. When we also take into consideration the uprooting effect of the wave action, we may assume that there is a positive correlation between the intensity of the wave action and the number of seedlings growing under fixed conditions. Figure 6 shows that towards the seaward side of the mangrove forests (where exposure to wave action is maximal) nearly 90% of the mangrove seedlings are fixed. In undisturbed mangrove forests this percentage decreases with higher positions in the intertidal gradient, coinciding with a gradient of decreasing wave action. Figure 6 shows as well that at the very landward site of the undisturbed mangrove forest (where exposure to wave action is minimal) only 25% of the seedlings grow under fixed conditions. In the over-exploited mangrove forests on the other hand the percentage of fixed seedlings does not decrease with higher positions in the intertidal gradient and even never drops beyond 75%. This can be explained (i) by an inadequate protection of the intertidal slope from wave action (more than 75% of the mangrove trees have been felled) and (ii) by the supremacy of the 'stranding' theory over the 'self-planting' theory in an over-exploited mangrove forest.

Inspection of the percentage covered seedlings in undisturbed and over-exploited mangrove forests (Table 1, Fig. 7), further supports the idea of coexistence of the 'stranding' and the 'self-planting' theory. Seedlings in undisturbed mangrove forests however are principally covered, indicating a preponderance of the 'self-planting' theory over the 'stranding' theory, whilst seedlings in over-exploited mangrove forests are principally not-covered, supporting here the idea of a preponderance of the 'stranding' theory.

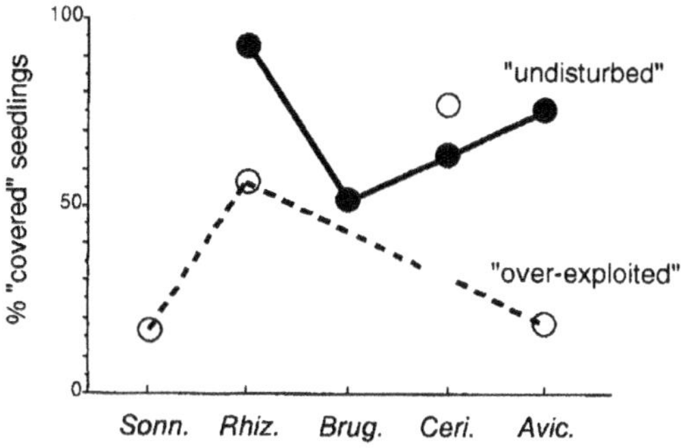

Fig. 7. Percentage seedlings growing under covered conditions in an over-exploited (Gazi 2; dashed line) and an undisturbed mangrove forest (Gazi 1, Mida 1 and Mida 2; full line). Mangrove species are listed according to their position on the intertidal slope, with *Sonneratia alba* (Sonn.) closest to the low water line, followed by *Rhizophora mucronata* (Rhiz.), *Bruguiera gymnorrhiza* (Brug.), *Ceriops tagal* (Ceri.) and *Avicennia marina* (Avic.). The *Ceriops* zone in the over-exploited mangrove forest behaves as in an undisturbed mangrove forest (it still has an overall tree cover of 70%), resulting in an untypical high percentage covered seedlings for an over-exploited mangrove forest.

Conclusion

Analysis of 449 vegetation quadrats along 36 line transects in mangrove forests at Gazi bay and Mida creek showed that establishment of mangrove propagules, once dispersed, takes place in species specific distribution zones in the intertidal zone, which we defined in terms of height above MLWS. The resulting zonation of mangrove seedlings starts with *Sonneratia alba* closest to the low water level, respectively followed by *Rhizophora mucronata*, *Bruguiera gymnorrhiza*, *Ceriops tagal*, *Avicennia marina*, *Lumnitzera racemosa* and *Xylocarpus granatum*. This zonation pattern perfectly corresponds with the zonation pattern displayed by their parent trees. This clear mangrove zonation pattern nevertheless allows the existence of mixed stands of mangrove trees in the intertidal zone.

Analysis of growing conditions of seedlings in undisturbed and over-exploited mangrove forests supported the idea that propagule dispersal follows both the 'self-planting' and the 'stranding' theory. The 'self-planting' theory being the major mechanism of propagule dispersal in undisturbed mangrove forests, where only the seaward margin is exposed to strong wave action and where the mangrove forest as such provides sufficient protection to prevent uprooting of dominantly 'self-planted' seedlings in the intertidal zone. The 'stranding' theory, on the other hand, becomes more important than the 'self-planting' theory in places without a direct input of 'self-planting' propagules such as in over-exploited or cleared mangrove forests.

Mangrove seedlings are more likely to be found under fixed conditions with an increasing potential of the wave action to uproot not-fixed seedlings. This is why fixed seedlings, as a rule, are found at the seaward margin of the mangrove forests, and at sites exposed to strong wave action, such as gaps in mangrove forests or over-exploited and totally cleared mangrove forests. Propagules dispersed by mechanisms belonging to the 'stranding' theory, as a result, rather grow under fixed conditions than under not-fixed conditions.

When considering mangrove re-afforestation projects along the Kenya coast, it is clear from the present study that seedlings should be planted under fixed conditions and in their species specific distribution zones.

Acknowledgements

This work was undertaken as part of the Nairobi University (Botany Department) undergraduate course SBT 203 – plant ecology – in June 1990. We are grateful to the university authorities for their overall support, and to Dr. M. J. Ntiba, Mr. A. O. Oluoch, Mr. C. A. Aketch, Mr. J. Muraya, Mr. S. Mathenge and 50 second year students for their help with the field work. We thank the Danish International Development Agency (DANIDA) at Nairobi, Kenya, for financing this study, as well as the African Wildlife Foundation of Washington D. C., African operations, for their contribution.

References

Beeckman, H., E. Gallin & E. Coppejans, 1990. Indirect gradient analysis of the mangal formation of Gazi Bay (Kenya). Silva Gandavensis 54: 57–72.

Chambers, J. M., W. S. Cleveland, B. Kleiner & P. A. Tukey, 1983. Graphical methods for data analysis. The Wadsworth statistics/probability series. Wadsworth International Group, Belmont, California. 395 pp.

Clarke, L. D. & N. J. Hannon, 1970. The mangrove swamp and salt marsh communities of the Sydney district. 3. Plant growth in relation to salinity and waterlogging. J. Ecol. 58: 351–369.

Coppejans, E. & E. Gallin, 1989. Macroalgae associated with the mangrove vegetation of Gazi Bay (Kenya). Bull. Soc. r. Bot. Belg. 122: 47–60.

Dawes, C. J., 1980. Marine Botany. John Wiley & Sons, New York. 628 pp.

Gallin, E., 1988. Vegetatie-ecologische studie van de mangrove langs de Westkust van Gazi Bay (Kenia). State University Ghent. M. Sc. Thesis, 157 pp.

Gallin, E., E. Coppejans & H. Beeckman, 1989. The mangrove vegetation of Gazi Bay. Bull. Soc. r. Bot. Belg. 122: 197–207.

Gang, P. O. & J. L. Agatsiva, 1990. The current status of mangroves along the Kenyan coast: a case study of the Mida Creek mangroves based on remote sensing. Paper presented at the International Symposium on Mangroves and related Ecosystems, 1990, Mombasa, Kenya.

Graham, R. M., 1929. Notes on the mangrove swamps of Kenya. J. E. Afr. U. Nat. Hist. Soc. 36: 157–164.

Hotelling, H., 1933. Analysis of a complex of statistical variables into principal components. J. Ed. Psych. 24: 417–441, 498–520.

Isaac, W. M. E. & F. M. N. Isaac, 1968. Marine botany of the Kenya coast. 3. General account of the environment, flora and vegetation. J. E. Afr. Nat. Hist. Soc. 27: 1–28.

Jongman, R. H. G., C. J. F. ter Braak & O. F. R. van Tongeren, 1987. Data analysis in community and landscape ecology. Pudoc, Wageningen. 299 pp.

Kokwaro, J. O., 1980. Economic importance and local use of the Kenyan mangroves. Proceedings of the Kenya National Seminar on Agroforestry, ICRAF, Nairobi: 377–386.

Kokwaro, J. O., 1985. The distribution and economic importance of the mangrove forests in Kenya. J. E. Afr. Nat. Hist. Soc. 75: 1–12.
MacNae, W. & M. Kalk, 1962. The ecology of the mangrove swamps at Inhaca island, Moçambique. J. Ecol. 50: 19–34.
Moorjani, S. A., 1977. The ecology of marine algae of the Kenya coast. University of Nairobi, Ph.D. Thesis. 285 pp.
Orlóci, L. & N. C. Kenkel, 1985. Introduction to data analysis. Statistical Ecology monographs 1. International Cooperative Publishing House, Maryland, USA. 339 pp.
Rabinowitz, D., 1978a. Dispersal properties of mangrove propagules. Biotropica 10: 47–57.
Rabinowitz, D., 1978b. Early growth of mangrove seedlings in Panama, and an hypothesis concerning the relationship of dispersal and zonation. J. Biogeogr. 5: 113–133.
Ruwa, R. K. & P. Polk, 1986. Additional information on mangrove distribution in Kenya: some observations and remarks. Kenya Journal of Science and Technology Series B 7 (2): 41–45.
Ruwa, R. K., 1990a. The effects of habitat complexities created by mangroves on macrofaunal composition in brackishwater intertidal zones at the Kenyan coast. Discovery and Innovation 2: 49–55.
Ruwa, R. K., 1990b. Impact on mangroves and crabs in creek and fringe mangrove vegetation due to disturbances at the Kenyan coast. Paper presented at the International Symposium on Mangroves and related Ecosystems, 1990, Mombasa, Kenya.
Tomlinson, P. B., 1986. The botany of mangroves. Cambridge university Press Tropical Biology Series. 413 pp.
Van Speybroeck, D., 1990. Field course to the Kenya coast. June 1990. Report on the activities. University of Nairobi, Botany Department, internal report. 65 pp.
van der Maarel, E., 1979. Transformations of cover-abundance values in phytosociology and its effects on community similarity. Vegetatio 39: 97–114.
Walter, H. & M. Steiner, 1936. The ecology of East African mangroves. Zeitschrift für Botanik 30.
Westhoff, V. & E. van der Maarel, 1978. The Braun-Blanquet approach. In: Whittaker, R. H. (ed.), Classification of plant communities. Dr W. Junk Publishers, The Hague: 287–399.

Hydrobiologia **247**: 253–259, 1992.
V. Jaccarini & E. Martens (eds), The Ecology of Mangrove and Related Ecosystems.

Impact of past management practices on the present status of the Muthupet mangrove ecosystem

Jayapaul Azariah, V. Selvam & S. Gunasekaran
Department of Zoology, University of Madras, Guindy Campus, Madras – 600 025, India

Key words: Mangrove ecosystem, species composition, zonation, community structure, past management practices

Abstract

Muthupet mangrove ecosystem (10° 25′ N and 79° 30′ E) is located at the southern end of the Cauvery delta on the east coast of India. The swamp is characterized by the presence of only 4 species of exclusive mangroves namely *Avicennia marina*, *Excoecaria agallocha*, *Aegiceras corniculatum* and *Acanthus ilicifolius*.

Community structure analysis carried out at four randomly selected quadrats showed that the relative density, relative dominance, relative frequency and importance value of *A. marina* was relatively higher than for the other three species. Hence, Muthupet mangrove ecosystem is characterized by the mono-specific dominance of *A. marina*. Earlier palynological studies indicate that species like *Rhizophora* were present in this swamp some 200 years ago. It is believed that the past management practices together with the present poor scientific policy of the decision makers and harsh environment conditions may be considered as a few factors responsible for the disappearance of *Rhizophora* from this mangrove swamp.

Introduction

The present status of the mangrove plants with respect to their ecological distribution and community structure, is poorly known in most of the Asian countries. Management policies formulated on the basis of this inadequate knowledge of the resource may result in unanticipated and irrevocable loss of valuable mangrove resources (Burbridge, 1984). The mangroves are essentially a forest formation and, like other forests, mangroves are also a renewable natural resource. It is thus possible to manage the mangrove ecosystem on a sustainable basis although, unfortunately, this is often not done. The concept of sustainable use involves either sustainable harvest or sustainable economic returns while at the same time the system can be maintained in as natural or close to its original status as possible (Saenger *et al.*, 1983). India is one of the few countries which realized the importance of the study and management of mangroves. As early as 1903 the flora of Sunderban including mangroves and the mangrove vegetation of Bombay Presidency were studied (Prain, 1905). In the second stage of mangrove management, two symposia were conducted during the years 1956–1957. They were the 9th Silvicultural Conference (1956) and the Mangrove Symposium (1957) which was organised by the Ministry of Food and Agriculture, Government of India. Later on significant contributions were made by the study of Sidhu (1963), Venkatesan (1966), Blasco (1975) and Untawale (1984). Along with scientific contributions, the

Government departments also contributed the working plan during the year 1918 and 1935 and a revised scheme of Forest management was published during the year 1935–1948 (Venkatesan, 1966). In India, at present, only the mangroves of Ganges delta and of Andaman and Nicobar islands are under the Management Plan (Untawale, 1987). In the present study, the status of the Muthupet mangrove ecosystem, with respect to its species composition, zonation and community structure are dealt with in detail and the impact of the past management practices and the present status of the mangrove plant community are discussed.

Description of the study area

Muthupet mangrove forest is located at the southern end of the Cauvery delta on the east coast of India. The Cauvery river is the main riverine system of the Thanjavur district. Many drainage arteries of the Cauvery basin namely, Paminiyar, Korayar, Kilaithangi, Marakakorayar flow through the Muthupet and adjacent villages and create a lagoon before they empty into the Palk Strait (Fig. 1). The northern as well as the western border of the lagoon are occupied by a dense mangrove community. The southern portion of the lagoon is formed by a sand spit which is devoid of mangroves.

Materials and methods

Species composition

In order to prepare a comprehensive list of plant species available in the study area, a number of surveys were made. Twigs and inflorescences were collected and an herbarium was made and the species were identified referring to appropriate taxonomic keys (Gamble & Fischer, 1957). The herbarium material was then counterchecked with the International Herbarium, Presidency College, Madras, India. In the final account, all the identified species were listed and placed in the relevant family.

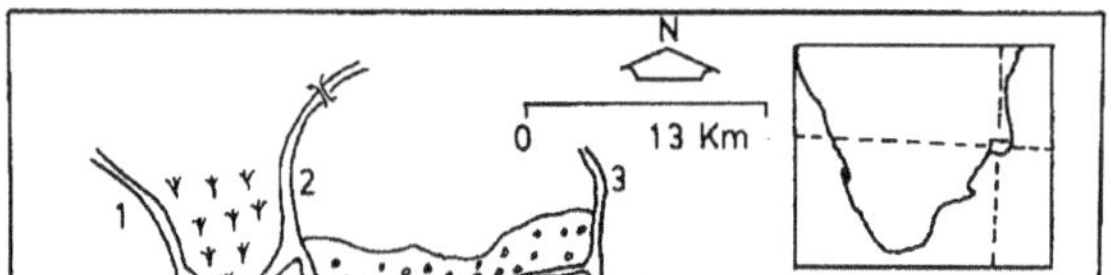

Fig. 1. Map showing the location of the Muthupet mangrove swamp and the zonation of the flora. 1. Pamini river 2. Korayar river 3. Kanthaperichan canal 4. Marakakorayar 5. Kilaithangi river. Y Y Y Prosopis zone; ∘∘∘∘ Suaeda zone; //// mangrove zone.

Zonation

The zonation pattern of the flora was studied by visual observation. A zone was demarcated clearly on the basis of a resident species which was predominant in it and the zone was named after that dominant species. Flora in each zone was collected and identified.

Community structure

Community structure of the flora was studied following the random quadrat method described by Cintron & Novelli (1984). A total number of 5 quadrats, each covering an area of 1 hectare, was laid at five different locations. Out of 5 quadrats, four were established in the pure thickets formed by exclusive mangroves and the remaining one was laid in the non-mangrove zone. From each quadrat the following data were collected: i) number of species of plants ii) number of individuals of each species iii) circumference of the trees at breast height. The circumference readings were divided by π to obtain the diameter at breast which were used for basal area computation, using the following formula.

$g(m^2) = 0.00007854\ dbh^2$, where 'g' is the basal area in m^2, dbh is the diameter at breast height.

The relative frequency (RF), relative dominance (RDo), relative density (RDe) and importance value (IV) were calculated using the standardized methodology (Cintron & Novelli, 1984). The importance value of a given species is the sum of RF, RDo and RDe of that species.

Past management practices

Information on the past management practices used to exploit the mangrove resources of Muthupet was collected from the working plans prepared by Chengappa (1918) and Ahmed (1937) and from the Government Gazette released in 1937.

Results

Species composition

The qualitative studies revealed that the Muthupet mangrove forest contains only four species of exclusive mangroves namely *Avicennia marina* (Forsk.) Vierh, *Excoecaria agallocha* L., *Acanthus ilicifolius* L. and *Aegiceras corniculatum* (L.) Blanco. The above species were recorded only on the border area of the lagoon.

Zonation

On the basis of distribution of flora, Muthupet mangrove ecosystem can be divided into 4 zones namely, Prosopis zone, Suaeda zone, mangrove zone and mud flat zone (Fig. 1). The Prosopis zone is present on the landward side of the lagoon and its length and breadth were approximately 1 to 1.2 km and 30 to 50 m respectively. This zone was completely occupied by the non-mangrove species *Prosopis spicigera* L. which is a thorny woody plant. A few individuals of *E. agallocha* were also present on the banks of the small tidal creeks of the Prosopis zone. Next to the Prosopis zone, a well marked Suaeda zone was present. This zone was occupied by the halophyte, *Suaeda maritima* is an undershrub. The mangrove zone was located next to the Suaeda zone and was restricted to the edges of the brackish water lagoon. The length of the mangrove zone varied from 0.2 km to 1.0 km and the breadth was approximately 0.05 km to 1.5 km. All the four exclusive species of the mangroves were present in this zone.

Community structure

The density of the different plant species in the randomly selected quadrats and the diameter of the each individual tree are given in Table 1. In the Prosopis zone, a total number of 513 individuals were present per hectare (Quadrat I), out of which 466 individuals belonged to the species, *P. spicigera*; 32 were identified as *E. agallocha* and 14 individuals were of the species *Opuntia monocantha* Haw. Nearly 75% of the *P. spicigera* had a diameter between 10.1 and 20 cm. In all other quadrats the density of *Av. marina* was high, the number of individuals per hectare being 302 (Quadrat II) 168 (Quadrat III) 274 (Quadrat IV) and 213 (Quadrat V) which represents 76.26%, 66.67%, 81.55% and 89.87%, of the total individuals recorded in the respective quadrats. In Quadrats II and III, *S. maritima* was the codominant species and its density was 94 per hectare in Quadrat II and 64 per hectare in Quadrat III. In Quadrat IV, *E. agallocha* was the codominant tree which constituted 18.45% of the total individuals present in that quadrat. *Ae. corniculatum* was the second dominant species in Quadrat V

Table 1. Plant density (no h^{-1}) and their diameter class.

Location	Species	Number of individuals per hectare	Diameter class (cm)				
			Less than 3.0	3.1–10.0	10.1–20.0	20.1–30.0	Greater than 30.0
Quadrat I	*Prosopis spicigera*	466 (91.02)	62 (13.40)	–	354 (75.97)	36 (7.73)	14 (3.00)
	Excoecaria agallocha	32 (6.25)	–	8 (25)	24 (75)	–	–
	Opuntia monocantha	14 (2.73)	14 (100)	–	–	–	–
	Total	512					
Quadrat II	*Avicennia marina*	302 (76.26)	–	–	26 (8.61)	272 (90.07)	4 (1.32)
	Suaeda maritima	94 (23.74)	9 (100)	–	–	–	–
	Total	396					
Quadrat III	*Avicennia marina*	168 (66.67)	–	–	14 (8.33)	126 (75.00)	28 (16.67)
	Suaeda maritima	84 (33.33)	–	–	–	–	–
	Total	252					
Quadrat IV	*Avicennia marina*	274 (81.55)	–	–	11 (4.01)	251 (91.61)	12 (4.38)
	Excoecaria agallocha	62 (18.45)	–	18 (29.03)	37 (59.68)	7 (11.29)	–
	Total	336					
Quadrat V	*Avicennia marina*	213 (89.87)	–	6 (2.81)	21 (9.86)	172 (80.75)	14 (6.57)
	Aegiceras corniculatum	24 (10.13)	–	20 (83.33)	4 (16.67)	–	–
		237					

Value in parenthesis indicates percentage.
– Indicates absence of individuals of species in that particular diameter class.

which constituted 10.13% of the total individuals.

The community structure indices such as relative dominance, relative frequency, relative density and importance value of each species of the five quadrats are given in Table 2. In Quadrat I, *P. spicigera* was the dominant species whose relative dominance, relative density, relative frequency and importance values were 93.54%, 91.01%, 33.34% and 217.89 respectively. In all

Table 2. Basal area (m^2), relative domainance (%) relative density (%) relative frequency (%) and importance value of flora of Muthupet mangrove swamp.

Species	Basal area (m^2)	Relative dominance (%)	Relative density (%)	Relative frequency (%)	Importance value
Quadrat I					
Prosopis spicigera	0.7321	93.54	91.01	33.34	217.89
Excoecaria agallocha	0.0506	6.36	6.25	33.33	46.94
Opuntia monocantha	–	–	2.74	33.33	–
Total	0.7827	100	100	100	
Quadrat II					
Avciennia marina	1.8324	100	75.26	50	226.26
Suaeda martima	–	–	23.74	40	–
	1.8324	100	100	100	
Quadrat III					
A. marina	1.2346	100	72.41	50	222.41
S. maritima	–	–	27.59	50	
	1.2346	100	100	100	
Quadrat IV					
A. marina	1.9080	87.35	81.55	50	218.90
E. agallocha	0.2766	12.65	18.45	50	81.01
	2.1846	100	100	100	
Quadrat V					
A. marina	1.5388	93.70	89.87	50	233.57
Aegiceras corniculatum	0.1035	6.30	10.13	50	66.43
	1.6423	100	100	100	

– indicates value could not be calculated as the stem diameter was less than 3 cm.

other quadrats *A. marina* was the dominant species. The relative dominance value of *A. marina* in Quadrats II and III was 100% whereas in Quadrat IV and V its value was 87.35% and 93.70% respectively. The relative density of *A. marina* in Quadrat II to V was 75.26%, 72.40%, 81.55% and 89.87% respectively. The importance value of *A. marina* was also high in all the last four quadrats the values being 226.20 (Q.II) 222.41 (Q.III) 218.90 (Q.IV) and 233.57 (Q.V). Thus the community structure analysis indicates that the landward portion of the swamp is dominated by the non-mangrove species *P. spicigera* whereas *A. marina* showed monospecific dominance in the mangrove zone.

Past management practices

According to Chengappa (1918), the Muthupet mangrove forest was under the control of the Chatram Department from 1853 to 1912. The Chatram Department, a branch of the then District Board of Tanjore, looked after the numerous Chatrams (free boarding and lodging houses) which were constructed by the Raja of Tanjore for the convenience of pilgrims and travellers. To maintain these Chatrams, the Chatram Department collected revenue from Muthupet mangrove forest. They harvested the mangrove wood and sold it in the market as timber and firewood. The revenue thus realized was spent to maintain the Chatrams. In 1912, the Revenue Department took charge and during this year the First Working Plan for Tanjore mangroves was proposed and advocated. According to this plan the coppice system of clear felling with 12 years rotation was introduced. The revenue collected from selling the mangrove was given to the Chatram Department. In 1923, once again the Chatram Department took over the responsibility of managing the mangrove forest and followed a 20 years rotation with clearfelling as advised by the Forest Department (Venkatesan, 1966). The Chatram Department exploited the mangrove resources of Muthupet till 1936. From 1923 to 1936, half of the revenue obtained from selling mangrove products was paid to the Revenue Department and the remaining half was spent to maintain the Chatrams. In February 1937, Muthupet mangrove forest was declared as reserve forest by the Government and accordingly the mangrove forest was handed over to the Forest Department of the Madras Presidency (Gazette, 1937). After 1937, the Forest Department managed the Muthupet mangroves following the 20 years rotation system. Based on the frequency and dominance of *Avicennia* which was found to be a coppicing species, 50 trees per acre were left as seedbearers. According to the study of Venkatesan (1966) it was found that the above rotation pattern was followed with the added prescription of retaining all advanced growth below 1.8 m (6 feet) in height.

No detailed information is available on the yield and revenue obtained by the above management bodies from Muthupet mangrove Forest. However, Venkatesan (1966) mentioned that from 1935 to 1948, the average yield was 16 tonnes/hectare in Tanjore district. In a neighbouring mangrove forest area, Cuddalore, South Division of Tamil Nadu, a cycling pattern of 40 years was followed leaving 38 mother trees/acre as seed bearers. The average yield was 25 tonnes/hectares.

Discussion

The species composition of Muthupet mangrove forest is very poor when compared to the number of exclusive mangrove species present in the Pitchavaram mangrove forest which is located just 140 km north of the Muthupet mangrove. Uncontrolled and unregulated exploitation by the authorities of the Chatram Department prior to 1912 and the unscientific management policies followed by the Revenue as well as Chatram Departments after 1912 and some environmental factors like poor inundation and damage due to cyclonic storm seem to be the reason for the poor species diversity of the Muthupet mangrove forest. The available records show, that prior to 1912, the woody plant species of Muthupet mangrove were exploited throughout the year for sales as timber as well as for fuelwood by the Chatram Department (Chengappa, 1918). After 1912, a coppice method with 12 years rotation period was followed for the harvest of mangrove wood. According to this method an area of mangrove forest was allotted for clear felling by the Revenue Department which was then brought to the notice of the Chatram Department. Since authorities of both the Revenue and Chatram Departments were not aware of the importance of leaving seed bearers in the clear felling area they removed all the trees which resulted in the poor regeneration of the mangrove plants in that particular area (Ahmed, 1937). To correct the above mistake, Ahmed (1937) introduced another system according to which all the trees, except 50 seed bearers/hectare were harvested. He also changed the 12 years clear felling rotation to 20 years. Ahmed (1937) also noticed the poor regeneration of the mangrove plants and prescribed artificial regeneration of *A. marina* by direct sowing. But this practice was withdrawn after a few trials due to unsatisfactory results. He also suggested the digging up of channels in areas where the ground level was raised due to fresh accretion of sand

and where *Avicennia* was showing signs of death due to insufficient inundation. This suggestion was discontinued after a few attempts due to high cost involved in it. The coppice system with 20 years rotation period was followed by the Forest Department till 1970. The management policy suggested by Chengappa (1918) was unscientific because it was based on the wrong assumption that all mangrove plant species could regenerate from the root stock. Blasco (1975) showed that only the species such as *Avicennia* and *Excoecaria* have the capacity to regenerate from root stocks.

The palynological studies carried out by Tissot (1987) indicate that the species of *Rhizophora* and *Sonneratia* were present in the Muthupet mangrove ecosystem some 200 years ago but they have disappeared completely now. The uncontrolled and unregulated exploitation carried out by the authorities of the Chatram and Revenue Departments, some unscientific policies introduced in 1920 and thereafter as well as the poor attention paid towards the replantation of the mangrove plants seem to be some of the causes responsible for the present poor species diversity of the Muthupet mangrove forest.

References

Ahmed, S., 1937. Working plan for the Tanjore Forest Division Revenue Department, District Board of Tanjore, 82 pp.

Blasco, F., 1975. Mangroves of India. Inst. Fr. Pondicherry, Trav. Sect. Sci. Tech., Vol. 14. 180 pp.

Burbridge, P. R., 1984. Management and Planning of Mangrove Resources of India. Proc. As. Symp. Mangr. Env. – Res. Manag., 27–42.

Chengappa, D., 1918. Working Plan for the Tanjore Forest Division. Revenue Department, District Board of Tanjore, 78 pp.

Cintron, G. & Y. S. Novelli, 1984. Methods for studying mangrove structure. In: Snedaker, S. C. & Snedaker J. G. (eds), The Mangrove Ecosystem: Research Methods. Monographs on Oceanographic Methodology. UNESCO, Paris, 91–113.

Gazette, 1937. Government of the Presidency of Madras, Madras, 318 pp.

Gamble, G. S. & C. E. S. Fischer, 1957. Flora of the Presidency of Madras Vol. I–III, Government of India, Calcutta, 1388 pp.

Prain, D., 1905. Flora of the Sunderbans. Rec. Bot. Sur. India 2: 237–370.

Saenger, P., E. J. Hegerl & J. D. S. Davis, 1983. Global status of mangrove ecosystems. Commission of Ecology paper No. 3 IUCN, Gland, 38 pp.

Sidhu, S. S., 1963. Studies on the mangroves of India. I-East Godavari region. Indian Forester 89: 337–351.

Tissot, C., 1987. Recent evolution of mangrove vegetation in the Cauvery Delta: A palynological study. J. mar. biol. Ass. India 29: 16–22.

Untawale, A. G., 1984. Present status of mangrove along the west coast of India. Proc. As. Symp. Mangr. Env. – Res. Manag: 57–74.

Untawale, A. G., 1987. Conservation of Indian mangroves – A national perspective. In: Rao, T. S. S. (ed.), Contribution in Marine Sciences. National Institute of Oceanography Goa: 85–104.

Venkatesan, K. R., 1966. The mangroves of Madras state. Indian Forester 92: 27–34.

Index

GPSR Compliance

The European Union's (EU) General Product Safety Regulation (GPSR) is a set of rules that requires consumer products to be safe and our obligations to ensure this.

If you have any concerns about our products, you can contact us on ProductSafety@springernature.com

In case Publisher is established outside the EU, the EU authorized representative is:

Springer Nature Customer Service Center GmbH
Europaplatz 3
69115 Heidelberg, Germany

Batch number: 10755809

Printed by Printforce, the Netherlands